TEXTS AND READINGS IN MATHEMATICS **65**

The Classical Groups

Their Invariants and Representations

Texts and Readings in Mathematics

The Classical Groups
Their Invariants and Representations

Hermann Weyl

HINDUSTAN
BOOK AGENCY

Published in India by
Hindustan Book Agency (India)
P 19 Green Park Extension
New Delhi 110 016
India

email: info@hindbook.com
http://www.hindbook.com

Indian edition published in 2013

ISBN 978-93-80250-35-9

In Memoriam

ISSAI SCHUR

PREFACE TO THE FIRST EDITION

Ever since the year 1925, when I succeeded in determining the characters of the semi-simple continuous groups by a combination of E. Cartan's infinitesimal methods and I. Schur's integral procedure, I have looked toward the goal of deriving the decisive results for the most important of these groups by direct algebraic construction, in particular for the full group of all non-singular linear transformations and for the orthogonal group. Owing mainly to R. Brauer's intervention and collaboration during the last few years, it now appears that I have in my hands all the tools necessary for this purpose. The task may be characterized precisely as follows: with respect to the assigned group of linear transformations in the underlying vector space, to decompose the space of tensors of given rank into its irreducible invariant subspaces. In other words, our concern is with the various kinds of "quantities" obeying a linear transformation law, which may be prepared under the reign of each group from the material of tensors. Such is the problem which forms one of the mainstays of this book, and in accordance with the algebraic approach its solution is sought for not only in the field of real numbers on which analysis and physics fight their battles, but in an arbitrary field of characteristic zero. However, I have made no attempt to include fields of prime characteristic.

The notion of an algebraic invariant of an abstract group γ cannot be formulated until we have before us the concept of a representation of γ by linear transformations, or the equivalent concept of a "quantity of type $\mathfrak{A}$." The problem of finding all representations or quantities of γ must therefore logically precede that of finding all algebraic invariants of γ. (For the notion of quantities and invariants of a more general character, and their close interdependence, the reader is referred to the restatement in Chapter I of Klein's Erlanger program in slightly more abstract terms.) My second aim, then, is to give a modern introduction to the theory of invariants. It is high time for a rejuvenation of the classic invariant theory, which has fallen into an almost petrified state. My vindication for having proceeded in a much more conservative manner than our young generation of algebraists would probably deem desirable, is the wish not to sacrifice the past; even so, I hope to have broken through to the modern concepts resolutely enough. I do not pretend to have written *the* book on modern invariant theory: A systematic handbook would have to include many things passed over in silence here.

As one sees from the above description, the subject of this book is rather special. Important though the general concepts and propositions may be with which the modern industrious passion for axiomatizing and generalizing has presented us, in algebra perhaps more than anywhere else, nevertheless I am convinced that the special problems in all their complexity constitute the stock and core of mathematics; and to master their difficulties requires on the whole the harder labor. The border line is of course vague and fluctuating. But quite intentionally scarcely more than two pages are devoted to the general theory of group representations, while the application of this theory to the particular groups that come under consideration occupies at least fifty times as much space. The general theories are shown here as springing forth from special problems whose analysis leads to them with almost inevitable necessity as the fitting tools for their solution; once developed, these theories spread their light over a wide region beyond their limited origin. In this spirit we shall treat among others the doctrine of associative algebras, which in the last decade has risen to a ruling position in mathematics.

The relations to other parts of mathematics are emphasized where occasion arises, and despite the fundamentally algebraic character of the book, neither the infinitesimal nor the topological methods have been omitted. My experience has seemed to indicate that to meet the danger of a too thorough specialization and technicalization of mathematical research is of particular importance in America. The stringent precision attainable for mathematical thought has led many authors to a mode of writing which must give the reader an impression of being shut up in a brightly illuminated cell where every detail sticks out with the same dazzling clarity, but without relief. I prefer the open landscape under a clear sky with its depth of perspective, where the wealth of sharply defined nearby details gradually fades away towards the horizon. In particular, the massif of topology lies for this book and its readers at the horizon, and hence what parts of it had to be taken into the picture are given in broad outline only. An adaptation of sight different from that required in the algebraic parts, and a sympathetic willingness to cooperate, are here expected from the reader.

The book is primarily meant for the humble who want to learn as new the things set forth therein, rather than for the proud and learned who are already familiar with the subject and merely look for quick and exact information about this or that detail. It is neither a monograph nor an elementary textbook. The references to the literature are handled accordingly.

The gods have imposed upon my writing the yoke of a foreign tongue that was not sung at my cradle.

"Was dies heissen will, weiss jeder,
Der im Traum pferdlos geritten,"

I am tempted to say with Gottfried Keller. Nobody is more aware than myself of the attendant loss in vigor, ease and lucidity of expression. If at least the worst blunders have been avoided, this relative accomplishment is to be ascribed solely to the devoted collaboration of my assistant, Dr. Alfred H. Clifford; and even more valuable for me than the linguistic, were his mathematical criticisms.

HERMANN WEYL

PRINCETON, N. J.,
September, 1938.

NOTE. A reference to formula (7.6) [or to (3.7.6)] indicates the formula 6 in section 7 labeled as (7.6) in the *same* chapter [or in Chapter III respectively].

PREFACE TO THE SECOND EDITION

The photostatic process employed for the reprinting ruled out any appreciable changes which otherwise might have been desirable. But a new chapter containing Supplements, a list of Errata and Addenda, and a short Bibliography for the years 1940–1945 have been added. Two of the supplements develop an alternate and more direct approach to some of the problems in the theory of the orthogonal and symplectic groups dealt with in Chapters II, V and VI. Supplement C describes a particularly straightforward and powerful process for the generation of invariants discovered by M. Schiffer, whereas supplement D applies the "matrix method" of Chapters III and IX to the splitting of a division algebra by extension of the ground field, without the limitation to normal algebras and finite extensions.

HERMANN WEYL

PRINCETON, N. J.,
March, 1946.

TABLE OF CONTENTS

CHAPTER I

INTRODUCTION

CHAPTER II

VECTOR INVARIANTS

CHAPTER III

MATRIC ALGEBRAS AND GROUP RINGS

Chapter VIII
GENERAL THEORY OF INVARIANTS
A. Algebraic Part

Chapter IX
MATRIC ALGEBRAS RESUMED

Chapter X
SUPPLEMENTS

CHAPTER I

INTRODUCTION

1. Fields, rings, ideals, polynomials

Before we can start talking algebra we must fix the *field* k of numbers wherein we operate. k is the closed universe in which all our actions take place. I should advise the reader at first to think of k as the continuum of the ordinary real or complex numbers. But generally speaking, k is any set of elements α, called *numbers*, closed with respect to the two binary operations: *addition* and *multiplication*. Addition and multiplication are supposed to be *commutative* and *associative*. Moreover, addition shall allow of a unique inversion (*subtraction*), i.e. there is a number o, called *zero*, such that

$$\alpha + o = \alpha$$

for every α, and each α has a *negative* $-\alpha$ satisfying $\alpha+(-\alpha) = o$. Multiplication shall fulfill the *distributive law* with respect to addition:

$$\alpha(\beta + \gamma) = (\alpha\beta) + (\alpha\gamma),$$

from which one readily deduces the universal equation

$$\alpha \cdot o = o. \tag{1.1}$$

Multiplication also is required to be invertible (*division*) with the one exception necessarily imposed by (1.1): there shall exist a *unit* ϵ or 1 satisfying

$$\alpha \cdot \epsilon = \alpha \tag{1.2}$$

for all α, and every α *except* o shall have an *inverse* α^{-1} or $1/\alpha$ such that $\alpha \cdot \alpha^{-1} = \epsilon$. Were $\epsilon = o$, all numbers α would be $= o$ according to (1.1) and (1.2); this degenerate case we once for all exclude by the axiom $\epsilon \neq o$.

Any number α gives rise to its *multiples*

$$\alpha = 1\alpha, \qquad \alpha + \alpha = 2\alpha, \qquad 2\alpha + \alpha = 3\alpha, \cdots ;$$

here the integers 1, 2, 3, $\cdots$ are symbols of "multipliers" rather than numbers in the reference field k. Two cases are possible: either all the multiples

$$n\epsilon \qquad\qquad (n = 1, 2, 3, \cdots)$$

of the unit ϵ are $\neq o$, or there is a least n for which $n\epsilon = o$. In the latter case the integer n must be a prime number p. Indeed for a composite number $n = n_1n_2$ (neither n_1 nor $n_2 = 1$) we should have

$$n\epsilon = n_1\epsilon \cdot n_2\epsilon = o,$$

and hence $n_1\epsilon$ or $n_2\epsilon$ would equal o in contradiction to n being the *least* vanishing multiple of ϵ. One distinguishes these two cases by ascribing the *characteristic* 0 *or* p to the field k. In a field of prime characteristic p the p-fold of any number α vanishes:

$$p\alpha = p(\epsilon\alpha) = (p\epsilon)\alpha = o.$$

In a field of characteristic zero we can form the aliquot part $\beta = \alpha/n$ of α with any integer n, i.e. a number β satisfying the equation $n\beta = \alpha$. Indeed this equation amounts to

$$n\epsilon \cdot \beta = \alpha,$$

and as the first factor $n\epsilon$ is $\neq o$ the equation is solvable according to the axiom of divisibility. Hence our field k contains the subfield of the rational multiples of ϵ:

$$m\epsilon/n \quad (n \text{ a positive integer } 1, 2, 3, \cdots,$$

$$m \text{ any integer } 0, \pm1, \pm2, \cdots),$$

which is isomorphic to the field of ordinary rational numbers m/n and may be identified with it. To this most primitive field of characteristic 0 we shall always refer as the *ground field* κ, and our remark thus asserts the fact that *any field k of characteristic* 0 *contains the ground field κ. From now on we shall assume the reference field k to be of characteristic* 0 without mentioning this restriction again and again; we shall not try to discuss any of our problems in a field of prime characteristic. So even when we use the phrase "in an arbitrary field" or something similar we mean "in an arbitrary field of characteristic 0".

If one omits the axiom requiring the existence of an inverse α^{-1} one obtains the general notion of a *ring* rather than a field; only addition, subtraction and multiplication are possible in a ring. The classical example is the set of all integers. If a product $\alpha\beta$ of two elements of the ring never vanishes unless at least one of the factors vanishes, the ring is *without null divisors.* Starting with a given ring R without null divisors, we may formally introduce fractions α/β as pairs of elements α, β in R of which the second term β is $\neq 0$, and then define equality, addition, and multiplication in accordance with the rules which we all learned in school. The fractions form a field, the *quotient field of R*; it contains R if we identify the fraction $\alpha/1$ with α.

With respect to a given ring R a set $\mathfrak{a}$ of its elements is called an *ideal* if

$$\alpha \pm \beta, \qquad \lambda\alpha$$

lie in $\mathfrak{a}$ for any α, β in $\mathfrak{a}$ and any number λ in R. The case where $\mathfrak{a}$ consists of the one element 0 only is expressly excluded. The classical example is provided by the integral multiples of a given integer. The ideals serve as modules for *congruences:*

$$\lambda \equiv \mu \pmod{\mathfrak{a}}$$

means that the difference $\lambda - \mu$ of the two numbers λ, μ of R lies in $\mathfrak{a}$. A finite number of elements $\alpha_1, \cdots, \alpha_r$ in $\mathfrak{a}$ constitute an (ideal) basis of $\mathfrak{a}$ if every element α in $\mathfrak{a}$ is of the form

$$\lambda_1\alpha_1 + \cdots + \lambda_r\alpha_r \qquad (\lambda_i \text{ in } R).$$

$\mathfrak{a}$ is then the ideal $(\alpha_1, \cdots, \alpha_r)$ with the basis $\alpha_1, \cdots, \alpha_r$. In a *field* k there is only one ideal, the field itself. For if α is a number $\neq 0$ in the given ideal $\mathfrak{a}$, the latter will contain all numbers of the form $\lambda\alpha$ and hence every number β whatsoever: $\lambda = \beta\alpha^{-1}$. In the ring of ordinary integers every ideal is a principal ideal (α). $\mathfrak{a}$ is a *prime ideal* if the congruence

$$\lambda\mu \equiv 0 \pmod{\mathfrak{a}}$$

never holds unless one of the factors λ, μ is $\equiv 0 \pmod{\mathfrak{a}}$.

A formal expression

$$f(x) = \sum_{i=0}^{n} \alpha_i x^i$$

involving the "indeterminate" (or variable) x, whose coefficients α_i are numbers in a field k, is called a (k-)polynomial of x of formal degree n. If $\alpha_n \neq 0$, n is its actual degree; 0 is the only polynomial not possessing an actual degree. Everybody knows how to add and multiply polynomials; they form a ring $k[x]$ without null-divisors. Indeed if a is of actual degree m, b of degree n:

$$a = \alpha_m x^m + \cdots, \quad b = \beta_n x^n + \cdots \qquad (\alpha_m \neq 0,\ \beta_n \neq 0),$$

then

$$ab = \alpha_m\beta_n x^{m+n} + \cdots$$

is of degree $m + n$ since $\alpha_m\beta_n \neq 0$. One sees that this proposition will still hold when the coefficients are taken from a ring without null-divisors rather than from a field k. This allows us to pass to polynomials of a new indeterminate y with coefficients taken from $k[x]$ or, what is the same, to k-polynomials of *two* indeterminates x, y, and so on. *The k-polynomials of several indeterminates $x, y, \cdots$ form a ring $k[x, y, \cdots]$ without null-divisors.*

In a given polynomial $F(u, v, \cdots)$ of certain indeterminates $u, v, \cdots$ one may carry out the *substitution*

$$u = f(x, y, \cdots), \qquad v = g(x, y, \cdots), \cdots$$

by means of certain polynomials $f, g, \cdots$ of other indeterminates $x, y, \cdots$; the result is a polynomial $\Phi(x, y, \cdots)$ of $x, y, \cdots$:

$$F(f(x, y, \cdots), g(x, y, \cdots), \cdots) = \Phi(x, y, \cdots).$$

In particular one may substitute *numbers* $\alpha, \beta, \cdots$ for the "arguments" $u, v, \cdots$ in F; the resulting number $F(\alpha, \beta, \cdots)$ is called the *value of F for the values $\alpha, \beta, \cdots$ of the arguments $u, v, \cdots$.*

$f(x)$ being a polynomial in x, α is a *zero* or a *root* of f if $f(\alpha) = 0$. A polynomial of degree n has at most n different zeros; this follows in the well-known way by proving that $f(x)$ contains the factors $(x - \alpha_1)(x - \alpha_2) \cdots$ if $\alpha_1, \alpha_2, \cdots$ are distinct zeros. Hence a polynomial $f(x) \neq 0$ does not vanish numerically for every value of x in k, provided the reference field k is of characteristic 0, because such a field contains infinitely many numbers. One can even find a *rational* value of x for which the value of f is $\neq 0$. Induction with respect to the number of indeterminates permits generalization of this proposition to polynomials with any number of arguments. If

$$F(x, y, \cdots); \qquad R_1(x, y, \cdots), \qquad R_2(x, y, \cdots), \cdots$$

are a number of non-vanishing k-polynomials then the product $FR_1R_2 \cdots$ is also $\neq 0$; and hence our statement can be sharpened to the following

LEMMA (1.1.A). (*Principle of the irrelevance of algebraic inequalities.*) *A k-polynomial $F(x, y, \cdots)$ vanishes identically if it vanishes numerically for all sets of rational values $x = \alpha$, $y = \beta$, $\cdots$ subject to a number of algebraic inequalities*

$$R_1(\alpha, \beta, \cdots) \neq 0, \qquad R_2(\alpha, \beta, \cdots) \neq 0, \cdots.$$

From the ring $k[x, y, \cdots]$ of k-polynomials in $x, y, \cdots$ one can pass to the *field* $k(x, y, \cdots)$ of the *rational functions* of $x, y, \cdots$ in k by forming the quotient field of $k[x, y, \cdots]$.

The *derivative* $f'(x)$ of a polynomial $f(x)$ is introduced as the coefficient of t in the expansion of $f(x + t)$ as a polynomial in t:

$$f(x + t) = f(x) + t \cdot f'(x) + \cdots \tag{1.3}$$

The familiar formal properties of derivation are immediate consequences thereof. One might restate the definition (1.3) as follows: there is a polynomial $g(x, y)$ satisfying the identity

$$f(y) - f(x) = (y - x) \cdot g(x, y); \tag{1.4}$$

$f'(x)$ is $= g(x, x)$. While in Calculus the unique determination of $g(x, x)$ is brought about by requiring $g(x, y)$ to be *continuous* even for $y = x$, Algebra attains the same by requiring g to be a *polynomial*. The derivative of

$$f(x) = \alpha_0 + \alpha_1 x + \alpha_2 x^2 + \cdots + \alpha_n x^n$$

is

$$f'(x) = \alpha_1 + 2\alpha_2 x + \cdots + n\alpha_n x^{n-1}.$$

Hence *the only polynomial $f(x)$ in a field of characteristic zero whose derivative $f'(x)$ vanishes is the constant*: $f(x) = \alpha_0$.

For a polynomial $f((x)) = f(x_1, \cdots, x_n)$ of n variables x_i one may form similarly to (1.3):

$$f((x + t \cdot y)) = f(x_1 + ty_1, \cdots, x_n + ty_n) = f((x)) + t \cdot f_1((x, y)) + \cdots. \tag{1.5}$$

The coefficient $f_1((x, y))$ of t in this expansion by t is called the ***polarized polynomial*** $D_{yx}f$ of f; it involves the new variables y_i in a homogeneous linear fashion:

$$D_{yx}f = \frac{\partial f}{\partial x_1} y_1 + \cdots + \frac{\partial f}{\partial x_n} y_n. \tag{1.6}$$

Sometimes the new variables y_i are designated by dx_i and then the polarized form is called the *total differential* df of f. The polar process has the formal properties of differentiation:

$$\begin{aligned} D(f + g) &= Df + Dg, \\ D(\alpha f) &= \alpha \cdot Df \quad (\alpha \text{ a number}), \\ D(f \cdot g) &= Df \cdot g + f \cdot Dg. \end{aligned} \tag{1.7}$$

The degree of a *monomial*

$$x_1^{r_1} x_2^{r_2} \cdots x_n^{r_n}$$

of our n variables $x_1, x_2, \cdots, x_n$ is the sum

$$r = r_1 + r_2 + \cdots + r_n$$

of the non-negative integral exponents $r_1, \cdots, r_n$. Each polynomial $f((x))$ is a linear combination of monomials; if all these monomials are of the same degree r:

$$f((x)) = \sum \alpha_{r_1 \cdots r_n} x_1^{r_1} \cdots x_n^{r_n}, \qquad (r_1 + \cdots + r_n = r) \tag{1.8}$$

the polynomial is called *homogeneous* or *a form of degree r*. In (1.8) the sum extends over all sets of non-negative integral exponents $r_1, \cdots, r_n$ with the sum r. Multiplication of all variables x_i with a numerical factor λ has the effect of changing

$$f((x)) \quad \text{into} \quad \lambda^r \cdot f((x)). \tag{1.9}$$

Another way of writing such a form is this:

$$f((x)) = \sum_{i=1}^{n} \beta(i_1, \cdots, i_r) x_{i_1} \cdots x_{i_r} \tag{1.10}$$

where each of the r indices i runs independently from 1 to n. In this expression the coefficients β are not uniquely determined; they become so, however, if one imposes the condition of symmetry upon the β:

$$\beta(i_{1'}, \cdots, i_{r'}) = \beta(i_1, \cdots, i_r)$$

provided $1', \cdots, r'$ is any permutation of $1, \cdots, r$. Then the β are obviously linked to the α's by the following relation:

$$\alpha_{r_1 \cdots r_n} = \frac{r!}{r_1! \cdots r_n!} \beta(i_1, \cdots, i_r) \tag{1.11}$$

if r_1 of the r indices i_α are $= 1$, r_2 of them $= 2, \cdots, r_n$ of them $= n$.

(1.10) suggests introducing the multilinear form

$$f((x, y, \cdots, z)) = \sum \beta(i_1 i_2 \cdots i_r) x_{i_1} y_{i_2} \cdots z_{i_r} \tag{1.12}$$

depending on r sets of n variables:

$$\begin{aligned} x &= (x_1, \cdots, x_n), \\ y &= (y_1, \cdots, y_n), \\ &\cdots\cdots\cdots\cdots\cdots \\ z &= (z_1, \cdots, z_n). \end{aligned} \tag{1.13}$$

From it we fall back upon the form (1.10) by identifying

$$x = y = \cdots = z \quad \text{with} \quad x. \tag{1.14}$$

Symmetry of the coefficients β with respect to the indices i_α is equivalent to the symmetry of the multilinear form $f(x, y, \cdots, z)$ with respect to permutation of the r sets $x, y, \cdots, z$. Hence our result may be expressed thus: there exists a uniquely determined symmetric multilinear form $f(x, y, \cdots, z)$ which by the identification (1.14) passes into a given form $f((x))$ of degree r.

On putting $\lambda = 1 + t$ in (1.9) one finds that the polarized form $D_{yx}f$ changes back into $r \cdot f$ if y is replaced by x:

$$\{D_{yx}f(x)\}_{y=x} = r \cdot f(x).$$

The same is clear from (1.10) which, under the assumption of symmetric β's, at once yields

$$D_{yx}f = r \cdot \sum_{i_1, \cdots, i_r} \beta(i_1, i_2, \cdots, i_r) y_{i_1} x_{i_2} \cdots x_{i_r}.$$

Hence the symmetric multilinear form $f(x, y, \cdots, z)$ corresponding to the given form $f(u)$ of degree r arises from $f = f(u)$ by *complete polarization*:

$$D_{xu} D_{yu} \cdots D_{zu} f(u) = r! \sum \beta(i_1, i_2, \cdots, i_r) x_{i_1} y_{i_2} \cdots z_{i_r}.$$

This again shows its uniqueness.

2. Vector space

The next fundamental concept on which we must come to a common understanding right at the beginning is that of *vector space (in k)*. A vector space P is a *k-linear set* of elements, called vectors; i.e. a domain in which addition of vectors and multiplication of a vector by a number in k are the permissible operations, satisfying the well-known rules of vector geometry.[1] n vectors $\mathfrak{e}_1, \cdots, \mathfrak{e}_n$ form a *coordinate system* or a *basis* if they are linearly independent, while enlargement of the sequence by any further vector would destroy this independence. Under these assumptions every vector $\mathfrak{x}$ is uniquely expressible in the form

$$\mathfrak{x} = x_1 \mathfrak{e}_1 + \cdots + x_n \mathfrak{e}_n \tag{2.1}$$

where the numbers x_i are the "*components*" of $\mathfrak{x}$. The number n, which does not depend on the choice of the coordinate system, is called the *dimensionality of the vector space* P or the *order of the linear set* P. Transition to another coordinate system $\mathfrak{e}_1', \cdots, \mathfrak{e}_n'$ is effected by a non-singular linear transformation A as described by the matrix $\| a_{ik} \|$ in the following manner:

$$\mathfrak{x} = x_1\mathfrak{e}_1 + \cdots + x_n\mathfrak{e}_n = x_1'\mathfrak{e}_1' - \cdots + x_n'\mathfrak{e}_n' ;$$

$$\mathfrak{e}_i' = \sum_k a_{ki}\mathfrak{e}_k , \qquad x_i = \sum_k a_{ik}x_k' \quad (i, k = 1, \cdots, n). \tag{2.2}$$

A non-singular matrix $A = \| a_{ik} \|$ is one whose determinant, det A or $| A |$, is different from 0; the inverse transformation A^{-1} sends the column of n numbers x' back into the column x. On writing the components in a column (matrix of n rows and one column), (2.2) lends itself to the abbreviation

$$x = Ax' \tag{2.3}$$

in terms of matrix calculus.

There is another interpretation of this, or rather of the modified equation $x' = Ax$, to the effect that it describes *a linear mapping* $\mathfrak{x} \rightarrow \mathfrak{x}'$ *of* P *upon itself* in terms of a fixed coordinate system. In that case we need not suppose A to be non-singular. A mapping $\mathfrak{x} \rightarrow \mathfrak{x}'$ carrying each vector $\mathfrak{x}$ into a vector $\mathfrak{x}'$ is linear if it sends

$$\mathfrak{x} + \mathfrak{y} \ \text{ into } \ \mathfrak{x}' + \mathfrak{y}' \ \text{ and } \ \alpha\mathfrak{x} \ \text{ into } \ \alpha\mathfrak{x}'$$

(α any number in k). If such a correspondence changes the basic vector $\mathfrak{e}_i$ of our coordinate system into

$$\mathfrak{e}_i' = \sum_k a_{ki}\mathfrak{e}_k ,$$

it will carry

$$\mathfrak{x} = \sum x_i\mathfrak{e}_i \ \text{ into } \ \mathfrak{x}' = \sum_i x_i\mathfrak{e}_i' = \sum x_i'\mathfrak{e}_i ,$$

where

$$x_i' = \sum_k a_{ik}x_k \quad \text{or} \quad x' = Ax. \tag{2.4}$$

The identical mapping $\mathfrak{x} \rightarrow \mathfrak{x}$ is represented by the *unit matrix*

$$E_n = E = \| \delta_{ik} \| .$$

When we express a given linear mapping $\mathfrak{x} \rightarrow \mathfrak{x}'$, (2.4), in another coordinate system in which the vector x has the components y given by

$$x = Uy, \tag{2.5}$$

U being the non-singular transformation matrix, the result will be

$$y' = (U^{-1}AU)y,$$

as one readily derives from (2.5) combined with

$$x' = Uy' \quad \text{or} \quad y' = U^{-1}x'.$$

Hence the matrix A changes into $U^{-1}AU$ which arises from A, as we shall say, by "transformation with U." Therefore the *characteristic polynomial*

$$| \lambda E - A | = \lambda^n - b_1\lambda^{n-1} + \cdots \pm b_n$$

of the indeterminate λ is independent of the coordinate system, in particular the *trace* b_1,

$$\text{tr}(A) = \sum_i a_{ii},$$

and the determinant

$$b_n = \det A.$$

A square matrix A of n rows and columns is said to have *degree* n; the same term applies to any *set* $\mathfrak{A} = \{A\}$ of n-rowed matrices A.

In the *algebraic model of the n-dimensional vector space* a vector simply means a sequence $\mathfrak{x}$ of n numbers:

$$\mathfrak{x} = (x_1, \cdots, x_n).$$

The numbers are the coordinates of $\mathfrak{x}$ with respect to the "*absolute coordinate system*":

$$\mathfrak{e}_1 = (1, 0, \cdots, 0),$$
$$\mathfrak{e}_2 = (0, 1, \cdots, 0),$$
$$\cdots\cdots\cdots\cdots\cdots$$
$$\mathfrak{e}_n = (0, 0, \cdots, 1).$$

What our considerations have shown is the simple fact that every n-dimensional vector space in the general abstract axiomatic sense is isomorphic to this unique algebraic model.

A *linear form* $f(\mathfrak{x})$ depending on an argument vector $\mathfrak{x}$ may be defined without reference to a coordinate system by the functional properties:

$$f(\mathfrak{x} + \mathfrak{x}') = f(\mathfrak{x}) + f(\mathfrak{x}'), \qquad f(\alpha\mathfrak{x}) = \alpha \cdot f(\mathfrak{x}) \qquad (\alpha \text{ a number}).$$

Its expression in terms of a coordinate system will be a linear form of the components x_i of $\mathfrak{x}$ in the algebraic sense:

$$f(\mathfrak{x}) = \alpha_1 x_1 + \cdots + \alpha_n x_n$$

with constant coefficients α_i. Hence we know what a *multilinear form* $f(\mathfrak{x}, \cdots, \mathfrak{z})$ is, depending on r argument vectors $\mathfrak{x}, \cdots, \mathfrak{z}$. By the identification $\mathfrak{x} = \mathfrak{y} = \cdots = \mathfrak{z}$ it leads to a form $f(\mathfrak{x})$ of degree r of a single argument vector $\mathfrak{x}$. In the same manner $f(\mathfrak{x})$ arises from each of the forms $sf(\mathfrak{x}, \cdots, \mathfrak{z})$ into which $f(\mathfrak{x}, \cdots, \mathfrak{z})$

changes by a permutation s of its r arguments $\mathfrak{x}, \mathfrak{y}, \cdots, \mathfrak{z}$, and hence in particular from the symmetric form

$$\frac{1}{r!} \sum_s sf(\mathfrak{x}, \cdots, \mathfrak{z});$$

the sum here extends to all $r!$ permutations s. It is clear how to tie these remarks up with the considerations at the end of the first section concerning algebraic forms. It is much easier to describe what a form of degree r is in a manner independent of the coordinate system, by passing through the corresponding multilinear forms. The definition of the polarized form by means of the expansion of $f(\mathfrak{x} + t \cdot \mathfrak{y})$ in terms of the parameter t shows that the polar process is invariant under any change of coordinates. A natural generalization is the study of forms $f(\mathfrak{x}, \mathfrak{y}, \cdots)$ depending on various argument vectors $\mathfrak{x}, \mathfrak{y}, \cdots$ with pre-assigned degrees $\mu, \nu, \cdots$. When we stick to the algebraic model of vector space our independent definitions prove that a form $f((x)) = f(x_1 \cdots x_n)$ of degree r changes into a similar form by any linear transformation (2.3); and the same is true of a form depending on various rows (1.13) each of which undergoes the same transformation (2.3) as x.

Within the n-dimensional vector space P there may be defined an m-dimensional linear subspace P_1, ($m \leqq n$). A coordinate system $\mathfrak{e}_1 \cdots \mathfrak{e}_m$ in P_1 can be supplemented by $n - m$ further vectors $\mathfrak{e}_{m+1}, \cdots, \mathfrak{e}_n$ to form a basis in P. With respect to this basis *adapted to* P_1 the vectors x in P_1 are those whose last $n - m$ components vanish:

$$x = (x_1, \cdots, x_m, 0, \cdots, 0).$$

Hence the universal algebraic model for this situation is described thus: the vectors in P are the n-uples $(x_1 \cdots x_n)$, the vectors of the subspace P_1 are the n-uples of the particular form $(x_1 \cdots x_m 0 \cdots 0)$. P mod P_1 is the $(n - m)$-dimensional vector space into which P turns if one identifies any two vectors $\mathfrak{x}$ and $\mathfrak{x}'$ which are congruent modulo P_1, i.e. whose difference lies in P_1.

P is *decomposed* into two linear subspaces

$$P = P_1 + P_2$$

if each vector $\mathfrak{x}$ splits into a sum $\mathfrak{x}_1 + \mathfrak{x}_2$ of a vector $\mathfrak{x}_1$ in P_1 and $\mathfrak{x}_2$ in P_2 in unambiguous fashion. Uniqueness is assured if the only decomposition of 0:

$$0 = \mathfrak{x}_1 + \mathfrak{x}_2 \qquad (\mathfrak{x}_1 \text{ in } P_1, \mathfrak{x}_2 \text{ in } P_2)$$

is $0 + 0$, or if the two spaces are linearly independent (have no common vector except 0). A basis for P_1 together with a basis for P_2 forms a coordinate system for the whole space P (*adaptation of the coordinates to a given decomposition*); hence the sum of the dimensionalities $n_1 + n_2$ of P_1 and P_2 equals n. Relative to the adapted coordinate system, the vectors of P_1 and P_2 have the form

$$(x_1 \cdots x_{n_1}, 0 \cdots 0), \qquad (0 \cdots 0, x_{n_1+1} \cdots x_n).$$

The word "sum" will occasionally be used (but never the word "decomposition") even if unicity or linear independence does not prevail. The process of summation of (independent) parts may easily be extended to more than two summands:

$$P = P_1 + \cdots + P_t ; \qquad \mathfrak{x} = \mathfrak{x}_1 + \cdots + \mathfrak{x}_t , \qquad \mathfrak{x}_\alpha \text{ in } P_\alpha ;$$

independence meaning that $0 + \cdots + 0$ is the only decomposition of 0 into components lying in the subspaces P_α.

In case a linear mapping A carries every vector in the subspace P_1 into a vector of the same subspace, we call P_1 *invariant under* A, and the mapping A of P_1 upon itself is called the transformation "*induced*" in P_1 by A. If the coordinate system is adapted to the subspace P_1 the matrix A has the form (2.6),

$$\text{(2.6)} \quad \begin{Vmatrix} A_1 & * \\ 0 & A_2 \end{Vmatrix} \qquad\qquad \text{(2.7)} \quad \begin{Vmatrix} A_1 & 0 \\ 0 & A_2 \end{Vmatrix}$$

where the matrix A_1 of degree m is that induced by A in the subspace, while A_2 may be interpreted as the corresponding transformation of the "projected" space $P \bmod P_1$. In case P breaks up into two subspaces $P_1 + P_2$ both invariant under A, the matrix A has the form (2.7) in terms of a basis of P adapted to that decomposition.

One knows how matrices of a given degree n may be added, multiplied by numbers and among each other; the multiplication is associative but not commutative. The *transposed matrix* of $A = \| a_{ik} \|$ shall be denoted by

$$A^* = \| a^*_{ik} \| , \qquad a^*_{ik} = a_{ki} .$$

It is the matrix of the substitution

$$\xi' = \xi A$$

where ξ denotes a *row* of numbers $(\xi_1 , \cdots , \xi_n)$. A column x of n numbers may be called a *covariant*, a row ξ a *contravariant vector*. From them we may form the *product*

$$\xi x = \xi_1 x_1 + \cdots + \xi_n x_n = (\xi x)$$

which is a one-rowed square matrix or a number. If under the influence of transition to a new coordinate system the x undergo a non-singular transformation $x = Uy$, the ξ shall be subjected to the contragredient transformation

$$\xi = \eta U^{-1}$$

so that ξx remains unchanged:

$$\xi x = \eta U^{-1} U y = \eta y.$$

We consider covariant and contravariant vectors as vectors in two different "*dual*" *spaces* P and P^*. A change of coordinates in one shall be automatically connected with the contragredient change in the other space, so that the product

ξx has an invariantive significance. The mapping $x \to x' = Ax$ is in invariant manner tied up with the mapping $\xi \to \xi' = \xi A$ in the dual space:

$$\xi' x = \xi A x = \xi x',$$

i.e. the product of ξ' with x equals the product of ξ with the image x' of x.

3. Orthogonal transformations, Euclidean vector geometry

An *orthogonal transformation* (2.4) is one leaving invariant the quadratic form

$$x^* x = x_1^2 + \cdots + x_n^2 \; (= x_1'^2 + \cdots + x_n'^2). \tag{3.1}$$

This amounts to the equation

$$A^* A = E \tag{3.2}$$

for A, from which follows at once

$$A A^* = E \tag{3.3}$$

since the relation of a matrix A and its inverse A^{-1} is mutual. Another way of putting it is to say that an orthogonal matrix is identical with its contragredient. As for the determinant, it follows from (3.2) that its square $= 1$, hence

$$\det A = +1 \text{ or } -1.$$

According to these two cases one speaks of a *proper* or an *improper orthogonal transformation.*

Let A_{ik} be $(-1)^{i-k}$ times the determinant of the matrix A after the ith row and k^{th} column have been cancelled. The familiar identities

$$\sum_{r=1}^{n} a_{ir} A_{kr} = \det A \quad \text{or} \quad 0$$

according as $i = k$ or $i \neq k$, show that for a non-singular A the quotient $A_{ki}/\det A$ is the (ik)-element of the inverse A^{-1}; hence the contragredient matrix is

$$\hat{A} = \| A_{ik}/\det A \|.$$

We therefore have

$$A_{ik} = \pm a_{ik} \tag{3.4}$$

according as A is a proper or an improper orthogonal transformation.

The *minor*

$$\begin{vmatrix} a_{i_1 k_1}, & \cdots, & a_{i_1 k_p} \\ \cdots & \cdots & \cdots \\ a_{i_p k_1}, & \cdots, & a_{i_p k_p} \end{vmatrix}$$

may be denoted by

$$a_{i_1 \cdots i_p, k_1 \cdots k_p}.$$

In the theory of determinants one proves the following identities between the minors of $\| a_{ik} \|$ and those of $\| A_{ik} \|$:

$$A_{i_1 \cdots i_\rho, k_1 \cdots k_\rho} = (\det A)^{\rho - 1} \cdot a_{\iota_1 \cdots \iota_\sigma, \kappa_1 \cdots \kappa_\sigma} . \tag{3.5}$$

Here $\rho + \sigma = n$ and

$$i_1 \cdots i_\rho \iota_1 \cdots \iota_\sigma , \qquad k_1 \cdots k_\rho \kappa_1 \cdots \kappa_\sigma$$

stand for two even permutations of the figures $1, 2, \cdots, n$. In particular

$$\det(A_{ik}) = \{\det(a_{ik})\}^{n-1}.$$

For an orthogonal matrix A we combine (3.5) with (3.4) and find:

$$a_{i_1 \cdots i_\rho, k_1 \cdots k_\rho} = \pm\, a_{\iota_1 \cdots \iota_\sigma, \kappa_1 \cdots \kappa_\sigma} , \tag{3.6}$$

the upper sign holding again for the proper and the lower for the improper transformations. These simple formal relations will later on be useful.

Everybody is familiar with the part the orthogonal transformations play in the most fundamental—the *Euclidean*—geometry, where, after the choice of the unit of length, foot or meter, each vector $\mathfrak{x}$ has a certain *length* the square of which is given by a positive definite quadratic form $(\mathfrak{x}\mathfrak{x})$ in $\mathfrak{x}$. The corresponding symmetric bilinear form is the *scalar product* $(\mathfrak{x}\mathfrak{y})$. The condition $(\mathfrak{x}\mathfrak{y}) = 0$ means that $\mathfrak{x}$ and $\mathfrak{y}$ are *perpendicular*. An *orthogonal* or *Cartesian coordinate system* $\mathfrak{e}_1, \cdots, \mathfrak{e}_n$ is one in which $(\mathfrak{x}\mathfrak{x})$ has the normal form

$$(\mathfrak{x}\mathfrak{x}) = x_1^2 + \cdots + x_n^2 ,$$

or, in other words, one such that

$$(\mathfrak{e}_i \mathfrak{e}_k) = \delta_{ik} .$$

All Cartesian coordinate systems are equivalent in Euclidean geometry; the transition from one Cartesian coordinate system to another is accomplished by means of an orthogonal transformation; according as it is proper or improper the two systems are of equal or opposite "*orientation.*" A linear mapping $\mathfrak{x} \rightarrow \mathfrak{x}' = A\mathfrak{x}$ leaving unchanged the lengths of vectors is expressed in terms of a Cartesian coordinate system by an orthogonal matrix A; in case A be proper we have to do with a "rotation."

Again and again in Euclidean geometry one has to construct a Cartesian coordinate system in the following inductive way. One first chooses (or is given) a vector $\mathfrak{a} \neq 0$. By the positive normalizing factor

$$\alpha = 1/\sqrt{(\mathfrak{a}\mathfrak{a})}$$

one changes $\mathfrak{a}$ into a vector $\mathfrak{e}_1 = \alpha\mathfrak{a}$ of unit length, and takes $\mathfrak{e}_1$ as the first basic vector. Thereupon one chooses an arbitrary vector $\mathfrak{x} \neq 0$ perpendicular to $\mathfrak{e}_1$:

$$(\mathfrak{x}\mathfrak{e}_1) = 0,$$

and then takes the "normalized" $\mathfrak{x}$ as the second element $\mathfrak{e}_2$ of our basis. The next step requires the solution of the two simultaneous homogeneous linear equations

$$(\mathfrak{x}\mathfrak{e}_1) = 0, \qquad (\mathfrak{x}\mathfrak{e}_2) = 0,$$

while at the last instant there are $n - 1$ such equations to solve:

$$(\mathfrak{x}\mathfrak{e}_1) = 0, \cdots, (\mathfrak{x}\mathfrak{e}_{n-1}) = 0.$$

According to the general theory of linear equations, our equations always have solutions $\mathfrak{x} \neq 0$ because throughout the whole process their number remains lower than the number n of the unknown components of $\mathfrak{x} = (x_1, \cdots, x_n)$. To this construction we sometimes refer as the *classical inductive construction of a Cartesian system of axes.*

In our previous remarks about Euclidean or orthogonal vector geometry we had in mind as reference field k that field which one ordinarily uses for all geometric and physical measurements: the *field* K *of all real numbers.* However on analyzing the last construction one realizes that it merely requires the field k to satisfy the two conditions:

1) A square sum $\alpha^2 + \beta^2 + \cdots + \gamma^2$ is never zero unless all the individual terms $\alpha, \beta, \cdots, \gamma$ are zero.

2) The Pythagorean equation

$$\alpha^2 + \beta^2 = \gamma^2$$

has a solution γ for any two given numbers α and β.

The second hypothesis allows us to express any square sum as a square; indeed a finite square sum

$$\alpha_1^2 + \alpha_2^2 + \alpha_3^2 + \cdots$$

may be added up step by step:

$$\alpha_1^2 + \alpha_2^2 = \alpha_{12}^2, \qquad \alpha_{12}^2 + \alpha_3^2 = \alpha_{123}^2, \cdots.$$

A field satisfying the first condition is called a *real field*; when the second condition also holds we speak of a *Pythagorean field.* The role of this condition in geometry is that it enables us to lay off a given segment on a given straight line.[2]

4. Groups. Klein's Erlanger program. Quantities

The set $GL(n)$ of all non-singular linear transformations in an n-dimensional vector space P, the set $SL(n)$ of all unimodular linear transformations (whose determinant $= 1$), the set $O(n)$ of all orthogonal transformations, and the set $O^+(n)$ of all proper orthogonal transformations are *groups.* And this notion of a group is the third pillar on which our edifice is to be erected. In any point field, to wit a given set of elements p called points, we can study *one-to-one corre-*

spondences $S: p \to p'$. E means the identity, S^{-1} the inverse correspondence $p' \to p$, and two correspondences

$$S: p \to p', \qquad T: p' \to p''$$

combine to form a *composite*

$$TS: p \to p''.$$

A *group* Γ is a set of correspondences containing the identity E, the inverse S^{-1} of any S in Γ and the composite TS of any two correspondences S and T in Γ. Considered as an *abstract group* γ, our set Γ consists of elements s (of irrelevant nature) for which a composition st is defined satisfying the three rules:

1) the associative law $(st)u = s(tu)$;
2) there is a unit element I such that $\mathsf{I}\, s = s\, \mathsf{I} = s$ for all s;
3) every element s has an inverse s^{-1}, $ss^{-1} = s^{-1}s = \mathsf{I}$.

When we turn to the abstract standpoint we shall always change the capitals like Γ, S, $\cdots$ into the corresponding lower case types γ, s, $\cdots$. The given transformation group Γ is a *faithful realization* of the abstract group scheme γ. A *realization of* γ is given if with every element s there is associated a one-to-one correspondence S: $s \to S$ such that

$$\mathsf{I} \to E, \qquad s^{-1} \to S^{-1}, \qquad ts \to TS;$$

it is *faithful* provided different elements s are associated with different S. *Every group* γ *in the abstract sense is capable of a faithful realization the point field of which is the group manifold* γ *itself*; this is accomplished by associating with the element a the "translation" in γ:

$$(a): s' = as \qquad \text{(with the inversion } s = a^{-1}s'\text{)} \tag{4.1}$$

(*regular realization*). A realization by means of linear substitutions in an n-dimensional vector space is called a *representation of degree n*.

This is not the place for repeating the string of elementary definitions and propositions concerning groups which fill the first pages of every treatise on group theory.[3] Following Klein's Erlanger program[4] (1872) we prefer to describe in general terms the significance of groups for the idea of *relativity*, in particular in geometry. Take Euclidean point space as an example. With respect to a Cartesian frame of reference $\mathfrak{f}$ each point p is represented by its coordinate $x = (x_1, x_2, x_3)$, a column of three real numbers. (On purpose I deviate from the common usage in calling the entire symbol (x_1, x_2, x_3) a coordinate, in the singular.) The coordinates are objectively individualized reproducible symbols, while the points are all alike. There is no distinguishing objective property by which one could tell apart one point from all the others; fixation of a point is possible only by a demonstrative act as indicated by terms like "this," "here." All Cartesian frames (of reference) are equally admissible; any objective geometric property possessed by one of them is shared by all others. The coordinates x, x' of the same arbitrary point p in two such frames are linked by a transformation S:

$$x_i' = a_i + \sum_k a_{ik}x_k \qquad (i, k = 1, 2, 3) \tag{4.2}$$

where the non-singular $A = \| a_{ik} \|$ satisfies a relation

$$A^*A = aE \tag{4.3}$$

with a number a (that accounts for the arbitrariness of the yardstick). *Each* such transformation (4.2) effects a transition from a given Cartesian frame to another one. At the same time S may be interpreted as the expression of a similarity mapping $p \rightarrow p' = \sigma p$ of the point space upon itself in terms of one given Cartesian frame. The group of all these transformations or automorphisms of space describes the exact kind of relativity inherent in the point space, its exact degree of homogeneity. For instance, the group characteristic of Euclidean geometry tells us that all points are alike, and at a given point all directions, and so on.

In affine point space the restriction (4.3) is replaced by the weaker one $\det A \neq 0$, or, if one takes the term affine in the sense in which it was first introduced by Euler where it included preservation of volume, $\det A = 1$. With these and other examples, in particular the projective geometry, before his eyes, Klein advanced the principle that any group of transformations may serve as the group of automorphisms and that it defines the kind of geometry we are dealing with. Let us distinguish two closely related ideas, (1) automorphism and (2) coordinatization (*sit venia verbo!*).

1) *Automorphisms.* Leibnitz declared: Two figures are similar or equivalent if they cannot be distinguished from each other when each is considered by itself, because they have every imaginable property of objective meaning in common.[5] Leibnitz thus exhibited the true general meaning of similitude. If geometry is based on a system of axioms one might describe the objective properties and relations as those logically defined in terms of the undefined fundamental geometrical concepts entering into the axioms. A similarity mapping or automorphism is a one-to-one correspondence $\sigma : p \rightarrow p' = \sigma p$ among the points p of the space which changes every figure into a similar one, or which does not destroy any objective point relation. (The mapping $p \rightarrow p'$ does not destroy the relation $R(p, q, \cdots)$ if the images p', q', $\cdots$ fulfill the relation R whenever the original points p, q, $\cdots$ do.) The automorphisms necessarily form a *group*, because each figure is equivalent to itself and equivalence is a symmetric and transitive relation. The group axioms are exactly the formal expression for these trivial facts. The mathematician unwilling to draw on any external truth will be inclined to take the view that any group whatsoever can be appointed as the group of automorphisms; he declares by this appointment or convention that he is going to study only such relations among points as are not destroyed by the mappings of his group.

2) *Coordinatization.* With respect to a frame $\mathfrak{f}$ the point space is mapped upon a field of reproducible symbols or coordinates:

$$p \to x \quad \text{or} \quad x = x(p).$$

(The word "field" is here used in the loose sense of a range of variability.) We suppose that the coordinatization sets up a one-to-one correspondence between the points p and the coordinates x. There is no objection to regarding as the frame of reference this coordinatization itself. By means of an automorphism

$$\sigma\colon p \to p' = \sigma p$$

we can define a new coordinatization

$$x'(p) = x(p') = x(\sigma p)$$

which is *equivalent* to the first one and in no way objectively distinguishable from it. Both are linked by the transformation S,

$$x' = S(x) \qquad \{x = x(p), \qquad x' = x(p')\},$$

which describes the automorphism σ in terms of the coordinate system x. The transformations S expressing the several automorphisms σ in terms of the given frame $\mathfrak{f}$ form a group isomorphic to the group of automorphisms. At the same time the group of the S describes the transitions between the various equivalent frames. The utmost we can hope for is to define objectively a *class of equally admissible frames* such that any two frames within that class will be equivalent. This is *the relativity problem*: *to fix objectively a class of equivalent coordinatizations and to ascertain the group of transformations S mediating between them.* (The individual transformation function S is, just like the coordinate x itself, a reproducible symbol.)

However, not only points are required to be represented by reproducible symbols, but also every other kind of geometric entity, and when passing to physics all sorts of physical quantities like velocities, forces, field strengths, wave functions, and what not, expect a similar symbolic treatment. One often acts as though once the points have been submitted to it by fixing a frame of reference for them, all these other things will follow suit without necessitating further provisions. This is certainly not quite true; at least further units of measurement have to be fixed at random so as to make the scheme of reference complete. Without prejudicing the situation beforehand we may then talk of a frame of reference which takes care of all sorts of entities, while the law of transformation for the symbols describing a given sort of entity (points or electromagnetic field strengths) relatively to the frames depends on the particular entity under consideration. The group of automorphisms will then be an abstract rather than a transformation group. This seems to be a natural step beyond Klein's own formulation of his program. The abstract group characterizes the "*geometry*" in Klein's sense while the *type* of a variable quantity in that geometry is characterized by its transformation law. Each element s of the abstract group describes the transition from one frame to another. The transformation law states how the symbol or coordinate representing any arbitrary value of the quantity under consideration with respect to a frame $\mathfrak{f}$ changes under transition to another frame $\mathfrak{f}'$ by means of s; it is therefore a realization of the abstract group through transformations in the field of coordinates. I now give the systematic axiomatic formulation in which (1) refers to the "geometry" or "space" as such, (2) to a particular quantity in it.

A. *The "symbolic" part* (dealing with group elements and coordinates).

(1) Let there be given a set γ of elements called *group elements*. Each pair s, t of group elements shall give rise to a composite element ts. There shall be a unit element I satisfying $\mathsf{I}s = s\mathsf{I} = s$ and an inverse s^{-1} for each group element s: $s^{-1}s = ss^{-1} = \mathsf{I}$. (The associative law is not explicitly required.)

(2) Let there be given a set (or "field") of elements called coordinates x, and a realization $\mathfrak{A}$: $s \to S$ of the group γ by means of one-to-one correspondences $x \to x' = Sx$ within that field.

B. *The "geometric" part* (dealing with frames and quantities).

(1) Any two *frames* $\mathfrak{f}$, $\mathfrak{f}'$ determine a group element s, called the *transition* from $\mathfrak{f}$ to $\mathfrak{f}'$. Vice versa, a group element s "carries" a frame $\mathfrak{f}$ into a uniquely determined frame $\mathfrak{f}' = s\mathfrak{f}$ such that the transition $(\mathfrak{f} \to \mathfrak{f}') = s$. The transition $\mathfrak{f} \to \mathfrak{f}$ is the unit element I, the transition $\mathfrak{f}' \to \mathfrak{f}$ the inverse element. If s, t are the transitions $\mathfrak{f} \to \mathfrak{f}'$, $\mathfrak{f}' \to \mathfrak{f}''$ respectively, then the composite ts is the transition $\mathfrak{f} \to \mathfrak{f}''$.

(2) A quantity q of the type $\mathfrak{A}$ is capable of different values. Relatively to an arbitrarily fixed frame $\mathfrak{f}$ each value of q determines a coordinate x such that $q \to x$ is a one-to-one mapping of the possible values of q on the field of coordinates. The coordinate x' corresponding to the same arbitrary value q in any other frame $\mathfrak{f}'$ is linked to x by the transformation $x' = Sx$ associated with the transition $(\mathfrak{f} \to \mathfrak{f}') = s$ by the given realization $\mathfrak{A}$.

For a better understanding we may add the following remarks. The connection between frames and group elements as established by B (1) is very similar to that between points and vectors in affine geometry.[6] The last axiom under B(1) entails the associative law for composition of group elements. The epistemologist will stress the fact that the objects under A, the group elements and the coordinates, are objectively individualized and reproducible symbols, while any two frames are, in Leibnitz's words, "indiscernable when each is considered by itself." They are introduced in order to make possible the fixation of the values of all sorts of quantities in our geometry by reproducible symbols. From a mathematical standpoint one ought to observe that the axioms B (1) involve in no way more than the axioms defining a group, so that the elements of *every* associative group may be considered as transitions between frames in an appropriate "geometry." Indeed, if a group γ is given, one may call each element s of γ at the same time a "frame" and define the transition from frame s to frame t as the group element ts^{-1}. Then our axioms linking group elements to frames are obviously fulfilled provided the group multiplication is associative. In the same manner, the mathematician will not hesitate to identify the values of the quantity q with their respective coordinates x, and the requirement that only such relations matter or have objective significance as stay unaltered when x is replaced by

$$x' = Sx \qquad (s \to S \text{ in } \mathfrak{A})$$

for every s will mean to him a mere convention by which he proclaims that he will study no other relations.

All this sounds general and abstract enough. Nevertheless our formulation B (2) is still too narrow for some important purposes since we have to consider the possibility that a single coordinate system will not be capable of covering the whole range of values of q. However, we are not going to dwell on such further generalizations; on the contrary, we shall from now on restrict ourselves to the particular case where the realization $\mathfrak{A}$ is a representation and the coordinate therefore any n-uple of numbers $(x_1, \cdots, x_n)$ in a given number field k (k-vector). The word "*quantity*" shall now be reserved for this case, and we once more repeat the definition under this limitation:[7]

A quantity q of type $\mathfrak{A}$ is characterized by a representation $\mathfrak{A}$ of γ in k: $s \to A(s)$ of a certain degree n. Each value of q relatively to a frame $\mathfrak{f}$ determines a k-vector $(x_1, \cdots, x_n)$ such that the "components" x_i of q transform under the transition s to another frame $\mathfrak{f}'$ according to $A(s)$.

Representations of degree 1 are representations by *numbers*:

$$s \to \lambda(s); \qquad \lambda(1) = 1, \qquad \lambda(st) = \lambda(s)\lambda(t).$$

The particular representation of degree 1 for which $\lambda(s) = 1$ identically in s: $s \to 1$, is called the *identical representation*; a quantity having this type is called a *scalar*.

We are now safely back in the waters of pure mathematics. The notions of *inequivalence, reduction,* and *decomposition* present themselves quite naturally in their application to a representation $\mathfrak{A}$ of γ or to a type $\mathfrak{A}$ of quantities. They are of even more general significance inasmuch as any set of matrices may replace the group $\mathfrak{A}$.

Let then $\mathfrak{A}$ be a set of linear transformations or matrices A in an n-dimensional vector space P. If one changes the basis of that space by means of a non-singular linear substitution U each A is changed into

$$A' = U^{-1}AU;$$

the A' form the *equivalent* set $\mathfrak{A}'(\sim\mathfrak{A})$.

$\mathfrak{A}$ is called *reducible* if P contains a linear subspace P′ invariant under all the transformations A of $\mathfrak{A}$ that is neither the whole space P nor the zero space consisting of the vector $0 = (0, \cdots, 0)$ only. When P is referred to a suitable coördinate system all matrices A are then of the form (2.6).

$\mathfrak{A}$ *decomposes* if P breaks up into two non-vanishing subspaces $P_1 + P_2$ invariant under all A's of $\mathfrak{A}$. In the coördinate system adapted to this decomposition $P_1 + P_2$ each matrix A has the form (2.7), which fact shall be indicated by

$$A = A_1 \dot{+} A_2 .$$

These definitions apply in particular to a group $\mathfrak{A}$: $s \to A(s)$ of matrices $A(s)$ homomorphic with the given group γ. If U is a fixed non-singular matrix the representation

$$s \to A'(s), \qquad A'(s) = U^{-1}A(s)U$$

is *equivalent* to the first one. We shall treat equivalent representations as one and the same, the difference lying only in the basis of the representation space in terms of which the linear operators are expressed in matrix form. The trace $\chi(s)$ of $A(s)$ is called the *character of the representation. Equivalent representations have the same character.* In case of *reduction* all $A(s)$ have, relatively to the adapted coördinate system, the form (2.6). The part $A_1(s)$ defines a representation $\mathfrak{A}_1$ of γ of degree m, and the part $(x_1, \cdots, x_m)$ of the quantity $(x_1, \cdots, x_n)$ is itself a quantity, of type $\mathfrak{A}_1$; we say that it is *contained* in the latter. If $\mathfrak{A}$ is *irreducible* the quantity itself is called *irreducible* or *primitive.* In case of *decomposition* (2.7) our quantity consists of the juxtaposition of two independent parts

$$(x_1, \cdots, x_{n_1}), \qquad (x_{n_1+1}, \cdots, x_n)$$

whose components transform only among themselves. Nothing prevents one from considering the electromagnetic four-potential together with the field strength as a single quantity of 10 components; but of course this is a rather artificial unity and it is much more natural to decompose it into its two independent parts of 4 and 6 components respectively, the potential and the field strength. It obviously is of paramount importance to know whether a quantity breaks up into a number of independent primitive partial quantities, i.e. whether a given representation $\mathfrak{A}$ may be split into irreducible constituents: is it true that a subspace P_1 of P invariant under the operators $A(s)$ of $\mathfrak{A}$ possesses a complementary invariant subspace P_2 such that the whole representation space P breaks up into the two linearly independent parts $P_1 + P_2$—and is this true for *any* representation $\mathfrak{A}$ of the given group γ? The answer is affirmative in the most important cases, in particular, as we shall see later, for all finite groups.

On our way we encountered the following process of *addition* by which two representations

$$\mathfrak{A}: s \to A(s) \quad \text{and} \quad \mathfrak{A}': s \to A'(s)$$

of the same group and of degrees m and n respectively give rise to the representation $\mathfrak{A} + \mathfrak{A}'$ of degree $m + n$:

$$s \to A(s) \dot{+} A'(s).$$

The quantity $(x_1, \cdots, x_m)$ of type $\mathfrak{A}$ is combined with the quantity $(y_1, \cdots, y_n)$ of type $\mathfrak{A}'$ to form the quantity

$$(x_1, \cdots, x_m, y_1, \cdots, y_n).$$

The character of $\mathfrak{A} + \mathfrak{A}'$ is the sum of the characters of $\mathfrak{A}$ and $\mathfrak{A}'$.

Another important procedure is that of *multiplication* $\mathfrak{A} \times \mathfrak{A}'$. If the vectors

$$x = (x_1, \cdots, x_m), \qquad y = (y_1, \cdots, y_n)$$

undergo the linear transformations

$$A = \| a_{ik} \|, \qquad A' = \| a'_{pq} \| \qquad \begin{pmatrix} i, k = 1, \cdots, m; \\ p, q = 1, \cdots, n \end{pmatrix}$$

respectively, then the mn products

$$x_i y_p \qquad (i = 1, \cdots, m; \quad p = 1, \cdots, n) \tag{4.4}$$

undergo a corresponding linear transformation $A \times A'$, called the *Kronecker product* of A and A'. In explicit form

$$C = A \times A' = \| c_{ip,kq} \|$$

is obviously given by

$$c_{ip,kq} = a_{ik} a'_{pq},$$

and our definition at once yields the law of composition

$$(A \times A')(B \times B') = (AB \times A'B').$$

$\mathfrak{A} \times \mathfrak{A}'$ is the representation

$$s \to (A(s) \times A'(s))$$

of degree mn. The same sign $\times$ will be applied to the corresponding quantities. The character of $\mathfrak{A} \times \mathfrak{A}'$ is the product of the character of $\mathfrak{A}$ with that of $\mathfrak{A}'$. The problem arises quite naturally to decompose the product of two primitive quantities into its primitive constituents, partial cases of which will be discussed later (Chapters IV and VII).

The numbers (4.4) may be considered as the components z_{ip} of a vector $z = x \times y$ in an mn-dimensional vector space PP′. When considering linear forms in that space it will often be convenient to replace the most general vector z with mn independent components by the vector $x \times y$ where x_i and y_p are independent variables; this procedure is called *the symbolic method* in invariant theory.

Let us call attention right here to some important representations of the full linear group $GL(n)$, by stressing the representation aspect of forms of n variables and their transformation under the influence of linear substitutions on the variables. Take any non-singular linear transformation A:

$$x_i' = \sum_k a_{ik} x_k . \tag{4.5}$$

Under its influence all monomials of given degree r,

$$x_1^{r_1} x_2^{r_2} \cdots x_n^{r_n} \qquad (r_1 + r_2 + \cdots + r_n = r),$$

undergo a linear transformation $(A)_r$, and this correspondence $A \to (A)_r$ is a representation of degree

$$n(n + 1) \cdots (n + r - 1)/1 \cdot 2 \cdots r.$$

On the other hand we consider an arbitrary form f of degree r depending on a contravariant argument vector $(\xi_1, \cdots, \xi_n)$ and write it as

$$f = \sum \frac{r!}{r_1! \cdots r_n!} u_{r_1 \cdots r_n} \xi_1^{r_1} \cdots \xi_n^{r_n}. \tag{4.6}$$

While ξ_i are transformed according to

$$\xi_i = \sum_k a_{ki} \xi_k' . \tag{4.7}$$

f changes into a form of the new variables ξ_i' whose coefficients $u'_{r_1 \cdots r_n}$ proceed from $u_{r_1 \cdots r_n}$ by the same substitution $(A)_r$ as encountered above. Indeed, the rth power of the invariant product

$$(\xi x) = \xi_1 x_1 + \cdots + \xi_n x_n$$

equals the special form (4.6) with

$$u_{r_1 \cdots r_n} = x_1^{r_1} \cdots x_n^{r_n}.$$

In the symbolic method one replaces the arbitrary form f by the specialized $(\xi x)^r$.

The products of the components of r vectors $x, y, \cdots, z$:

$$x_{i_1} y_{i_2} \cdots z_{i_r}$$

which are cogrediently transformed into vectors $x', y', \cdots, z'$ by the same A, (4.5), undergo the transformation $A \times A \times \cdots \times A$. A quantity F of this type with the n^r components $F(i_1, i_2, \cdots, i_r)$ is called *tensor* of rank r. According to (1.10) we write our form (4.6) as

$$\sum v(i_1 i_2 \cdots i_r) \xi_{i_1} \xi_{i_2} \cdots \xi_{i_r}$$

with symmetric coefficients $v(i_1 i_2 \cdots i_r)$, where

$$u_{r_1 \cdots r_n} = v(i_1 i_2 \cdots i_r)$$

if r_1 of the r indices i_α are $= 1$, r_2 of them $= 2$, and so on. Hence the manifold of forms of degree r is nothing else than the set of all *symmetric* tensors F of rank r—where symmetry means that $F(i_1 i_2 \cdots i_r)$ does not change its value under arbitrary permutations of the indices or arguments $i_1 i_2 \cdots i_r$. Within the tensor space of rank r the symmetric tensors form a linear subspace which is invariant under all (non-singular) transformations A. Another such invariant subspace consists of all skew-symmetric tensors whose components change sign under the influence of transposition of two arguments, e.g. i_1 and i_2:

$$F(i_2 i_1 i_3 \cdots i_r) = -F(i_1 i_2 i_3 \cdots i_r).$$

A special case of the problem mentioned above of decomposing the product of several primitive quantities into its primitive components is the splitting of tensor space into irreducible invariant subspaces. This will be treated in Chapter IV for the full linear, in Chapters V and VI for some other groups. Indeed, the $\times$-product of r (covariant) vectors is the quantity called arbitrary tensor of rank r.

Before closing this section I want to touch on a question concerning *relativity in Euclidean space* although it is but loosely connected with our interests here, on which however there has arisen much confusion among mathematicians and philosophers; I mean the question of the relation of *congruence* to the group of automorphisms of Euclidean space. If one bases geometry on a number of fundamental relations as "lies on", "between", "congruent", an automorphism is a correspondence not affecting any such relation. Leaving out first the relation of congruence one finds that an automorphism, as far as vectors are concerned, must be a *linear* vector transformation S. Supposing that congruence is established by means of proper orthogonal transformations A, the further condition on S, viz. not to destroy congruence, amounts to the requirement that S *commutes* with the whole group O^+ of congruences:

$$S^{-1}AS = A'$$

must be a proper orthogonal matrix whenever A is such. All linear transformations S satisfying this condition form a group, the so-called *normalizer* of O^+. The normalizer of a group comprises that group; in our case it is actually larger because it contains the "reflections" (improper orthogonal transformations) and dilatations, besides the "rotations". The group O^+ plays its intrinsic part in Euclidean geometry long before the "extrinsic" question of all automorphisms arises; its normalizer rather than O^+ itself is the group of automorphisms. What shall we say then to Kant's discussion of the distinction of "left" and "right" in §13 of the Prolegomena[8] where he claims that "by no single concept, but only by pointing to our left and right hand, and thus depending directly on intuition (Anschauung), can we make comprehensible the difference between similar yet incongruent objects (such as oppositely winding snails)." No doubt the meaning of congruence in space is based on intuition, but so is similitude. Kant seems to aim at some subtler point; but just this point is one which can be subsumed under the general "concept" of a group and its normalizer. While a group is, generally speaking, not derivable from its normalizer, there is nothing mysterious in the possibility that the normalizer may be actually larger than the group itself.

5. Invariants and covariants

Let there be given a group Γ of linear transformations A in an n-dimensional vector space P. A function $f(x, y, \cdots)$ depending on a number of argument vectors

$$x = (x_1, \cdots, x_n), \qquad y = (y_1, \cdots, y_n), \cdots \tag{5.1}$$

in P will change into a transform $f' = Af$ if $x, y, \cdots$ are sent by the linear transformation A into $x' = Ax, y' = Ay, \cdots$:

$$f'(x', y', \cdots) = f(x, y, \cdots).$$

We then have $B(Af) = (BA)f$, as it should be. If $Af = f$ for all substitutions A of our Γ, the function f is called an *invariant* of Γ. In this sense the scalar products (xx), (xy), $\cdots$ are orthogonal vector invariants. We shall be concerned with the *algebraic case* exclusively where f is a polynomial, homogeneous with respect to the components of each argument vector, and therefore is called an *invariant form*. The degrees $\mu, \nu, \cdots$ of f in $x, y, \cdots$ may coincide or not.

With this elementary notion we contrast the *general notion of invariant*. While the former is concerned with a given *group* Γ *of linear transformations* A, the latter is relative to a given *abstract* group $\gamma = \{s\}$ and a number of representations of γ of degrees $m, n, \cdots$ respectively:

$$\mathfrak{A}: s \to A(s), \qquad \mathfrak{B}: s \to B(s), \cdots. \tag{5.2}$$

A function $\varphi(x, y, \cdots)$ depending on an arbitrary quantity x of type $\mathfrak{A}$, another y of type $\mathfrak{B}$, $\cdots$, will be a certain function f of the numerical vectors

$$x = (x_1, \cdots, x_m), \qquad y = (y_1, \cdots, y_n), \cdots \tag{5.3}$$

in terms of a given frame of reference, and will be a certain function f' in another frame of reference in which the same argument quantities have the components

$$x' = (x'_1, \cdots, x'_m), \qquad y' = (y'_1, \cdots, y'_n), \cdots :$$

$$\varphi = f(x, y, \cdots) = f'(x', y', \cdots).$$

If s is the transition from the first to the second frame:

$$x' = A(s)x, \qquad y' = B(s)y, \cdots,$$

we denote the transform f' by sf. Our function φ is an *invariant* provided its algebraic expression f does not depend on the frame: $sf = f$ for all elements s of the group γ. If one is a mathematical purist and therefore wants to eliminate from the definition the "frames" and "quantities" in favor of the group elements and the numerical vectors, one will call the function $f(x, y, \cdots)$ of the numerical vectors $x, y, \cdots$ an invariant with respect to the given representations $\mathfrak{A}$, $\mathfrak{B}, \cdots$ of γ if

$$f(A(s)x, B(s)y, \cdots) = f(x, y, \cdots).$$

It is evident how to subsume the elementary case, the case of "vector invariants", under this more general scheme.

In the classical theory of invariants, γ is the special (or unimodular) linear group $SL(n)$ and the argument quantities $x, y, \cdots$ are arbitrary forms in n variables. We explained near the end of the preceding section how such forms are to be interpreted as quantities of a particular type. The usual viewpoint of the classical theory is slightly different in that it looks upon the variables and the coefficients of the forms as numbers whose *ratios only* matter because the values of the n variables are considered as homogeneous coordinates of a point in *projective* $(n - 1)$*-space* rather than as components of a vector in an affine n-space. The vanishing of a given form defines an algebraic spread of $n - 2$ dimensions; the vanishing of an invariant J depending on a number of arbitrary such forms defines a projectively invariant algebraic relationship between the corresponding spreads.

One particular extension of the elementary concept toward the general deserves special mention, the case when the arguments are a number of *covariant vectors* $x, y, \cdots$ and a number of *contravariant vectors* $\xi, \eta, \cdots$. While $x, y, \cdots$ are transformed cearediently according to any substitution A in the group Γ each of the vectors $\xi, \eta, \cdots$ undergoes the contragredient transformation $\hat{A}$. The product (ξx) is the most fundamental invariant of this type, both for the full linear group $GL(n)$ and for any subgroup Γ of it. The study of such functions $f(xy \cdots, \xi\eta \cdots)$ shall be included under the title of vector invariants, because here, as in the strictly elementary case, we refer to *one* given group Γ of *linear substitutions* rather than to an *abstract* group γ and a number of its representations. (If $x, y, \cdots$ are point coordinates in a projective $(n - 1)$-space, then $\xi, \eta, \ldots$ are plane coordinates and vice versa.)

In returning to the general case, let us fix our attention on the forms $f(x, y, \cdots)$ which are *of pre-assigned degrees* $\mu, \nu, \cdots$ in the argument quantities $x, y, \cdots$. These forms are linear combinations of the

$$(5.4) \qquad N = \frac{m(m+1)\cdots(m+\mu-1)}{1\cdot 2\cdots\mu}\cdot\frac{n(n+1)\cdots(n+\nu-1)}{1\cdot 2\cdots\nu}\cdots$$

monomials of the prescribed degrees,

$$z(\alpha_1 \cdots \alpha_m\beta_1 \cdots \beta_n \cdots) = x_1^{\alpha_1} \cdots x_m^{\alpha_m} y_1^{\beta_1} \cdots y_n^{\beta_n} \cdots$$
$$(\alpha_1 + \cdots + \alpha_m = \mu, \beta_1 + \cdots + \beta_n = \nu, \cdots).$$

The exponents $\alpha, \beta, \cdots$ of course are non-negative integers. While the n components x_i undergo the linear transformation $A(s)$, the y_k the transformation $B(s), \cdots$, our monomials undergo a certain compound transformation $U(s)$

which establishes a compound representation $\mathfrak{U}$: $s \rightarrow U(s)$ of degree N. The invariant forms $f(x, y, \cdots)$ are thus turned into *linear invariants of a single quantity* z of type $\mathfrak{U}$. However, one should emphasize the fact that this linearization of the problem of invariants is possible only if we study invariant forms of pre-assigned degrees $\mu, \nu, \cdots$.

The *linear invariants* $L(x)$ of a quantity $x = (x_1, \cdots, x_n)$ of given type $\mathfrak{A}$: $s \rightarrow A(s)$ form a subspace of the n-dimensional space of all linear forms of x. l being its dimensionality, we have exactly l linearly independent linear invariants. On taking them as the first l coordinates of an arbitrary x in a new coordinate system, we obtain a reduction of $\mathfrak{A}$ to the form

$$A(s) = \begin{Vmatrix} E_l & 0 \\ * & * \end{Vmatrix}.$$

l is the maximum degree with which the unit representation $s \rightarrow E_l$ is contained in $\mathfrak{A}$. When in particular the theorem of full reduciblity holds, we may describe l as the maximum number of times the identical representation $s \rightarrow 1$ is contained in $\mathfrak{A}$.

An invariant may be described as *a scalar depending on a number of arbitrary quantities* $x, y, \cdots$ *of prescribed types*. If the transform sf differs from f by a constant factor $\lambda(s)$,

$$sf = \lambda(s) \cdot f, \tag{5.5}$$

f is called a *relative invariant with the multiplier* $\lambda(s)$. $f = 0$ is still an invariant relation between the variable quantities $x, y, \cdots$. The invariants in the original sense are then to be distinguished as *absolute invariants*. The multiplier is a representation $s \rightarrow \lambda(s)$ of degree 1. Still more generally, a *covariant of type* $\mathfrak{H}$: $s \rightarrow H(s)$ is a quantity f of that type depending on arguments $x, y, \cdots$ which are quantities of given types $\mathfrak{A}, \mathfrak{B}, \cdots$ respectively. With regard to a given frame of reference, f will have h components

$$f_1 = f_1(x, y, \cdots), \cdots, f_h = f_h(x, y, \cdots),$$

just as $x, y, \cdots$ have the components (5.3). After transition s to another frame, the new components which arise from the old ones by the linear substitution $H(s)$ shall be the transforms $f'_1 = sf_1, \cdots, f'_h = sf_h$; hence on putting

$$f'_i(x', y', \cdots) = f_i(x, y, \cdots) \qquad (i = 1, \cdots, h)$$

with

$$x' = A(s)x, \qquad y' = B(s)y, \cdots,$$

the equation

$$f' = sf = H(s)f$$

is to hold. The system of simultaneous equations

$$f_1 = 0, \cdots, f_h = 0$$

then has an invariant significance, independent of the frame of reference.

As an illustration of relative invariants let us consider *the classical case* where γ is the full linear group $GL(n)$, consisting of the linear transformations

$$A: x'_i = \sum_k a_{ik} x_k ,$$

and where the quantities which appear as arguments in f are arbitrary forms of given degrees in the n variables ξ_i. Under these circumstances $\lambda(A)$ will be a homogeneous polynomial of the n^2 variables a_{ik} ; hence for the special transformation

$$x'_i = a x_i \tag{5.6}$$

the multiplier $\lambda(A)$ will equal a^G with a non-negative integral exponent G. On applying the relation

$$\lambda(A)\lambda(B) = \lambda(AB)$$

to the transformation A and that one $B = \| A_{ki} \|$ whose elements consist of the minors A_{ki} of A:

$$AB = BA = \Delta \cdot E$$

where $\Delta = | A |$ is the determinant of A, one finds

$$\lambda(A)\lambda(B) = \Delta^G \tag{5.7}$$

But since Δ is an *irreducible* polynomial of the n^2 variables a_{ik} and $\lambda(B)$ is a polynomial no less than $\lambda(A)$, (5.7) forces $\lambda(A)$ to be a power of Δ:

$$\lambda(A) = \Delta^g \tag{5.8}$$

The integral exponent g is called the *weight* of the relative invariant. On account of formula (5.8) relative invariants of the full linear group $GL(n)$ are absolute invariants of the unimodular group $SL(n)$.

On the background of these general notions concerning fields, vectors, groups, representations we now set out to study the algebraic vector invariants of the most important groups, in particular of the full and of the unimodular linear group, $GL(n)$ and $SL(n)$, and of the orthogonal group, $O(n)$ or $O^+(n)$, in n dimensions.

CHAPTER II

VECTOR INVARIANTS

1. Remembrance of things past

The *theory of invariants* originated in England about the middle of the nineteenth century as the genuine analytic instrument for describing configurations and their inner geometric relations in projective geometry. The functions and algebraic relations expressing them in terms of projective coördinates are to be invariant under all homogeneous linear transformations. *Cayley* first passed from the consideration of determinants to more general invariants. This procedure accounts for the title of his paper, Mémoire sur les Hyperdéterminants[1] (1846), which one may look upon as the birth certificate of invariant theory. In his later nine famous Memoirs on Quantics[2] (1854–1859) he succeeds, among other things, in obtaining a complete set of invariants for cubic and biquadratic forms. His work was taken up in England by Sylvester and Salmon. Sylvester taught at Johns Hopkins University for some years, and there founded the first mathematical journal on this continent: The American Journal of Mathematics. The pages of its first volumes are filled with papers on invariant theory from Sylvester's prolific pen. In Germany, Aronhold, Clebsch and Gordan became adherents and promoters of the new discipline. In Italy, Brioschi, Cremona, Beltrami, and Capelli were attracted to the subject. This early period has a formal character throughout: the development of formal processes and the actual computation of invariants stand to the fore. Almost all papers refer to one group, the continuous group of all homogeneous linear transformations.

Another impulse, in a somewhat different direction, came from number theory, more particularly from the arithmetic theory of binary quadratic forms. Here one had been led to consider not a continuous but a discrete group, the group of unimodular linear substitutions with integral coefficients. Gauss, in his Disquisitiones arithmeticae, studied equivalence of quadratic forms with respect to this group. Besides and after Gauss, we have Jacobi in Germany and Hermite in France as outstanding men in this line of investigation.

The formal period of classic invariant theory is followed by a more critical and conceptual one which solves the general problems of finiteness less by explicit computations than by developing suitable general notions and their general properties along such abstract lines as have lately come into fashion all over the whole field of algebra. Here there is only one man to mention—Hilbert. His papers (1890/92) mark a turning point in the history of invariant theory.[3] He solves the main problems and thus almost kills the whole subject. But its life lingers on, however flickering, during the next decades. A. Hurwitz makes a new and important contribution by introducing integral processes extending over the group manifold (1897); in England A. Young, working more or less alone in this field, obtains far-reaching results on the representations of the symmetric group and uses them for invariant-theoretic purposes (1900 and later). In recent times the tree of invariant theory has shown new life, and has begun to blossom again, chiefly as a consequence of the interest in invariant-theoretic questions awakened by the revolutionary developments in mathematical physics (relativity theory and quantum mechanics), but also due to the connection of invariant theory with the extension of the theory of representations to continuous groups and algebras.

The rise of projective geometry made such an overwhelming impression on the geometers of the first half of the nineteenth century that they tried to fit all geometric considerations into the projective scheme. The narrowing down of the projective group to the affine group or to the group of Euclidean motions of metric geometry was accordingly effected by adjoining some so-called "absolute" entities: the plane at infinity, the absolute involution. The same attitude is expressed when one treats metric geometry in vector space by allowing arbitrary affine coördinate systems and arbitrary linear transformations, while adding the fundamental metric form $x_1^2 + x_2^2 + \cdots + x_n^2$ as something absolute, instead of sticking to the metrically equivalent Cartesian coördinate systems only and the corresponding group of orthogonal transformations. As this procedure easily admits extension into infinitesimal geometry, it has remained in use with great success, particularly for the purpose of general relativity theory. In group theory it amounts to considering each group of linear transformations as a subgroup of and in relation to the total linear group. The dictatorial regime of the projective idea in geometry was first successfully broken by the German astronomer and geometer Möbius, but the classical document of the democratic platform in geometry establishing the group of transformations as the ruling principle in any kind of geometry and yielding equal rights of independent consideration to each and any such group, is F. Klein's Erlangen program. The adjustment of invariant theory to this standpoint has been slow; it could not be made without recognizing that the study of the groups themselves and their representations necessarily has to precede the study of their invariants.

Decisive for the development of the *theory of groups* was the use E. Galois (1832) made of groups of permutations for the investigation of algebraic equations; he recognized that the relation of an algebraic extension K to its ground field k is to a large extent determined by the group of automorphisms. His theory may be described as the algebraic relativity theory of a finite set of numbers which are given as the roots of an algebraic equation.[4] Galois's brief allusions remained for a long time a book of seven seals. Only by C. Jordan's Traité des Substitutions (1870) was the newly gained field opened up to a wider circle of mathematicians. The algebraic problems connected with the elliptic and modular functions—partition, transformation, complex multiplication—furnished the most important material for the new concepts. Going ahead in this direction F. Klein and H. Poincaré created the theory of automorphic functions. While Galois's theory deals with finite groups, infinite discrete groups here come to the fore. Crystallography became the motive for a detailed study of infinite discrete groups of motions.[5] S. Lie founded a general theory of continuous groups from the infinitesimal standpoint, and showed its importance by many applications to geometric questions and differential equations.[6]

The theory of *representations of groups by linear transformations* was created above all by G. Frobenius[7] during the years 1896–1903. Burnside, independent of him, and I. Schur in continuance of his work, found an essentially simpler approach by emphasizing the representing matrix itself rather than its trace, the Frobenius character. For Lie's infinitesimal groups E. Cartan demonstrated the fundamental propositions concerning structure and representations.[8] The matter is closely connected with hypercomplex number systems or algebras. After Hamilton's foundation of the quaternion calculus (1843), and a long period of more or less fomal work in which the name of B. Peirce is outstanding, Molien (1892) was really the first to win some general and profound results in this direction.[9] Of paramount importance for the modern development is Wedderburn's paper[10] of the year 1908, where he investigates associative algebras in an arbitrary number field k; also I. Schur's study of irreducible representations in an arbitrary number field (1909) should be mentioned as fundamental.[11] Since then the development has been pushed ahead, in America chiefly by L. Dickson's and A. A. Albert's efforts, in Germany through E. Noether and R. Brauer.

This brief enumeration of names must suffice here in place of a real history, as our link to the past.[12] The bibliography will help to round out the picture with respect to modern times.

2. The main propositions of the theory of invariants

It will be convenient, before going on, to illustrate the notion of vector invariant (Chapter I, §5) by two familiar examples, the symmetric group and the orthogonal group.

In the theory of algebraic equations one is led to consider symmetric functions $f(x_1, x_2, \cdots, x_n)$ of n arguments $x_1, \cdots, x_n$, i.e. functions invariant under the group π_n of all $n!$ possible permutations of the n arguments. These permutations are obviously linear transformations of the n-dimensional vector $x = (x_1, x_2, \cdots, x_n)$. The elementary symmetric functions $\varphi_1, \varphi_2, \cdots, \varphi_n$ are the coefficients of the polynomial $\Phi(t)$ of the indeterminate t,

$$(2.1) \qquad \Phi(t) = (t - x_1)(t - x_2) \cdots (t - x_n) = t^n - \varphi_1(x)t^{n-1} + \varphi_2(x)t^{n-2} - \cdots \pm \varphi_n(x),$$

whose roots are $x_1, x_2, \cdots, x_n$. Thus

$$\varphi_1(x) = \sum_i x_i,$$

$$\varphi_2(x) = \sum_{i<k} x_i x_k,$$

$$(2.2) \qquad \varphi_3(x) = \sum_{i<k<l} x_i x_k x_l,$$

$$\cdots\cdots\cdots\cdots\cdots$$

$$\varphi_n(x) = x_1 x_2 \cdots x_n.$$

The main fact concerning symmetric functions is that they can be expressed in terms of the elementary symmetric functions $\varphi_i(x)$. More explicitly, if $f(x_1, x_2, \cdots, x_n)$ is any symmetric function of n arguments $x_1, \cdots, x_n$, there exists a function $F(\xi_1, \xi_2, \cdots, \xi_n)$ of n arguments $\xi_1, \cdots, \xi_n$ such that

$$(2.3) \qquad f(x) = F(\varphi_1(x), \varphi_2(x), \cdots, \varphi_n(x)).$$

We say that the functions $\varphi_i(x)$, $(i = 1, 2, \cdots, n)$, form a functional basis for the symmetric functions. This is almost trivial if we take the notion of function in its widest scope; for then it simply states the fact that the values of the elementary symmetric functions $\varphi_1(x), \cdots, \varphi_n(x)$ determine the values of the arguments $x_1, \cdots, x_n$ uniquely but for their order. Indeed, the equation $\Phi(t) = 0$ uniquely determines the set of its roots. But if $f(x_1, \cdots, x_n)$ is a *polynomial* in $x_1, \cdots, x_n$ the question arises whether or not f is expressible in terms of the functions $\varphi_i(x)$, $(i = 1, 2, \cdots, n)$, in the same algebraic fashion; that is, does there exist a *polynomial* F for which (2.3) holds? The truth of this is asserted by the so-called fundamental theorem of symmetric functions: the functions $\varphi_i(x)$, $(i = 1, 2, \cdots, n)$, constitute an *integral rational basis*, or *integrity basis*, for the symmetric forms. The restriction of the hypothesis, namely that the given symmetric function is integral-rational, is thus counterbalanced by a corresponding narrowing of the inference: the functional expression F of f by means of the basis φ is also of integral rational nature. Thus the "algebraic theorem" referring to forms f is not a particular case of the "functional theorem" in which the functional dependence in f and F is understood in the widest possible sense; on the contrary, it is the algebraic theorem alone that needs an elaborate proof.

A similar situation prevails in many cases. "All invariants are expressible in terms of a finite number among them": this so-called first main theorem of invariant theory is suggested by our present example. We cannot claim its validity for every group γ; rather, it will be our chief task to investigate for each particular group whether a finite integrity basis exists or not; the answer, to be sure, will turn out affirmative in the most important cases. In those cases, and here is the point I wish to emphasize, one will find the purely functional part—asserting that the values of all invariants are determined by the values of the basic invariants—almost trivial; the essential difficulties lie in the algebraic part only.

I choose the group which rules the classic Euclidean geometry, the group $\Gamma = O(n)$ of orthogonal transformations, as a further instance to throw more light on this point. Let us consider functions of two arbitrary vectors x, y which are invariant under all (proper and improper) orthogonal transformations. The first fundamental theorem asserts that the scalar products which may be constructed for these two vectors, namely the three products

$$(xx), \quad (xy) = (yx), \quad (yy), \tag{2.4}$$

form a basis. The functional part of this statement is nothing else than the fundamental proposition about the congruence of triangles: "The triangles ABC and $A'B'C'$ are congruent when two sides and the included angle of one triangle coincide with the corresponding elements of the other," or "Two figures each consisting of a couple of vectors x, y and x', y' are congruent, i.e. are changeable into each other by an appropriate orthogonal transformation, if and only if

$$(xx) = (x'x'), \qquad (xy) = (x'y'), \qquad (yy) = (y'y').\text{"}$$

Deeper lying but still true is the algebraic proposition that every orthogonally invariant *form* $f(x, y)$ is expressible as a *polynomial* of the three scalar products (2.4). Let us see if we can prove this by the methods used in demonstrating the congruence theorem in analytic n-dimensional geometry, where the coördinates vary in the field K of all real numbers.

Let the two vectors x, y be numerically fixed. By the "classic inductive construction" (Chapter I, §3) one may choose a new Cartesian coördinate system $e^1, e^2, \cdots, e^n$ such that x lies in the direction of the first fundamental vector e^1 and y lies in the plane (e^1, e^2):

$$x = \alpha e^1,$$
$$y = \beta e^1 + \gamma e^2.$$

The integral rational invariant $f(x, y)$ must then be equal to $f(x', y')$, where

$$x' = (\alpha, 0, 0, \cdots, 0),$$
$$y' = (\beta, \gamma, 0, \cdots, 0).$$

Thus $f(x', y')$ is a polynomial in the three quantities α, β, γ. But

$$\alpha^2 = (x'x') = (xx),$$
$$\alpha\beta = (x'y') = (xy),$$
$$\beta^2 + \gamma^2 = (y'y') = (yy);$$

hence

$$\alpha = \sqrt{(xx)}, \qquad \beta = \frac{(xy)}{\sqrt{(xx)}}, \qquad \gamma = \sqrt{\frac{(xx)(yy) - (xy)^2}{(xx)}}.$$

In this way, square roots and the denominator (xx) creep in.

It is fairly easy, however, to get rid of the square roots. We found $f(x, y)$ to be equal to a certain polynomial F of the quantities α, β, γ. Invariance of f for the particular orthogonal transformations which consist in changing the direction of the first or the second fundamental axis shows that F remains unaltered under the two substitutions

$$(1)\ \gamma \to -\gamma; \qquad (2)\ \alpha \to -\alpha, \quad \beta \to -\beta.$$

The polynomial F is a linear combination of monomials

$$M = \alpha^a \beta^b \gamma^c;$$

because of the invariance just mentioned, the exponent c must be even in all terms of F and the two exponents a and b of equal parity, i.e. either both even, or both odd. According to these two cases, M is either a monomial of the squares α^2, β^2, γ^2 or $\alpha\beta$ times such a monomial. Hence F can be written as a polynomial in α^2, β^2, γ^2 and $\alpha\beta$, that is, in

$$(xx),\ (xy),\ (yy),\ (xy)^2/(xx)\ .$$

f is thus *rationally* expressible by the scalar products with a power of (xx) as denominator.

In a similar manner one may find a rational expression of $f(x, y)$ in terms of the scalar products containing a power of (yy) as its denominator. By combining both results one can get rid of the denominators also, as we shall soon see. But the whole method is too clumsy to encourage its generalization to more than two arguments.

The problem of orthogonal invariants of an arbitrary number of argument vectors $x, y, \cdots, z$ will be solved in §9 along another line of approach. The result is analogous. The symmetric matrix of all the scalar products

$$\left\| \begin{matrix} (xx) & (xy) & \cdots & (xz) \\ (yx) & (yy) & \cdots & (yz) \\ \cdots & \cdots & \cdots & \cdots \\ (zx) & (zy) & \cdots & (zz) \end{matrix} \right\| \tag{2.5}$$

is a complete table of basic invariants.

This suggests the possibility of assigning to a given group Γ of linear transformations a finite number of *typical* basic invariants independent of the number of argument vectors to be considered. Such a table would consist of certain invariants depending on some "typical" argument vectors $u, v, \cdots$; and it would yield an integrity basis for the invariants of an arbitrary number of argument vectors $x, y, z, \cdots$ if one substitutes these argument vectors $x, y, z, \cdots$ in all possible combinations (repetitions not excluded) for the typical ones $u, v, \cdots$. In this sense the orthogonal group possesses the scalar product (uv) as its only typical basic invariant. For one gets a basis of invariants of h independent vectors $x, y, \cdots, z$, whatever this number h may be, in forming all the scalar products (2.5).

When one is called upon to express certain functions like the invariants in terms of given quantities like the basic invariants, it is essential to know whether these quantities are dependent or not. Thus the basic invariants for the symmetric group, the elementary symmetric functions $\varphi_i(x)$, are independent in the strict functional sense that they can simultaneously assume arbitrarily assigned values $a_i : \varphi_i(x) = a_i$. Indeed, the components x_i of the vector x are taken as the roots of the equation

$$t^n - a_1 t^{n-1} + a_2 t^{n-2} - \cdots \pm a_n = 0.$$

The algebraic independence of the polynomials $\varphi_i(x)$, the fact that there exists no rational relation among them, is an immediate consequence of this strict functional independence. It is desirable, however, to have a purely algebraic proof for this purely algebraic proposition: that a polynomial $F(\xi_1, \xi_2, \cdots, \xi_n)$ of n independent variables $\xi_1, \xi_2, \cdots, \xi_n$ vanishes identically in the latter if it vanishes identically in $x_1, x_2, \cdots, x_n$ after the ξ_i are replaced by the elementary symmetric functions $\varphi_i(x)$. Such a proof becomes indispensable when one operates in an arbitrary number field rather than in the domain of complex numbers.

We give a demonstration by means of a double induction, on the number n and on the degree of F. Observe that

$$\begin{aligned} \varphi_1(x) &= x_1 + \varphi_1'(x) \\ \varphi_2(x) &= x_1\varphi_1'(x) + \varphi_2'(x) \\ &\cdots\cdots\cdots\cdots \\ \varphi_n(x) &= x_1\varphi_{n-1}'(x), \end{aligned} \tag{2.6}$$

where $\varphi_1', \cdots, \varphi_{n-1}'$ are the elementary symmetric functions in $x_2, \cdots, x_n$. Hence if we put $x_1 = 0$ in the identity

$$F(\varphi_1, \varphi_2, \cdots, \varphi_n) = 0,$$

we obtain

$$F(\varphi_1', \varphi_2', \cdots, \varphi_{n-1}', 0) = 0.$$

Since we assume as hypothesis for induction that $\varphi_1', \cdots, \varphi_{n-1}'$ are algebraically independent, it follows that

$$F(\xi_1, \xi_2, \cdots, \xi_{n-1}, 0) = 0,$$

whence F has the form

$$F = \xi_n G(\xi_1, \cdots, \xi_n).$$

Since $\varphi_n(x) \neq 0$, $G(\varphi_1, \cdots, \varphi_n) = 0$; but G is of lower degree than F and so must vanish identically.

One cannot expect that this same situation, which we encounter here in the case of the symmetric group, will prevail in general; there may exist algebraic relations among the basic invariants even though their set is not redundant.

This happens for instance if Γ is the alternating group, consisting of the even permutations of the arguments $x_1, x_2, \cdots, x_n$. A set of basic invariants for the alternating group consists of the elementary symmetric functions $\varphi_1(x), \cdots, \varphi_n(x)$ together with the "difference product"

$$\Delta(x) = \prod_{i<k} (x_i - x_k).$$

The square Δ^2, the "discriminant," is a symmetric function and is therefore expressible in terms of the $\varphi_i(x)$. But we shall see that this relation is in a sense the "only" one to which our set of invariants is bound.

We prove first that the invariants $\varphi_1(x), \cdots, \varphi_n(x); \Delta(x)$ form an integrity basis for the alternating group. A form $f(x_1, \cdots, x_n)$ that is invariant with respect to the even permutations is changed into one and the same form f' by all odd permutations. The sum $f + f' = F$ is symmetric whereas $f - f' = g$ is alternating, i.e. changes its sign under the influence of transposition of two variables. g therefore vanishes if we identify two of its variables, say x_i and x_k, and so must be divisible by $x_i - x_k$. Being divisible by each of the prime polynomials $x_i - x_k$, it must be divisible by their product $\Delta(x)$:

$$g = \Delta G.$$

The polynomial G is evidently symmetric. After expressing the symmetric forms F and G in terms of the elementary symmetric functions, one gets an expression for f in terms of the $\varphi_i(x)$ and $\Delta(x)$ by means of the equation

$$f = \tfrac{1}{2}(F + \Delta G).$$

(It is not surprising that Δ appears in the first power only, since Δ^2 can be expressed as a polynomial D of the $\varphi_i(x)$.)

In turning to the second part of our statement, we observe that in two senses, the "functional" and the "algebraic" sense, it can be maintained that the quadratic equation

$$\Delta^2 - D(\varphi_1, \cdots, \varphi_n) = 0 \tag{2.7}$$

mentioned above is the only one holding among our basic invariants $\varphi_1(x), \cdots, \varphi_n(x); \Delta(x)$. As for the *functional* aspect, we remark that the invariants may take on arbitrary values $\varphi_1 = a_1, \cdots, \varphi_n = a_n$, $\Delta = b$ subject only to

$$b^2 = D(a_1, \cdots, a_n). \tag{2.8}$$

Indeed, the coefficients $a_1, \cdots, a_n$ determine the roots $x_1, \cdots, x_n$ but for their order, and according to this order $\Delta(x)$ may take on either of the values $\pm b$ allowed by (2.8). As for the *algebraic* aspect, every polynomial $H(\xi_1, \cdots, \xi_n; \eta)$ of the independent variables $\xi_1, \cdots, \xi_n; \eta$ which vanishes identically in $x_1, \cdots, x_n$ after the substitution $\xi_i = \varphi_i(x)$, $\eta = \Delta(x)$, is a multiple of the left side of (2.7):

$$H(\xi_1, \cdots, \xi_n; \eta) = L(\xi_1, \cdots, \xi_n; \eta)\{\eta^2 - D(\xi_1, \cdots, \xi_n)\},$$

L being a polynomial again. To prove this, consider H as a polynomial in η and divide by $\eta^2 - D(\xi_1, \cdots, \xi_n)$. The remainder is linear: $A(\xi_1, \cdots, \xi_n) + \eta B(\xi_1, \cdots, \xi_n)$. The above substitution yields both equations $A(\varphi) \pm \Delta B(\varphi) = 0$, according to the arrangement of the variables $x_1, \cdots, x_n$. Therefore $A(\varphi)$ and $B(\varphi)$ vanish individually, whence $A(\xi) = B(\xi) = 0$.

No less instructive regarding the problem of relations among the basic invariants than this example of a finite group is the continuous group of all orthogonal transformations. We considered above invariants depending on two vectors x, y. Their basic invariants (xx), (xy), (yy)—at least if the number of dimensions is $\geqq 2$—are capable of all numerical values satisfying the inequality

$$(xy)^2 \leqq (xx) \cdot (yy); \tag{2.9}$$

for the lengths of two sides of the triangle and the angle included may be assigned arbitrarily. The inequality (2.9) is surely to be counted as a relation from the general functional standpoint; from the algebraic standpoint, however, (xx), (xy), (yy) are independent since they are not bound by any algebraic equation.

What is the behavior in this respect of an arbitrary number h of independent vectors $x, y, \cdots, z$ and their table of scalar products (2.5)? The scalar products are algebraically independent as long as h is less than or equal to the dimensionality n, but not so if $h > n$. The scalar products of $n + 1$ vectors $x, y, \cdots, z$ satisfy, for instance, the equation

$$\begin{vmatrix} (xx) & (xy) & \cdots & (xz) \\ (yx) & (yy) & \cdots & (yz) \\ \cdots & \cdots & \cdots & \cdots \\ (zx) & (zy) & \cdots & (zz) \end{vmatrix} = 0.$$

In case $h \leqq n$ the problem of determining h vectors $x, y, \cdots, z$ such that the matrix (2.5) of their scalar products coincides with a given symmetric matrix $\| a_{ik} \|$ of h rows and columns has a solution in the real field K if and only if the quadratic form with the coefficients a_{ik} is positive definite. Our statement is merely a different formulation of the well-known fact that such a form may be linearly transformed into the square sum of the h independent variables. One sees that the table (2.5) of the basic invariants is bound only by inequalities when $h \leqq n$; algebraic equations, however, appear as soon as the number h of vectors surpasses the dimensionality n. A purely algebraic proof valid in any reference field will be given in Part C, §17.

We proved in the case $h = 2$ that every invariant form $f(x, y)$ depending on two vectors x, y is expressible either as

$$F((xx), (xy), (yy))/(xx)^{\alpha} \quad \text{or as} \quad G((xx), (xy), (yy))/(yy)^{\beta},$$

where F and G are polynomials. The polynomial

$$\zeta^{\beta} F(\xi, \eta, \zeta) - \xi^{\alpha} G(\xi, \eta, \zeta)$$

vanishes if we make the substitution $\xi = (xx)$, $\eta = (xy)$, $\zeta = (yy)$. Since (xx), (xy), and (yy) are algebraically independent, it must vanish identically in ξ, η, ζ. Thus F is divisible by ξ^α, say $F = \xi^\alpha F_1$, whence

$$f(x, y) = F_1((xx), (xy), (yy)).$$

In spite of our success in proving the first fundamental theorem for orthogonal invariants of two vectors, we should get into serious trouble if we attempted to deal in the same manner with $h > 2$ independent vectors. The procedure becomes entirely hopeless if h surpasses the dimensionality n and the scalar products are no longer algebraically independent. To overcome these difficulties a new formal apparatus is needed.

The first main problem in the theory of vector invariants of a given group Γ is the determination of a set of basic invariants, and the first main theorem (which we cannot, however, assert for all groups Γ) states the finiteness of such a basis. The second main problem consists in determining "all" algebraic relations holding among the basic invariants

$$\varphi_1(x, y, \cdots), \qquad \varphi_2(x, y, \cdots), \cdots, \varphi_r(x, y, \cdots),$$

or rather, to find a number of such relations of which all others are algebraic consequences. The finiteness of this number is averred by the second main theorem. It holds for any group for which the first main theorem is true; for it is a special case of Hilbert's general theorem asserting that every polynomial ideal has a finite ideal basis. It was first developed by Hilbert exactly in this context of the theory of invariants. All polynomials $R(\xi_1, \xi_2, \cdots, \xi_r)$ ("relations") of r independent variables $\xi_1, \xi_2, \cdots, \xi_r$ that vanish after the substitution $\xi_i = \varphi_i(x, y, \cdots)$ identically in $x, y, \cdots$ form an ideal $\mathfrak{J}$ within the ring of polynomials of the indeterminates $\xi_1, \cdots, \xi_r$. The ideal $\mathfrak{J}$ has a finite ideal basis $R_1, \cdots, R_t$ according to Hilbert's theorem. All relations $R = 0$ holding among the r basic invariants $\varphi_1(x, y, \cdots)$ are thus consequences of the t relations

$$R_1 = 0, \cdots, R_t = 0.$$

This general solution of our problem does not free us from the duty of ascertaining an actual basis $R_1, \cdots, R_t$ for the relations in each particular case that may come under our consideration.

A. First Main Theorem

3. First example: the symmetric group

After all the planning of our journey, the moment of departure has finally come; we now embark on a systematic investigation of vector invariants. Our first concern will be the first main theorem, which we shall prove by explicit construction of a finite integrity basis for the most important groups. In this section we study the group π_n of all $n!$ permutations of the n components $x_1, \cdots, x_n$ of our generic vector x.

It is an immediate consequence of the definition of the polar process (Chapter I, §1) that it changes an invariant into an invariant, whatever may be the underlying group Γ of linear transformations in our n-dimensional vector space. More precisely, if $f(x, y, \cdots)$ is a form depending on several vectors $x, y, \cdots$ and invariant under cogredient transformation of $x, y, \cdots$ by any element of Γ, then the polarized form $D_{ux}f(x, y, \cdots)$ is also invariant; here u is a new vector transforming cogrediently with the others. On this remark rests the importance of polarization for invariant theory.

We mentioned in §2 the fundamental algebraic theorem that the elementary symmetric functions (2.2) form an integrity basis for invariants $f(x)$ of π_n depending on a single vector $x = (x_1, \cdots, x_n)$. Our task now is to solve the same problem for invariants $f(x^{(1)}, x^{(2)}, \cdots, x^{(m)})$ depending on an arbitrary number m of independent vectors $x^{(1)}, x^{(2)}, \cdots, x^{(m)}$. The conjecture which offers itself at once is that *full polarization of* (2.2) *will yield a complete table of typical basic invariants.* If one adjoins the factor $i!$ to φ_i, the polarized table reads as follows:

$$
\begin{aligned}
\varphi_1(u) &= \sum u_i \\
\varphi_2(u, v) &= \sum u_i v_k && (i \neq k) \\
\varphi_3(u, v, w) &= \sum u_i v_k w_l && (i, k, l \text{ all } \neq) \\
&\cdots\cdots\cdots\cdots\cdots\cdots \\
\varphi_n(u, v, \cdots, w) &= \sum u_i v_k \cdots w_l && (i, k, \cdots, l \text{ all } \neq).
\end{aligned} \tag{3.1}
$$

Our statement means (§2) that one obtains an integrity basis for the invariants $f(x^{(1)}, x^{(2)}, \cdots, x^{(m)})$ depending on m argument vectors $x^{(1)}, x^{(2)}, \cdots, x^{(m)}$ by substituting these arguments for the "typical" arguments $u, v, w, \cdots$ in all possible combinations (repetitions included) in the forms (3.1):

$$\varphi_i(x^{(\alpha_1)}, x^{(\alpha_2)}, \cdots, x^{(\alpha_i)}), \qquad (\alpha_1, \alpha_2, \cdots, \alpha_i = 1, 2, \cdots, m; i = 1, 2, \cdots, n).$$

In order to avoid the crowding of indices, the m argument vectors will be called $x, y, \cdots, z$. The proof will be given by complete induction with respect to the dimensionality n. For this purpose each vector $x = (x_1, x_2, \cdots, x_n)$ is to be considered as combining a number x_1 with an $(n-1)$-dimensional vector $x' = (x_2, \cdots, x_n)$. The elementary symmetric functions of the $(n-1)$-dimensional vectors (unpolarized or polarized) are designated by $\varphi_1', \varphi_2', \cdots, \varphi_{n-1}'$. Every form $f(x, y, \cdots, z)$ is an aggregate (linear combination) of terms

$$x_1^\alpha y_1^\beta \cdots z_1^\gamma \cdot f_{\alpha\beta\cdots\gamma}(x', y', \cdots, z').$$

If f is symmetric, so also is $f_{\alpha\beta\cdots\gamma}$, and, according to our assumption of the truth of our theorem for $n-1$ dimensions, $f_{\alpha\beta\cdots\gamma}$ is expressible (in an integral rational way) by the polarized elementary symmetric forms $\varphi_1', \cdots, \varphi_{n-1}'$. The given $f(x, y, \cdots, z)$ appears thus as an aggregate of terms

$$x_1^\alpha y_1^\beta \cdots z_1^\gamma \cdot (\varphi_1')^{\rho_1}(\varphi_2')^{\rho_2} \cdots (\varphi_{n-1}')^{\rho_{n-1}},$$

where $(\varphi_i')^{\rho_i}$ indicates a product of ρ_i factors φ_i' each of which may depend on different arguments.

We now make use of the equations (2.6) in their polarized form

$$\varphi_i(x^{(1)}, x^{(2)}, \cdots, x^{(i)}) = \varphi_i'(x^{(1)}, x^{(2)}, \cdots, x^{(i)}) + \sum_{\alpha=1}^{i} x_1^{(\alpha)} \varphi_{i-1}'(x^{(1)}, \cdots, x^{(\alpha-1)}, x^{(\alpha+1)}, \cdots, x^{(i)}) \quad (i = 1, \cdots, n-1; \varphi_0' = 1),$$

and thus express φ_i' in terms of φ_i, φ_{i-1}' and the variables $x_1, y_1, \cdots, z_1$. In this way we eliminate the quantities

$$\varphi_{n-1}', \varphi_{n-2}', \cdots, \varphi_1'$$

one after the other, replacing them by $\varphi_{n-1}, \varphi_{n-2}, \cdots, \varphi_1$ and $x_1, y_1, \cdots, z_1$. $f(x, y, \cdots, z)$ is then expressed as an aggregate of terms

$$x_1^\alpha y_1^\beta \cdots z_1^\gamma \cdot (\varphi_1)^{\rho_1} \cdots (\varphi_{n-1})^{\rho_{n-1}}. \tag{3.2}$$

The last part of this term is symmetric even when the first component x_1 of each vector x is interchanged with the components $x_2, \cdots, x_n$ of x. As the whole functon f is symmetric in all n components, the term (3.2) may be replaced by

$$(\varphi_1)^{\rho_1} \cdots (\varphi_{n-1})^{\rho_{n-1}} \cdot \frac{1}{n} \sum_{i=1}^{n} x_i^\alpha y_i^\beta \cdots z_i^\gamma.$$

The sum appearing here arises by consecutive polarizations from the power sum

$$\sigma_\nu(x) = \sum_{i=1}^{n} x_i^\nu \qquad (\nu = \alpha + \beta + \cdots + \gamma),$$

and Newton's well-known formulae show how to express these in terms of the elementary symmetric functions $\varphi_1, \varphi_2, \cdots, \varphi_n$.

To leave no gap we add their simplest derivation. The polynomial

$$\psi(\lambda) = \prod_{i=1}^{n} (1 - \lambda x_i) = 1 - \varphi_1(x)\lambda + \varphi_2(x)\lambda^2 - \cdots \pm \varphi_n(x)\lambda^n$$

has the logarithmic derivative

$$-\frac{\psi'(\lambda)}{\psi(\lambda)} = \sum_{i=1}^{n} \frac{x_i}{1 - \lambda x_i} = \sigma_1(x) + \sigma_2(x)\lambda + \sigma_3(x)\lambda^2 + \cdots.$$

The Taylor expansion on the right side is to be understood in the formal sense such that no questions of convergence arise:

$$-\psi'(\lambda) \equiv \psi(\lambda) \cdot \sum_{\nu=1}^{N} \sigma_\nu(x)\lambda^{\nu-1} \qquad (\text{mod } \lambda^N).$$

From this follow the recursive formulae

$$\begin{aligned}
\varphi_1(x) &= \sigma_1(x) \\
-2\varphi_2(x) &= -\varphi_1(x)\sigma_1(x) + \sigma_2(x) \\
3\varphi_3(x) &= \varphi_2(x)\sigma_1(x) - \varphi_1(x)\sigma_2(x) + \sigma_3(x) \\
&\cdots\cdots\cdots\cdots\cdots\cdots\cdots\cdots\cdots\cdots
\end{aligned}$$

with the convention $\varphi_\nu(x) = 0$ for $\nu > n$.

It should be observed that the case $n = 1$ is itself covered in our inductive process, and hence the theorem has been proved in full detail.

4. Capelli's identity

We have shown in the preceding section that an integrity basis for invariants of the symmetric group depending on one argument vector becomes one for invariants depending on an arbitrary number of argument vectors by complete polarization. This, however, is definitely not so for every linear group Γ. None the less it is a remarkable fact that an integrity basis for invariants depending on n arguments, where n is the degree of Γ, does yield a basis for $m > n$ arguments by complete polarization. To show this we need a certain powerful formal instrument, Capelli's identity.[13] It is concerned with the result of successive polarizations.

Let $x, y, z, \cdots$ be a row of independent vectors in an n-dimensional vector space, $x', y', z', \cdots$ the same vectors in (the same or) a different arrangement. $\Delta_{x'x}$ as well as $D_{x'x}$ may symbolize the polar process. By performing several polarizations like $D_{x'x}$, $D_{y'y}$, $D_{z'z}$ in succession on a form $f(x, y, z, \cdots)$ one will get

$$D_{z'z}D_{y'y}D_{x'x}f = \sum_{i,k,l} x'_i y'_k z'_l \frac{\partial^3 f}{\partial x_i \partial y_k \partial z_l},$$

provided x' does not coincide either with y or z, nor y' with z. The auxiliary symbols $\Delta_{x'x}$, $\Delta_{y'y}$, $\cdots$ may be used instead of $D_{x'x}$, $D_{y'y}$, $\cdots$ to indicate by their composition this result regardless of whether the coincidences just mentioned occur or not:

$$\Delta_{z'z}\Delta_{y'y}\Delta_{x'x}f = \sum_{i,k,l} x'_i y'_k z'_l \frac{\partial^3 f}{\partial x_i \partial y_k \partial z_l}.$$

We propose to compute how the composite operator $D_{z'z}D_{y'y}D_{x'x}$ differs from this "pseudo-composition." For this purpose we introduce the symbol $\delta_{x'x}$ defined by

$$\delta_{x'x} = \begin{cases} 1 & \text{if } x' = x, \\ 0 & \text{if } x' \neq x. \end{cases}$$

We then obtain directly from these definitions

$$\begin{aligned}
D_{z'z}\Delta_{y'y}\Delta_{x'x}f &= \sum_{l} z_l' \frac{\partial}{\partial z_l}\left(\sum_{i,k} x_i' y_k' \frac{\partial^2 f}{\partial x_i\, \partial y_k}\right) \\
&= \sum_{i,k,l} x_i' y_k' z_l' \frac{\partial^3 f}{\partial x_i\, \partial y_k\, \partial z_l} + \delta_{x'z} \sum_{i,k} z_i' y_k' \frac{\partial^2 f}{\partial x_i\, \partial y_k} + \delta_{y'z} \sum_{i,k} x_i' z_k' \frac{\partial^2 f}{\partial x_i\, \partial y_k} \qquad (4.1)\\
&= \Delta_{z'z}\Delta_{y'y}\Delta_{x'x}f + \delta_{x'z}\Delta_{y'y}\Delta_{z'x}f + \delta_{y'z}\Delta_{z'y}\Delta_{x'x}f.
\end{aligned}$$

Our chief interest is in the alternating sum

$$\sum \pm D_{z'z}\,\Delta_{y'y}\,\Delta_{x'x} = \begin{vmatrix} D_{z'z} & \Delta_{z'y} & \Delta_{z'x} \\ D_{y'z} & \Delta_{y'y} & \Delta_{y'x} \\ D_{x'z} & \Delta_{x'y} & \Delta_{x'x} \end{vmatrix}$$

extending to the 3! permutations of x', y', z'. The individual terms in the expansion of this determinant of operators are to be written in such a way that the factors follow each other from left to right as they stand in the determinant itself: the first factor is taken from the first column, the second factor from the second column, etc.; the same rule is to be observed throughout the following. After performing alternation on (4.1) we may exchange x' and z' in the second term on the right, and y' and z' in the third term, provided we change the signs of these terms. We thus obtain

$$\sum_{(x',y',z')} \pm D_{z'z}\,\Delta_{y'y}\,\Delta_{x'x} = \sum \pm \Delta_{z'z}\,\Delta_{y'y}\,\Delta_{x'x} - 2\sum \pm \delta_{z'z}\,\Delta_{y'y}\,\Delta_{x'x},$$

or

$$\sum \pm (D_{z'z} + 2\delta_{z'z})\,\Delta_{y'y}\,\Delta_{x'x} = \sum \pm \Delta_{z'z}\,\Delta_{y'y}\,\Delta_{x'x}.$$

Written in determinant form, this result reads:

$$(D_3) \qquad \begin{vmatrix} \vdots & \vdots & \vdots \\ D_{zz} + 2 & \Delta_{zy} & \Delta_{zx} \\ D_{yz} & \Delta_{yy} & \Delta_{yx} \\ D_{xz} & \Delta_{xy} & \Delta_{xx} \end{vmatrix} = \begin{vmatrix} \vdots & \vdots & \vdots \\ \Delta_{zz} & \Delta_{zy} & \Delta_{zx} \\ \Delta_{yz} & \Delta_{yy} & \Delta_{yx} \\ \Delta_{xz} & \Delta_{xy} & \Delta_{xx} \end{vmatrix},$$

the meaning of this equation being that corresponding minors of order three are equal.

This equation for three consecutive polarizations should have been preceded by those holding for one or two such operators:

$$(D_1) \quad \begin{vmatrix} \vdots \\ D_{yx} \\ D_{xx} \end{vmatrix} = \begin{vmatrix} \vdots \\ \Delta_{yx} \\ \Delta_{xx} \end{vmatrix}; \qquad (D_2) \quad \begin{vmatrix} \vdots & \vdots \\ D_{yy} + 1 & \Delta_{yx} \\ D_{xy} & \Delta_{xx} \end{vmatrix} = \begin{vmatrix} \vdots & \vdots \\ \Delta_{yy} & \Delta_{yx} \\ \Delta_{xy} & \Delta_{xx} \end{vmatrix}.$$

Their extension to four or more consecutive polarizations is obvious. Using first (D_2) for the last two columns of the determinant on the left side of (D_3) and then (D_1) for its last column, one gets the fundamental equation we were aiming at:

$$\begin{vmatrix} \cdot & \cdot & \cdot \\ \cdot & \cdot & \cdot \\ \cdot & \cdot & \cdot \\ D_{zz}+2 & D_{zy} & D_{zx} \\ D_{yz} & D_{yy}+1 & D_{yx} \\ D_{xz} & D_{xy} & D_{xx} \end{vmatrix} = \begin{vmatrix} \cdot & \cdot & \cdot \\ \cdot & \cdot & \cdot \\ \cdot & \cdot & \cdot \\ \Delta_{zz} & \Delta_{zy} & \Delta_{zx} \\ \Delta_{yz} & \Delta_{yy} & \Delta_{yx} \\ \Delta_{xz} & \Delta_{xy} & \Delta_{xx} \end{vmatrix}.$$

Similarly, for m independent vectors $x, y, \cdots, z$ instead of three, we obtain

$$\begin{vmatrix} D_{zz}+(m-1) & \cdots & D_{zy} & D_{zx} \\ \cdots & \cdots & \cdots & \cdots \\ D_{yz} & \cdots & D_{yy}+1 & D_{yx} \\ D_{xz} & \cdots & D_{xy} & D_{xx} \end{vmatrix} = \begin{vmatrix} \Delta_{zz} & \cdots & \Delta_{zy} & \Delta_{zx} \\ \cdots & \cdots & \cdots & \cdots \\ \Delta_{yz} & \cdots & \Delta_{yy} & \Delta_{yx} \\ \Delta_{xz} & \cdots & \Delta_{xy} & \Delta_{xx} \end{vmatrix}.$$

(Apologies should be made that $x, y, \cdots, z$ run upwards and backwards; this is a consequence of the bad habit of reading operators from right to left.)

The operator determinant on the right side transforms $f(x, y, \cdots, z)$ into

$$\sum_{i,k,\cdots,l}\left(\sum \pm x'_i y'_k \cdots z'_l\right)\cdot\frac{\partial^m f}{\partial x_i\, \partial y_k \cdots \partial z_l},$$

with the inner sum extending alternatingly over all permutations $x', y', \cdots, z'$ of $x, y, \cdots, z$. This sum is zero unless all m indices $i, k, \cdots, l$ are different. As this is impossible when $m > n$ the result will in this case be zero. When $m = n$ the inner sum is $\pm\,[xy \cdots z]$ or 0, according to whether the sequence $i, k, \cdots, l$ is an even or odd permutation of $1, 2, \cdots, n$ or contains equal indices. Here the "bracket" $[xy \cdots z]$ designates the determinant

$$\begin{vmatrix} x_1 & x_2 & \cdots & x_n \\ y_1 & y_2 & \cdots & y_n \\ \cdots & \cdots & \cdots & \cdots \\ z_1 & z_2 & \cdots & z_n \end{vmatrix},$$

which is invariant under the unimodular group $SL(n)$; as a matter of historical interest, it is this particular invariant from which the whole development of invariant theory started. Hence for $m = n$ one obtains

$$[xy \cdots z]\cdot\Omega f,$$

where

$$\Omega f = \sum_{(i,k,\cdots,l)} \pm \frac{\partial^n f}{\partial x_i \partial y_k \cdots \partial z_l} = \begin{vmatrix} \frac{\partial}{\partial x_1} & \frac{\partial}{\partial x_2} & \cdots & \frac{\partial}{\partial x_n} \\ \frac{\partial}{\partial y_1} & \frac{\partial}{\partial y_2} & \cdots & \frac{\partial}{\partial y_n} \\ \cdots & \cdots & \cdots & \cdots \\ \frac{\partial}{\partial z_1} & \frac{\partial}{\partial z_2} & \cdots & \frac{\partial}{\partial z_n} \end{vmatrix} f$$

is derived from f by the so-called Cayley Ω-process. The sum extends alternatingly over all permutations $i, k, \cdots, l$ of $1, 2, \cdots, n$.

We thus arrive at Capelli's identity. In its final form the notation $x, y, \cdots, z$ may be replaced by the more pedantic $x^1, x^2, \cdots, x^m$, and the symbols $D_{x^\beta x^\alpha}$ by $D_{\beta\alpha}$.

THEOREM (2.4. A).

$$\begin{vmatrix} D_{mm} + (m-1) & \cdots & D_{m2} & D_{m1} \\ \cdots & \cdots & \cdots & \cdots \\ D_{2m} & \cdots & D_{22} + 1 & D_{21} \\ D_{1m} & \cdots & D_{12} & D_{11} \end{vmatrix} f = \begin{cases} 0 & \text{if } m > n, \\ [x^1 x^2 \cdots x^n] \cdot \Omega f & \text{if } m = n. \end{cases}$$

We shall refer to the two cases $m > n$, $m = n$ of this formula as Capelli's general and special identity, respectively.

5. Reduction of the first main problem by means of Capelli's identities

The way in which Capelli's identity is used for the investigation of invariants depending on m argument vectors $x^1, x^2, \cdots, x^m$ can be described as follows for an arbitrary group Γ of linear transformations A. Polarization $D_{\beta\alpha}$ carries an (absolute or relative) invariant $f(x^1, x^2, \cdots, x^m)$ into an invariant $D_{\beta\alpha}f$ (of the same multiplier). The special Capelli identity shows that, in the case $m = n$, Ωf is a relative invariant whose multiplier equals the multiplier of f divided by the transformation determinant. In particular the operator Ω applied to an absolute invariant f will yield an absolute invariant provided Γ consists of unimodular transformations; in the following we have preferably this case in mind.

A form $f(x^1, x^2, \cdots, x^m)$ is of a certain degree $r_1, r_2, \cdots, r_m$ in each of its arguments $x^1, x^2, \cdots, x^m$. The sum $r = r_1 + r_2 + \cdots + r_m$ is its total degree. We arrange forms $f(x^1, x^2, \cdots, x^m)$ first according to their total degree, i.e. f is of lower rank than f^* in the hierarchy of forms if the total degree of f is lower than that of f^*. For instance Ωf is lower than f (in the case $m = n$) since the Ω-process diminishes the total degree by n. Within the set of all forms of given total degree, lexicographic arrangement according to the individual degrees $r_1, \cdots, r_m$ is followed; thus f is lower than f^* if the first of the degrees $r_1, \cdots, r_m$ in which f and f^* differ has a lower value for f. Forms coinciding in all their m degrees r_i are considered of equal rank; we abstain from imposing an order of precedence upon them.

The main term in the operator determinant of Capelli's identity,

$$(D_{mm} + m - 1) \cdots (D_{22} + 1)D_{11},$$

transforms f into

$$r_1(r_2 + 1) \cdots (r_m + m - 1)f = \rho f.$$

The numerical factor ρ multiplying f is $\neq 0$ when $r_1 > 0$, i.e. when f actually contains the first variable x^1. For any other term we observe that we may drop the diagonal factors $D_{\alpha\alpha} + \alpha - 1$; their effect is merely to multiply f by certain constants which we collect into a factor ρ^*. Such a term is then of the form

$$\rho^* D_{\beta_r\alpha_r} \cdots D_{\beta_2\alpha_2} D_{\beta_1\alpha_1};$$

where $\alpha_r > \alpha_{r-1} > \cdots > \alpha_2 > \alpha_1$, $\beta_i \neq \alpha_i$, and $\beta_1, \beta_2, \cdots, \beta_r$ is a permutation of $\alpha_1, \alpha_2, \cdots, \alpha_r$. In particular. $\beta_1 > \alpha_1$ and $r \geqq 2$. The main term being the only one in which no factor $D_{\beta\alpha}$ with different indices α, β appears, it follows that if we set

$$\mathcal{P} = D_{\beta_r\alpha_r} \cdots D_{\beta_2\alpha_2},$$

$$f^* = -\rho^* D_{\beta_1\alpha_1} f \qquad (\beta_1 > \alpha_1),$$

the left side of Capelli's identity takes the form

$$\rho f - \sum \mathcal{P} f^*.$$

The degree of $D_{\beta_1\alpha_1} f$ in x^{α_1} is one less than that of f, while its degree in x^{β_1} is one greater. The former is decisive; since $\alpha_1 < \beta_1$, f^* is lower than f.

Capelli's identities may now be written as follows:

$$\rho f = \sum \mathcal{P} f^* \qquad (m > n), \tag{5.1}$$

$$\rho f = \sum \mathcal{P} f^* + [x^1 x^2 \cdots x^n]\Omega f \qquad (m = n), \tag{5.2}$$

where (1) f^* and Ωf are of lower rank than f and are invariants if f is an invariant, (2) $\mathcal{P}$ is a succession of polar operations, and (3) $\rho > 0$ if f actually contains x^1.

Let us now suppose we have chosen from among the (absolute) invariants of the group Γ depending on m argument vectors $x^1, \cdots, x^m$ a finite set

$$\varphi_1, \varphi_2, \cdots, \varphi_l \tag{5.3}$$

for which we want to prove that they form an integrity basis for all these invariants. We assume that the table (5.3) is closed under polarization:

ASSUMPTION I: *Every $D_{\beta\alpha}\varphi_j$ is itself one of the φ's or at least is expressible in an integral rational manner by the set of the φ's.*

I contend: this assumption once granted, (5.3) is an integrity basis for the $m > n$ arguments $x^1, \cdots, x^m$ provided those φ_j not depending on x^1 form an integrity basis for the $m - 1$ arguments $x^2, \cdots, x^m$.

Indeed, owing to (5.1), f is expressible by the set (5.3) if the invariants f^* occurring on the right side are so expressible; for in view of Assumption I, and the formal properties (1.1.7) of polar operators, such an expression of $f^* = F^*(\varphi_1, \varphi_2, \cdots, \varphi_l)$ leads to a similar expression of $\mathcal{P}f^*$, as one is able to turn the polar processes $D_{\beta\alpha}$ of which $\mathcal{P}$ is composed, upon the arguments $\varphi_1, \cdots, \varphi_l$ of F^*. Our reasoning presupposes that f actually contains x^1, for then ρ does not vanish. The f^* are of lower rank than f. In using complete induction with respect to the rank, we shall be stopped only when the invariant f under consideration becomes of degree $r_1 = 0$ with respect to x^1.

One can go one step further and, even in the case $m = n$, cut the number of arguments down to $n - 1$ by means of Capelli's special identity (5.2) if one adds the

ASSUMPTION II (concerning the case $m = n$ only): *The determinant* $[x^1x^2 \cdots x^n]$ *occurs among the* φ*'s or is expressible in terms of them.*

Since we vary the number m of arguments our result is more conveniently expressed in terms of typical basic invariants. We once more formulate this notion as follows: Let there be given a number of invariants

$$\varphi_1^*((u)), \qquad \varphi_2^*((u)), \cdots \tag{5.4}$$

depending *in linear fashion* on some variable vectors $u^1, u^2, \cdots$ (not necessarily the same for each function); they constitute a complete table of typical basic invariants for m arguments if (5.4) changes into an integrity basis for invariants of m arguments $x^1, \cdots, x^m$ by substituting for $u^1, u^2, \cdots$ these arguments in all possible combinations (repetitions included). As such substitution results in a set (5.3) satisfying the requirement I, we infer by using induction with respect to m:

THEOREM (2.5.A). *A finite table of typical basic invariants of a linear group of degree n will be a complete set for any number m of arguments if this can be shown to be true for n arguments; even $n - 1$ arguments will suffice provided the determinant $[u^1u^2 \cdots u^n]$ appears in the table or is at least expressible by the invariants of the table.*

Even if the elements A of Γ are not unimodular, the first part of the theorem holds for relative invariants whose multipliers belong to a given group, i.e. to a set containing $\mu_1(A)/\mu_2(A)$ together with any two multipliers $\mu_1(A)$, $\mu_2(A)$; the second part requires that group to include the transformation determinant $|A|$.

In almost all cases the proof of the first main theorem consists of two parts: a formal part in which the reduction to n or $n - 1$ arguments takes place by dint of Capelli's identities, and another more conceptual part in which the proof for this restricted number of argument vectors is accomplished by considerations similar to those we used in proving the theorem of congruence in §2. The formal part can be carried through for any group, whereas the conceptual part cannot be brought down to such a general mechanized procedure and remains specific for each particular group. How this combination works out shall now be shown for the groups $SL(n)$ and $O(n)$.

6. Second example: unimodular group $SL(n)$

We shall take up the question of *invariants of the unimodular group* $SL(n)$ at once for any number of covariant or Latin vectors $x, y, \cdots$ and any number of contravariant or Greek vectors $\xi, \eta, \cdots$.

THEOREM (2.6.A).

$$[xy \cdots z], \qquad (\xi x), \qquad [\xi\eta \cdots \zeta]$$

is a complete table of typical basic invariants for the unimodular group.

By means of Capelli's identities the proof is reduced to the case where only $n - 1$ Latin and $n - 1$ Greek vectors

$$x^1, \cdots, x^{n-1}; \qquad \xi^1, \cdots, \xi^{n-1} \tag{6.1}$$

are present. We must show that an invariant form f depending on them is expressible in terms of the $(n - 1)^2$ products

$$(\xi^i x^k) \qquad (i, k = 1, \cdots, n - 1).$$

Let us suppose that the vectors (6.1) are numerically given in arbitrary fashion but so that the determinant

$$\Delta = \underset{i,k=1,\cdots,n-1}{\det.} (\xi^i x^k) \neq 0. \tag{6.2}$$

LEMMA (2.6.B). *Under the assumption* (6.2) *we can introduce a new coordinate system by unimodular transformation such that* $x^1, \cdots, x^{n-1}$ *coincide with the first* $n - 1$ *basic vectors* $e^1, \cdots, e^{n-1}$ *and such that the* n^{th} *component of each of the* $n - 1$ *contravariant vectors* $\xi^1, \cdots, \xi^{n-1}$ *vanishes.*

Taking this lemma for granted we proceed as follows. *Before* the transformation we have

$$\begin{array}{l|l} x^1 = (x_1^1, x_2^1, \cdots, x_n^1) & \xi^1 = (\xi_1^1, \xi_2^1, \cdots, \xi_n^1) \\ \cdots\cdots\cdots\cdots\cdots\cdots & \cdots\cdots\cdots\cdots\cdots\cdots \\ x^{n-1} = (x_1^{n-1}, x_2^{n-1}, \cdots, x_n^{n-1}) & \xi^{n-1} = (\xi_1^{n-1}, \xi_2^{n-1}, \cdots, \xi_n^{n-1}); \end{array} \tag{6.3}$$

after the transformation,

$$\begin{array}{l|l} x^1 = (1, 0, \cdots, 0, 0) & \xi^1 = (\bar\xi_1^1, \bar\xi_2^1, \cdots, \bar\xi_{n-1}^1, 0) \\ \cdots\cdots\cdots\cdots\cdots\cdots & \cdots\cdots\cdots\cdots\cdots\cdots \\ x^{n-1} = (0, 0, \cdots, 1, 0) & \xi^{n-1} = (\bar\xi_1^{n-1}, \bar\xi_2^{n-1}, \cdots, \bar\xi_{n-1}^{n-1}, 0). \end{array} \tag{6.4}$$

Because of the invariance of $(\xi^i x^k)$,

$$(\xi^i x^k) = (\bar\xi^i e^k) = \bar\xi_k^i ;$$

moreover, the invariant f depending on the arguments (6.3) is equal to the same function f of the arguments (6.4). We introduce the polynomial $\Phi\{\xi^i_k\}$ of $(n-1)^2$ variables ξ^i_k by

$$f\begin{pmatrix} 1 & 0 & \cdots & 0 & 0 & \xi^1_1 & \cdots & \xi^1_{n-1} & 0 \\ \cdots & \cdots & \cdots & \cdots & \cdots & \cdots & \cdots & \cdots & \cdots \\ 0 & 0 & \cdots & 1 & 0 & \xi^{n-1}_1 & \cdots & \xi^{n-1}_{n-1} & 0 \end{pmatrix} = \Phi\{\xi^i_k\}.$$

The last two observations then lead to the equation

$$f(x^1, \cdots, x^{n-1} \mid \xi^1, \cdots, \xi^{n-1}) = \Phi\{(\xi^i x^k)\}, \tag{6.5}$$

holding numerically for any vectors x and ξ satisfying the algebraic inequality (6.2). Δ is not identically zero, as is shown by taking $x^1 \cdots x^{n-1}$ and $\xi^1 \cdots \xi^{n-1}$ both to be the first $n-1$ basic vectors $e^1 \cdots e^{n-1}$ of the absolute coordinate system. Hence, on account of the principle of irrelevance of algebraic inequalities, (6.5) is an identity.

The proof of the lemma follows. Passing from the absolute coordinate system e^i to a new one $\bar{e}^i$ by means of

$$\begin{aligned} \bar{e}^1 &= x^1_1 e^1 + \cdots + x^1_n e^n, \\ &\cdots\cdots\cdots\cdots\cdots\cdots \\ \bar{e}^{n-1} &= x^{n-1}_1 e^1 + \cdots + x^{n-1}_n e^n, \\ \bar{e}^n &= z_1 e^1 + \cdots + z_n e^n, \end{aligned}$$

will produce the effect that $x^1, \cdots, x^{n-1}$ coincide with the new axes $\bar{e}^1, \cdots, \bar{e}^{n-1}$. In denoting by D_k the minors of the component matrix

$$\begin{Vmatrix} x^1_1, & \cdots, & x^1_n \\ \cdots & \cdots & \cdots \\ x^{n-1}_1, & \cdots, & x^{n-1}_n \end{Vmatrix}, \tag{6.6}$$

the condition that our transformation be unimodular is expressed by the following linear equation for the unknowns z_k:

$$\sum_{k=1}^{n} D_k z_k = 1. \tag{6.7}$$

An arbitrary contravariant vector ξ transforms according to

$$\begin{aligned} \bar{\xi}_i &= x^i_1 \xi_1 + \cdots + x^i_n \xi_n \qquad (i = 1, \cdots, n-1), \\ \bar{\xi}_n &= z_1 \xi_1 + \cdots + z_n \xi_n . \end{aligned}$$

If we require that the new n^{th} component $\bar{\xi}_n$ of each of our $n-1$ given contravariant vectors $\xi^1, \cdots, \xi^{n-1}$ vanish, we have to add the following $n-1$ equations to (6.7):

$$\sum_{k=1}^{n} \xi^i_k z_k = 0 \qquad (i = 1, \cdots, n-1). \tag{6.8}$$

The system (6.7), (6.8) of n equations for the n unknowns z_k has a uniquely determined solution provided its determinant

$$(6.9) \qquad \begin{vmatrix} \xi_1^1, & \cdots, & \xi_n^1 \\ \cdots & \cdots & \cdots \\ \xi_1^{n-1}, & \cdots, & \xi_n^{n-1} \\ D_1, & \cdots, & D_n \end{vmatrix}$$

is different from zero. This would suffice to prove our main theorem because we could well use this determinant instead of Δ. However, by a simple formal calculation Δ can be shown to coincide with (6.9).

The theorem and its demonstration are valid in any reference field k.

7. Extension theorem. Third example: group of step transformations

In the present section we deal with covariant vectors only. From a given group Γ of linear transformations on n variables $x_1, \cdots, x_n$ we construct the "*extended group*" Γ^ν in $n + \nu$ variables

$$(7.1) \qquad x_1 \cdots x_n \mid \bar{x}_1 \cdots \bar{x}_\nu$$

consisting of all matrices

$$(7.2) \qquad \begin{Vmatrix} A & 0 \\ B & C \end{Vmatrix}$$

which satisfy the following conditions:

$$(7.3) \qquad A \text{ in } \Gamma, \qquad \det C = 1.$$

The partition in (7.2) corresponds to the partition of the $n + \nu$ variables into *block* $(x_1 \cdots x_n)$ and *rim* $(\bar{x}_1 \cdots \bar{x}_\nu)$. In terms of homogeneous plane coordinates the extension Γ^1 $(\nu = 1)$ represents:

1) the group of *translations*, if Γ consists of the identity only;
2) the group characteristic for *affine n-space* if Γ is $GL(n)$ or $SL(n)$;
3) the group characteristic for *Euclidean geometry* if Γ is the orthogonal group $O(n)$ or $O^+(n)$.

In view of such important applications it is gratifying that we are able to master the consequences of extension upon the invariants by the following simple and general[14]

THEOREM (2.7.A). *A complete list of typical basic invariants for* Γ *becomes one for the extension* Γ^ν *by adding the bracket factor* $[xy \cdots z]$ *depending on* $n + \nu$ *typical arguments* $xy \cdots z$.

PROOF. Capelli's identities reduce the burden of our proof to showing that an invariant of Γ^ν,

$$f\begin{pmatrix} x_1 & \cdots & x_n \bar{x}_1 & \cdots & \bar{x}_{\nu-1} \bar{x}_\nu \\ \cdots & \cdots & \cdots & \cdots & \cdots \\ y_1 & \cdots & y_n \bar{y}_1 & \cdots & \bar{y}_{\nu-1} \bar{y}_\nu \end{pmatrix}, \tag{7.4}$$

depending on $n + \nu - 1$ arguments $x, \cdots, y$, does not involve the rim variables and hence is an invariant in n-space with respect to Γ. We assume $x, \cdots, y$ to be numerically given vectors for which the determinant

$$\begin{vmatrix} x_1 & \cdots & x_n \bar{x}_1 & \cdots & \bar{x}_{\nu-1} \\ \cdots & \cdots & \cdots & \cdots & \cdots \\ y_1 & \cdots & y_n \bar{y}_1 & \cdots & \bar{y}_{\nu-1} \end{vmatrix} \neq 0, \tag{7.5}$$

and we make the following substitution belonging to Γ^ν:

$$x'_i = x_i \ (i = 1, \cdots, n); \qquad \bar{x}'_\alpha = \bar{x}_\alpha \ (\alpha = 1, \cdots, \nu - 1);$$

$$\bar{x}'_\nu = (b_1x_1 + \cdots + b_nx_n) + (c_1\bar{x}_1 + \cdots + c_{\nu-1}\bar{x}_{\nu-1} + \bar{x}_\nu).$$

We can determine the $n + \nu - 1$ coefficients $b_1, \cdots, b_n; c_1, \cdots, c_{\nu-1}$ such that the last transformed component $\bar{x}'_\nu$ becomes 0 for each of the numerically given vectors $x, \cdots, y$: the determinant of these $n + \nu - 1$ linear equations for the $n + \nu - 1$ unknown coefficients is exactly (7.5). Because of the invariance of f we find that (7.4) equals

$$f\begin{pmatrix} x_1 & \cdots & x_n \bar{x}_1 & \cdots & \bar{x}_{\nu-1} & 0 \\ \cdots & \cdots & \cdots & \cdots & \cdots & \cdots \\ y_1 & \cdots & y_n \bar{y}_1 & \cdots & \bar{y}_{\nu-1} & 0 \end{pmatrix}.$$

As this equation holds numerically for any arguments satisfying the algebraic inequality (7.5), it must be a formal identity. Therefore f is free of the last component $\bar{x}_\nu$ of its argument vectors. In the same way one shows that the other rim components $\bar{x}_1, \cdots, \bar{x}_{\nu-1}$ are missing.

By direct application of the extension theorem we determine a complete table of typical basic invariants depending on covariant vectors only for the group of "step transformations," i.e. of all transformations (7.2) which satisfy the conditions:

$$\det A = 1, \qquad \det C = 1.$$

We again suppose the widths of the steps to be n and ν and use the notation (7.1) for the block and rim components.

THEOREM (2.7.B). *All purely covariant vector invariants of the group of step transformations are expressible in terms of two types: the (total) bracket* $[x \cdots yz]_{\text{tot.}}$ *depending on* $n + \nu$ *arguments, and the block bracket of* n *arguments:*

$$[x \cdots y]_{\text{block}} = \begin{vmatrix} x_1 & \cdots & x_n \\ \cdots & \cdots & \cdots \\ y_1 & \cdots & y_n \end{vmatrix}$$

Generalization to a staircase consisting of more than two steps is immediate. In particular one may consider a staircase of n steps of width 1 in n-space, i.e. the group of all "recurrent matrices"

$$\begin{Vmatrix} 1 & 0 & 0 & \cdots & 0 & 0 \\ a_{21} & 1 & 0 & \cdots & 0 & 0 \\ \cdots & \cdots & \cdots & \cdots & \cdots & \cdots \\ a_{n1} & a_{n2} & & \cdots & a_{n,n-1} & 1 \end{Vmatrix}$$

with 1's along the main diagonal. Invariants of this group are called *semi-invariants.*

8. A general method for including contravariant arguments

The extension theorem of the last section works only for invariants of covariant vectors. There is, however, a general method for deriving a table of typical basic invariants for both kinds of vectors from a table for covariant vectors. Let $x^1, \cdots, x^{n-1}$ be $n - 1$ covariant or Latin vectors in n-space; the minors (of order $n - 1$) of the component matrix (6.6), when arranged in proper order and fitted with alternating signs, are the components of a contravariant vector

$$\xi = [x^1 \cdots x^{n-1}],$$

as is readily seen from the identity in the covariant vector x:

$$\begin{vmatrix} x_1^1 & \cdots & x_n^1 \\ \cdots & \cdots & \cdots \\ x_1^{n-1} & \cdots & x_n^{n-1} \\ x_1 & \cdots & x_n \end{vmatrix} = \xi_1 x_1 + \cdots \xi_n x_n = (\xi x).$$

Our statement holds with respect to any group Γ of *unimodular* linear transformations. On interpreting the ξ as homogeneous point coördinates in projective $(n - 1)$-space this is the well-known process of generating a point by intersection of $n - 1$ planes.

Let $F(xy \cdots ; \xi\eta \cdots)$ be an invariant form depending on a number of Latin and Greek arguments; in particular it may be of degree $h \geqq 1$ in ξ. We carry F into an invariant G whose degree in ξ is $h - 1$ by first polarizing with respect to ξ:

$$\sum_{i=1}^{n} \frac{\partial F}{\partial \xi_i} \bar{\xi}_i,$$

and then taking for ξ the vector $[x^1 \cdots x^{n-1}]$. The new Latin arguments $x^1, \cdots, x^{n-1}$ introduced by this process we shall call the *symbolic vectors*. The result is the invariant

$$\frac{1}{h}\begin{vmatrix} x_1^1, & \cdots, & x_n^1 \\ \cdots & \cdots & \cdots \\ x_1^{n-1}, & \cdots, & x_n^{n-1} \\ \frac{\partial F}{\partial \xi_i}, & \cdots, & \frac{\partial F}{\partial \xi_n} \end{vmatrix} = G(x^1 \cdots x^{n-1}; \cdots).$$

We added the factor $1/h$ so as to make certain that by the *restitution* of ξ for $[x^1 \cdots x^{n-1}]$ our G changes back to F. ξ is called the *restituent*. The process of restitution applies to any invariant G depending linearly on the $n - 1$ symbolic vectors $x^1, \cdots, x^{n-1}$:

$$G(x^1, \cdots, x^{n-1}) = \sum b(i_1 \cdots i_{n-1}) x_{i_1}^1 \cdots x_{i_{n-1}}^{n-1}.$$

One first makes G skew-symmetric by alternation:

$$\frac{1}{(n-1)!}\sum \pm G(x^1, \cdots, x^{n-1}),$$

the sum extending alternatingly over the permutations of $x^1, \cdots, x^{n-1}$, and then performs the restitution $[x^1 \cdots x^{n-1}] \to \xi$. The result, which we indicate by $G(x^1 \cdots x^{n-1}) \to F$, is given by the formula

$$F = \frac{1}{(n-1)!}\sum \pm b(i_1 \cdots i_{n-1})\xi_{i_n},$$

where the sum runs alternatingly over all permutations $i_1 \cdots i_{n-1} i_n$ of $1, 2, \cdots, n$.

Here are the two simplest examples of restitution:

$$[x^1 \cdots x^{n-1} x] \to (\xi x). \tag{8.1}$$

$$(\xi^1 x^1) \cdots (\xi^{n-1} x^{n-1}) \to \frac{1}{(n-1)!}[\xi^1 \cdots \xi^{n-1} \xi]. \tag{8.2}$$

By repeating the process which led from F to G one can lower the degree in ξ until it becomes zero; after having thus eliminated ξ one may in the same manner do away with the other Greek arguments. To be sure, a new set of symbolic Latin vectors is introduced at every step, and it is for this reason that it is so important that we possess a table of typical basic invariants sufficient for *any number* of Latin arguments. Our considerations evidently lead to the following

LEMMA (2.8.A). *In order to show that a table of typical basic invariants is complete it suffices to convince one's self:*

1) *that it contains a complete table of such invariants for covariant arguments only;*

2) *that a "term" (a product of basic invariants) containing in linear fashion*

the $n - 1$ symbolic vectors changes by restitution into an invariant expressible in terms of the basic invariants of our table. (In applying restitution, such factors of the term as contain no symbolic arguments, in particular "purely Greek" factors, can be disregarded.)

The method was worked out in detail by R. Weitzenböck;[15] he uses what he calls a "complex symbol" $p = (p_1, \cdots, p_n)$ for the symbolic vectors $x^1, \cdots, x^{n-1}$, the same for each of them, and automatically takes care of alternation by the multiplication rule $p_k p_i = -p_i p_k$.

When applying the method one has to have frequent recourse to a certain formal identity concerning bracket factors. $L(x)$ being a linear form, one has

$$\text{(8.3)} \qquad \begin{vmatrix} L(x^1), & x_1^1, & \cdots, & x_n^1 \\ \cdots & \cdots & \cdots & \cdots \\ L(x^{n+1}), & x_1^{n+1}, & \cdots, & x_n^{n+1} \end{vmatrix} = 0.$$

We write $x\, y \cdots z$ for $x^{i+2}, \cdots, x^{n+1}$, develop the determinant by the first column, and break it between the $(i + 1)^{\text{th}}$ and $(i + 2)^{\text{th}}$ term:

$$\text{(8.4)} \quad [x^1 \cdots x^i xy \cdots z]\cdot L(x^{i+1}) - + \cdots = L(x)\cdot[x^1 \cdots x^{i+1} y \cdots z] - + \cdots.$$

The sum on the left side consists of $i + 1$ alternating terms which arise from the one written down by selecting one vector at a time from the sequence $x^1, \cdots, x^i, x^{i+1}$ as the argument of L, leaving the remaining ones in the bracket factor in their natural order; the same holds for the right side with regard to the sequence $x, y, \cdots, z$.

When we are given a term containing a bracket factor, the identity (8.4) obviously enables us to draw one symbolic vector after the other into this bracket factor, which then lends itself readily to restitution:

$$[x^1 \cdots x^{n-1} x] \to (\xi x).$$

If our table contains the invariant (ξx) we may therefore in Lemma (2.8.A) assume that the term destined for restitution involves no bracket factor. Had we proved in Theorem (2.6.A) merely the fact that the bracket factor $[xy \cdots z]$ is the one basic "Latin" type for the group $SL(n)$, then this reasoning together with the formula (8.2) would have provided a new demonstration of the complete Theorem (2.6.A) dealing with both kinds of vectors.

By induction one derives from (8.4) the identity

$$\text{(8.5)} \quad \sum \pm [x^1 \cdots x^i x \cdots yz \cdots u]\cdot L(x^{i+1}, \cdots, x^k)$$
$$= \sum \pm L(x, \cdots, y)\cdot[x^1 \cdots x^k z \cdots u].$$

L is a skew-symmetric multilinear form of $k - i$ arguments. The sum on the left extends alternatingly to all "mixtures" of

$$x^1, \cdots, x^i \quad \text{with} \quad x^{i+1}, \cdots, x^k,$$

i.e. to all permutations of $x^1 \cdots x^k$ which preserve the order within both sections, while the right-hand sum runs over all mixtures of

$$x, \cdots, y \quad \text{with} \quad z, \cdots, u.$$

This identity simultaneously draws $k - i$ further symbolic vectors into the bracket. A special case is the formula

$$(8.6) \qquad [x^1 \cdots x^i x \cdots yz](\xi x^{i+1}) \cdots (\eta x^{n-1}) \to \begin{vmatrix} (\xi x), & \cdots, & (\xi z) \\ \cdots & \cdots & \cdots \\ (\zeta x), & \cdots, & (\zeta z) \end{vmatrix},$$

where ζ is the restituent. As an application we prove

THEOREM (2.8.B). *By adding to the table for $SL(n + \nu)$ given in Theorem* (2.6.A) *the block bracket $[x \cdots y]_{\text{bl}}$ of n covariant arguments $x, \cdots, y$ and the rim bracket*

$$(8.7) \qquad [\xi \cdots \eta]_{\text{rim}} = \begin{vmatrix} \bar{\xi}_1 & \cdots & \bar{\xi}_\nu \\ \cdots & \cdots & \cdots \\ \bar{\eta}_1 & \cdots & \bar{\eta}_\nu \end{vmatrix}$$

of ν contravariant arguments $\xi, \cdots, \eta$, one obtains a complete table for the group of step transformations dealt with in Theorem (2.7.B).

According to a previous remark and to formula (8.2) one needs bother merely about terms containing no total Latin bracket and at least one Latin block bracket. We then split the term into a product of factors of type $[x \cdots y]_{\text{bl}}$. and a product of factors of type (ξx). The first partial product is concerned with the n-space of block components only, and by applying to it our identity we may gather all symbolic vectors $x^1, \cdots, x^i$ present in that part into one such factor $[x \cdots y]_{\text{bl}}$. We are then left with a term of the form

$$[x^1 \cdots x^i x \cdots y]_{\text{bl.}} \cdot (\xi x^{i+1}) \cdots (\eta x^{n+\nu-1}).$$

If the absolute coördinate system be denoted by $e_1 \cdots e_n$, $\bar{e}_1 \cdots \bar{e}_\nu$, the block bracket may be written as a total bracket factor:

$$[x^1 \cdots x^i x \cdots y \bar{e}_1 \cdots \bar{e}_\nu].$$

We apply (8.6) and thus find after restitution with ζ as the restituent

$$\begin{vmatrix} (\xi x), & \cdots, & (\xi y), & (\xi \bar{e}_1), & \cdots, & (\xi \bar{e}_\nu) \\ \cdots & \cdots & \cdots & \cdots & \cdots & \cdots \\ (\zeta x), & \cdots, & (\zeta y), & (\zeta \bar{e}_1), & \cdots, & (\zeta \bar{e}_\nu) \end{vmatrix} = \begin{vmatrix} (\xi x), & \cdots, & (\xi y), & \bar{\xi}_1, & \cdots, & \bar{\xi}_\nu \\ \cdots & \cdots & \cdots & \cdots & \cdots & \cdots \\ (\zeta x), & \cdots, & (\zeta y), & \bar{\zeta}_1, & \cdots, & \bar{\zeta}_\nu \end{vmatrix}.$$

Simultaneous expansion with respect to the last ν columns shows that this is an aggregate of terms of the form (ξx) and (8.7).

9. Fourth example: orthogonal group

In investigating the group $O(n)$ of all proper and improper orthogonal transformations[16] A it is convenient to take into consideration (besides the absolute or "even" invariants) the special kind of relative invariant (called "odd") whose multiplier $\mu(A)$ is $+1$ for proper and -1 for improper rotations. The determinant $[x^1 \cdots x^n]$ of n vectors is an odd invariant. Capelli's special identity shows that Ωf is an odd or even invariant according as the invariant f is even or odd. For the time being, the reference field is supposed to be the field K of all real numbers. In the case of the orthogonal group no distinction need be made between covariant and contravariant vectors.

The odd as well as the even invariants are absolute invariants for the proper orthogonal group $O^+(n)$. *Vice versa,* an absolute invariant f of $O^+(n)$ is carried by all proper rotations into itself, by all improper rotations into one and the same new form f', and f is therefore the sum of an even and an odd invariant with respect to the total group $O(n)$:

$$f = \tfrac{1}{2}(f + f') + \tfrac{1}{2}(f - f').$$

THEOREM (2.9.A). *A complete table of typical basic invariants of the orthogonal group consists of* (1) *the scalar product* (uv) *and* (2) *the bracket factor* $[u^1u^2 \cdots u^n]$.

The product of two bracket factors can be expressed by the scalar products owing to the well-known relation

$$(9.1) \qquad [x^1x^2 \cdots x^n]\,[y^1y^2 \cdots y^n] = \begin{vmatrix} (x^1y^1) & (x^1y^2) & \cdots & (x^1y^n) \\ (x^2y^1) & (x^2y^2) & \cdots & (x^2y^n) \\ \cdots & \cdots & \cdots & \cdots \\ (x^ny^1) & (x^ny^2) & \cdots & (x^ny^n) \end{vmatrix}$$

On taking this into account one may state our theorem thus:

T_n^m. *Every even orthogonal invariant depending on* m *vectors* $x^1, x^2, \cdots, x^m$ *in* n*-dimensional vector space is expressible in terms of the* m^2 *scalar products* $(x^\alpha x^\beta)$. *Every odd invariant is a sum of terms*

$$[u^1u^2 \cdots u^n]\cdot f^*(x^1, \cdots, x^m),$$

where $u^1, \cdots, u^n$ *are selected from the row* $x^1, \cdots, x^m$ *and* f^* *is an even invariant.*

The proof follows the scheme laid out in §6; it becomes, however, a little more involved because the starting point for the application of the Capelli argument as set forth in §5 must be secured by an inductive conclusion from $n - 1$ to n. By using Capelli's general and special identities, the theorem T_n^m $(m \geqq n - 1)$ is reduced to the theorem T_n^{n-1} concerning $n - 1$ argument vectors. When $n - 1$ vectors $x^1, \cdots, x^{n-1}$ are numerically given and linearly independent, one may introduce a new orthogonal coördinate system such that they lie in the $(n - 1)$-dimensional space spanned by the first $n - 1$ fundamental vectors ("non-formal" part). Thus one has reduced the question to the study of orthogonal invariants in $n - 1$ dimensions, or more precisely, since they depend on exactly $n - 1$ vectors, to T_{n-1}^{n-1}. In view of this situation, it seems best first to pass from

$$(9.2) \qquad T_{n-1}^{n-1} \to T_n^{n-1} \to T_n^n,$$

and then to generalize T_n^n to T_n^m. The two steps into which the transition $T_{n-1}^{n-1} \to T_n^n$ breaks up according to (9.2) are performed by the "non-formal" argument and Capelli's special identity respectively, whereas the transition $T_n^n \to T_n^m$ $(m > n)$ rests on Capelli's general identity. As it is obvious how to carry out the second part, we turn to the inductive proof of T_n^n according to the scheme (9.2). Let us first restate

T_n^n. *An even invariant depending on n vectors* $x^1, \cdots, x^n$ *in n-dimensional space is expressible in terms of their* n^2 *scalar products; every odd invariant arises from an even one by multiplication with the bracket factor* $[x^1 \cdots x^n]$.

We prove first the step $T_{n-1}^{n-1} \to T_n^{n-1}$. Let

$$f(x, \cdots, y) = f\begin{pmatrix} x_1, & \cdots, & x_{n-1}, & x_n \\ \cdots & \cdots & \cdots & \cdots \\ y_1, & \cdots, & y_{n-1}, & y_n \end{pmatrix}$$

be an even invariant depending on $n - 1$ vectors $x, \cdots, y$ in n dimensions. The function

$$f_0\begin{pmatrix} x_1, & \cdots, & x_{n-1} \\ \cdots & \cdots & \cdots \\ y_1, & \cdots, & y_{n-1} \end{pmatrix} = f\begin{pmatrix} x_1, & \cdots, & x_{n-1}, & 0 \\ \cdots & \cdots & \cdots & \cdots \\ y_1, & \cdots, & y_{n-1}, & 0 \end{pmatrix}$$

is an even orthogonal invariant in $n - 1$ dimensions, and hence according to T_{n-1}^{n-1} is expressible as a polynomial F in the $(n-1)^2$ scalar products

$$\begin{matrix} (xx)^*, \cdots, (xy)^*, \\ \cdots\cdots\cdots \\ (yx)^*, \cdots, (yy)^*, \end{matrix}$$

where

$$(xy)^* = x_1y_1 + \cdots + x_{n-1}y_{n-1}\,.$$

If f were odd, the improper orthogonal substitution

$$J_n : x'_1 = x_1\,, \cdots, x'_{n-1} = x_{n-1}\,, x'_n = -x_n \tag{9.3}$$

would show that $f_0 = -f_0$, hence $f_0 = 0$.

If $x, \cdots, y$ are numerically given one can find a vector $z \neq 0$ perpendicular to all of them and then apply the classical inductive construction of a Cartesian coördinate system in such a way that the last axis e^n has the direction of z. In terms of this new coördinate system $e^1, \cdots, e^n$ the last component of each of the vectors $x, \cdots, y$ will vanish:

$$\begin{matrix} x = \bar{x}_1e^1 + \cdots + \bar{x}_{n-1}e^{n-1}, \\ \cdots\cdots\cdots \\ y = \bar{y}_1e^1 + \cdots + \bar{y}_{n-1}e^{n-1}. \end{matrix}$$

Invariance of f with respect to the proper orthogonal transformation which we have thus performed results in the equation

$$f(x, \cdots, y) = f_0(\bar{x}, \cdots, \bar{y}),$$

where $\bar{x}, \cdots, \bar{y}$ are the $(n - 1)$-dimensional vectors with the components

$$\begin{matrix} \bar{x}_1, \cdots, \bar{x}_{n-1}, \\ \cdots\cdots\cdots \\ \bar{y}_1, \cdots, \bar{y}_{n-1}. \end{matrix}$$

If f be odd we obtain at once

$$f = 0; \tag{9.4}$$

if f be even we apply T_{n-1}^{n-1} to the even orthogonal $(n - 1)$-dimensional invariant f_0 as mentioned above and thus find

$$f_0(\bar{x}, \cdots, \bar{y}) = F\begin{pmatrix} (\bar{x}\bar{x}), & \cdots, & (\bar{x}\bar{y}) \\ \cdots & \cdots & \cdots \\ (\bar{y}\bar{x}), & \cdots, & (\bar{y}\bar{y}) \end{pmatrix}.$$

Since our transformation was orthogonal,

$$(xy) = (\bar{x}\bar{y}),$$

and therefore, as we claimed,

$$f(x, \cdots, y) = F\begin{pmatrix} (xx), & \cdots, & (xy) \\ \cdots & \cdots & \cdots \\ (yx), & \cdots, & (yy) \end{pmatrix}. \tag{9.5}$$

The equations (9.4) and (9.5), one for the odd and the other for the even invariants, hold numerically irrespective of the values of the vectors $x, \cdots, y$ and consequently also as identities in the formal sense. Our result is

T_n^{n-1}. *There does not exist any odd invariant form of $n - 1$ vectors in n dimensions, while every even invariant of $n - 1$ vectors is expressible by their scalar products.*

The other step $T_n^{n-1} \to T_n^n$ is taken care of by Capelli's special identity applied to invariants $f(x^1, \cdots, x^n)$ depending on n vectors. Its right side,

$$[x^1 \cdots x^n] \cdot \Omega f, \tag{9.6}$$

contains the factor Ωf of lower rank than f. If f is even, Ωf is odd, and by hypothesis for induction can be expressed as the product of $[x^1 \cdots x^n]$ with a polynomial of scalar products. One then resorts to the equation

$$[x^1 \cdots x^n]^2 = \det (x^\alpha x^\beta) \qquad (\alpha, \beta = 1, \cdots, n),$$

in order to express the even invariant (9.6) in terms of scalar products only. It should be noticed that merely this special case of the equation (9.1) enters into our proof.

The reader is asked to compare the proof thus finished with the preliminary clumsy attempts at achieving the same end in §2; the promise given there is now redeemed. The method goes through, as it stands, in any Pythagorean reference field k.

By the Extension Theorem (2.7.A) we can determine a complete table of typical invariants depending only on covariant vectors for the group that is characteristic for Euclidean geometry of rank ν in an $(n + \nu - 1)$-dimensional space, i.e. for the extension Γ^ν of the proper orthogonal group $\Gamma = O^+(n)$. The table consists of: scalar product, block bracket and total bracket. By means of Weitzenböck's method of complex symbols contravariant arguments may be included.[17]

B. A Close-up of the Orthogonal Group

10. Cayley's rational parametrization of the orthogonal group

At first glance our hypothesis requiring the underlying field to be of Pythagorean nature seems necessary in the case of Euclidean geometry, where laying off a given segment on a given line is the fundamental metric construction. It is not without surprise, therefore, that one finds the decisive results to hold for any number field k (of characteristic 0); this is essentially due to the possibility of a rational parametrization of the orthogonal group which was first discovered by Cayley.[18] Unfortunately Cayley's parametric representation leaves out some of the orthogonal matrices, and a good deal of our efforts will be spent in rendering these exceptions ineffective. Our algebraic ambition is aroused; we interrupt the main trend and enter into a closer investigation of the orthogonal group in an arbitrary reference field k.

A matrix A may be called *non-exceptional* if

$$\det(E + A) \neq 0.$$

We then introduce a matrix S by

$$E + S = 2(E + A)^{-1},$$

with the inversion

$$E + A = 2(E + S)^{-1}.$$

S is likewise non-exceptional, and we have the mutual relations

$$S = (E - A)(E + A)^{-1} = (E + A)^{-1}(E - A), \tag{10.1}$$

$$A = (E - S)(E + S)^{-1} = (E + S)^{-1}(E - S). \tag{10.2}$$

In view of the commutability of both factors,

$$A = \frac{E - S}{E + S}$$

would not be an inappropriate notation for the latter.

Let now $G = \| g_{ik} \|$ be an arbitrary matrix. The equation

$$A^*GA = G \tag{10.3}$$

expresses the condition that the substitution A leaves invariant the bilinear form

$$\sum g_{ik}x_iy_k.$$

LEMMA (2.10.A). *If the non-exceptional matrices A and S are connected by (10.1) and (10.2), and G is any matrix, then $A^*GA = G$ if and only if*

$$S^*G + GS = 0. \tag{10.4}$$

Taking the transpose of (10.1) one gets

$$E - A^* = S^*(E + A^*).$$

Multiplying on the right by GA and taking account of (10.3) one finds

$$G(A - E) = S^*G(A + E),$$

and hence, on dividing by $A + E$ on the right,

$$-GS = S^*G.$$

Vice versa, if we assume (10.4) and multiply the transposed equation

$$A^*(E + S^*) = E - S^*$$

of (10.2) on the right by G, we find

$$A^*G(E - S) = G(E + S),$$

which yields (10.3) on right-multiplication by $(E + S)^{-1}$.

The usefulness of the substitution (10.2) lies in its changing the quadratic relations (10.3) for A into the linear relations (10.4) for S. We shall make use of this lemma in the case when G is symmetric and non-degenerate, in particular $G = E$, and also when G is anti-symmetric and non-degenerate. Let us repeat our result when applied to the case $G = E$:

THEOREM (2.10.B). *Every non-exceptional orthogonal matrix A is expressible in the form (10.2) where S is a non-exceptional skew-symmetric matrix. Conversely, if S is a non-exceptional skew-symmetric matrix, then the matrix A defined by (10.2) is non-exceptional and orthogonal.*

Throughout this and the following sections, $S = \| s_{ik} \|$ stands for a skew-symmetric matrix. Notice the equation holding for such matrices:

$$\det (E - S) = \det (E + S^*) = \det (E + S). \tag{10.5}$$

A queer observation is urged upon us: A as given by (10.2) has the determinant $+1$ since numerator and denominator have the same determinant $\neq 0$ according to (10.5). Hence this representation and the hypothesis

$\det(E + A) \neq 0$ on which it was based must be impossible for improper orthogonal transformations:

COROLLARY 1.

$$\det(E + A) = 0$$

for every improper orthogonal matrix A.

$-A$ is improper if A be proper, provided our space is of odd dimensionality n. Hence

$$\det(E - A) = 0$$

for every proper orthogonal matrix A in such a space. This relation implies that the homogeneous linear equations

$$\sum_{k=1}^{n} (\delta_{ik} - a_{ik})x_k = 0 \quad \text{or} \quad x_i = \sum_k a_{ik} x_k$$

have a solution $(x_1, \cdots, x_n)$ different from zero.

COROLLARY 2. *A proper rotation in a space of odd dimension has an "axis" through the origin whose points are fixed under the rotation.*

There must be a more direct reason for the fact established in Corollary 1 than our argument reveals. It is simple enough. On taking the determinant of both sides of the equation

$$A^*(E + A) = A^* + E,$$

one finds for an improper A,

$$-\det(E + A) = \det(A^* + E) = \det(E + A).$$

We shall now turn to a closer study of proper orthogonal matrices, in particular to those *exceptional* matrices A satisfying the equation

$$\det(E + A) = 0.$$

If A is exceptional, so is any conjugate $A' = U^{-1}AU$, U being an arbitrary orthogonal matrix; for

$$E + A' = U^{-1}(E + A)U,$$

and hence

$$\det(E + A') = \det(E + A).$$

If the non-exceptional A is represented in the form

$$A = (E - S)/(E + S),$$

then the conjugate $A' = U^{-1}AU$ is given by

$$A' = (E - S')/(E + S'),$$

where

$$S' = U^{-1}SU \qquad \text{and} \qquad \det (E + S') = \det (E + S) \neq 0.$$

For an exceptional A the homogeneous linear equations

$$\sum_{k=1}^{n} (\delta_{ik} + a_{ik})x_k = 0 \qquad (i = 1, \cdots, n) \tag{10.6}$$

have a non-vanishing solution x_k. The linear subspace of all vectors x satisfying these equations may be called P^0. To P^0 we are going to apply the following

LEMMA (2.10.C). *P^0 being an m-dimensional linear subspace of P one can introduce an orthogonal coordinate system $e_1, e_2, \cdots, e_n$ such that the first m fundamental vectors $e_1, \cdots, e_m$ lie in (and span) P^0. This is true provided the reference field k is real and Pythagorean. If k is real, the construction may require several consecutive adjunctions of square roots of square sums.*

The proof of the first part consists in the classic inductive construction of a Cartesian coördinate system, which is feasible under the hypothesis that the field k is real and Pythagorean. Otherwise it requires the adjunction of square roots of square sums. A few words about this process!

Let k be a given real field and c a number given in k as a square sum,

$$c = c_1^2 + \cdots + c_h^2,$$

but not itself a square in k. The polynomial $\gamma^2 - c$ of the indeterminate γ is then irreducible in k. The field $(k, \sqrt{c})$ arising from k by the adjunction of the square root $\sqrt{c}$ consists of all k-polynomials of the indeterminate γ modulo $\gamma^2 - c$; i.e., the vanishing of such an element $f(\gamma)$ means its divisibility by $\gamma^2 - c$. All elements of $(k, \sqrt{c})$ are uniquely expressible in the form

$$a + b\gamma \qquad (a, b \text{ in } k).$$

The one observation we wish to make here is that $(k, \sqrt{c})$ is real provided k itself is real. In other words, it never comes to pass that

$$(a_1 + b_1\gamma)^2 + \cdots + (a_\nu + b_\nu\gamma)^2 \equiv 0 \qquad (\text{mod } \gamma^2 - c) \tag{10.7}$$

unless all the numbers a_ρ, b_ρ in k vanish. (10.7) is equivalent to the following two equations in k:

$$(a_1^2 + \cdots + a_\nu^2) + (b_1^2 + \cdots + b_\nu^2)c = 0,$$

$$a_1b_1 + \cdots + a_\nu b_\nu = 0.$$

The first equation may be written

$$\sum_\rho a_\rho^2 + \sum_{\rho, i} (b_\rho c_i)^2 = 0 \qquad (\rho = 1, \cdots, \nu; i = 1, \cdots, h).$$

Because k is real one infers from this

$$a_\rho = 0, \qquad b_\rho c_i = 0,$$

or indeed the vanishing of all a_ρ and b_ρ.

The process of adjoining the square root of a square sum may be called "*Pythagorean adjunction.*" Our observation makes possible successive Pythagorean adjunctions without bringing our construction to a breakdown, since the field stays real throughout the extensions.

Let us now proceed, as we promised before, to apply our lemma to the m-dimensional subspace P^0 of all vectors x satisfying (10.6). After introducing the new Cartesian coördinate system, the first m axes $e_1, \cdots, e_m$ of which span P^0, our matrix A takes on the form

$$\begin{Vmatrix} -E_m & * \\ 0 & B \end{Vmatrix}$$

since (10.6) now has the solutions

$$\begin{aligned} e_1 &= (1, 0, \cdots, 0 \mid 0 \cdots 0), \\ &\cdots\cdots\cdots\cdots\cdots\cdots \\ e_m &= (0, 0, \cdots, 1 \mid 0 \cdots 0). \end{aligned}$$

The square sum of the elements in the first row is $= 1$,

$$(-1)^2 + \sum_{k=m+1}^{n} a_{1k}^2 = 1, \quad \text{hence} \quad \sum_{k=m+1}^{n} a_{1k}^2 = 0.$$

On account of the reality of the field this has the consequence that all elements a_{1k} ($k = m + 1, \cdots, n$) vanish. The same applies to each of the first m rows, proving that the starred upper right rectangle is empty as well as the lower left one. The det $(E + B)$ must not vanish; otherwise we should have a solution

$$(0, \cdots, 0 \mid x_{m+1}, \cdots, x_n)$$

of (10.6) outside P^0. Hence B is necessarily a proper orthogonal transformation, and as A was supposed to be proper, the determinant of our m-dimensional $-E_m$ must equal $+1$; thus m has to be even: $m = 2p$, $n - 2p = q$. The result is this:

After a suitable orthogonal transformation U of A,

$$A = U^{-1}\tilde{A}U,$$

$\tilde{A}$ breaks up according to

$$\tilde{A} = (-E_{2p}) \dotplus B,$$

where B is a non-exceptional proper orthogonal matrix in q dimensions.

From this we easily deduce

LEMMA (2.10.D). *In a real reference field k any proper orthogonal matrix A*

may be written as the product of two commuting non-exceptional proper orthogonal matrices A_1, A_2.

PROOF. Let us first assume our field k to be real and Pythagorean. Making use of our transformation U we put

$$\tilde{A}_1 = \begin{pmatrix} 0, & -1 \\ 1, & 0 \end{pmatrix} \dotplus \cdots \dotplus \begin{pmatrix} 0, & -1 \\ 1, & 0 \end{pmatrix} \dotplus E_q,$$

(p summands)

$$\tilde{A}_2 = \begin{pmatrix} 0, & -1 \\ 1, & 0 \end{pmatrix} \dotplus \cdots \dotplus \begin{pmatrix} 0, & -1 \\ 1, & 0 \end{pmatrix} \dotplus B;$$

$$A_1 = U^{-1}\tilde{A}_1U, \qquad A_2 = U^{-1}\tilde{A}_2U.$$

Then all conditions desired prevail for A_1, A_2 with regard to A, as they prevail for the transformed $\tilde{A}_1$, $\tilde{A}_2$ with regard to $\tilde{A}$.

If k is real, the construction goes through after a suitable chain of Pythagorean adjunctions extending k to a larger real field K. Because of

$$\det (E + A_1) \neq 0$$

we may write, according to Theorem (2.10.B),

$$A_1 = (E - S)(E + S)^{-1}$$

The skew-symmetric $S = \| s_{ik} \|$ will commute with A, and

$$\det (E + S) = \det (E - S) \neq 0. \tag{10.8}$$

Incidentally $S = U^{-1}\tilde{S}U$, where

$$\tilde{S} = \begin{pmatrix} 0, & 1 \\ -1, & 0 \end{pmatrix} \dotplus \cdots \dotplus \begin{pmatrix} 0, & 1 \\ -1, & 0 \end{pmatrix} \dotplus 0_q,$$

and 0_q is the q-rowed zero matrix.

$$A_2 \text{ is } = A_1^{-1}A = (E + S)(E - S)^{-1}A = (E + S)A(E - S)^{-1};$$

consequently the inequality $\det (E + A_2) \neq 0$ amounts to

$$\det \{(E - S) + (E + S)A\} \neq 0 . \tag{10.9}$$

Commutability with A,

$$SA = AS,$$

imposes a certain number of linear conditions on the $\frac{1}{2}n(n - 1)$ unknowns s_{ik} $(i < k)$. As the coefficients of these equations lie in k we are able to find basic solutions $S^{(1)}, \cdots, S^{(N)}$ in k such that every solution (in K or in any field over k) is a linear combination of them:

$$S = \tau_1S^{(1)} + \cdots + \tau_NS^{(N)}. \tag{10.10}$$

Both determinants (10.8) and (10.9) become polynomials in $\tau_1, \cdots, \tau_N$ after introducing S by (10.10). They cannot vanish identically in the formal sense since we know values of the parameters τ in K for which both are $\neq 0$. Hence we can find such values in k or even rational values for which neither vanishes numerically.

Although this finishes the proof of our lemma, it still may be worth while to follow up the prevailing situation a little more closely. In multiplying two commuting non-exceptional proper orthogonal matrices

$$A_1 = \frac{E - S_1}{E + S_1}, \qquad A_2 = \frac{E - S_2}{E + S_2},$$

the result will be

$$(10.11)\quad A = A_1 A_2 = \frac{(E - S_1)(E - S_2)}{(E + S_1)(E + S_2)} = \frac{(E + S_1 S_2) - (S_1 + S_2)}{(E + S_1 S_2) + (S_1 + S_2)},$$

and this again is of the same form (10.2) with the skew-symmetric

$$S = \frac{S_1 + S_2}{E + S_1 S_2}.$$

One encounters here what in the scalar sphere is known as the law of addition for tangents. The rule breaks down of course if

$$\det (E + S_1 S_2) = 0,$$

and the wider range of binary products (10.11) by which they embrace the exceptional besides the non-exceptional transformations, is due to this occurrence.

11. Formal orthogonal invariants

Cayley's parametrization would show at once that the proper orthogonal matrices form a rational irreducible algebraic manifold within the n^2-dimensional space of all matrices—were it not for the exceptional elements left out in the cold by the parametrization. It is a sound guess that eventually they will not matter. Acting on that hope we try the following modified definition of an orthogonal invariant whose advantage lies in its more formally algebraic character. Consider an arbitrary form $f(x, y, \cdots)$ depending on several vectors $x, y, \cdots$ in our n-space and homogeneous of prescribed degrees $\mu, \nu, \cdots$ in these arguments. Perform the substitution

$$(E - S)(E + S)^{-1}$$

cogrediently upon x and y and $\cdots$, where $S = \| s_{ik} \|$ is skew-symmetric and the $\frac{1}{2}n(n - 1)$ quantities s_{ik} $(i < k)$ are treated as indeterminates. The result will be a rational function of those indeterminates whose denominator is a certain power $h = \mu + \nu + \cdots$ of $\det (E + S)$:

$$\frac{f(x, y, \cdots ; s_{ik})}{| E + S |^h}.$$

We postulate the identity

$$f(x, y, \cdots ; s_{ik}) = | E + S |^h \cdot f(x, y, \cdots)$$

in $x, y, \cdots$ and the $\frac{1}{2}n(n - 1)$ variables s_{ik} : such a function f is called a *formal orthogonal invariant.*

This postulate submits the coefficients of f to a number of homogeneous linear equations with rational coefficients. Hence we can find a set of linearly independent basic invariants $f_1, \cdots, f_N$ of the prescribed degrees *with rational coefficients* such that every invariant f (in k) is a linear combination $a_1f_1 + \cdots + a_Nf_N$ with constant coefficients a (in k). Irrespective of the underlying number field k of characteristic 0, our problem has now been brought down to one concerning the ground field κ of rational numbers only.

The main issue will be to confirm Theorem (2.9.A) for *formal* orthogonal invariants. We shall simply have to overhaul our former procedure so as to remove what clogs the "exceptions" may put on its wheels. Lemma (2.10.D) was devised to meet that purpose. Indeed, a function invariant with respect to the transformations A_1 and A_2 is also invariant for $A = A_1A_2$, and thus invariance can be extended from the non-exceptional to *all* orthogonal transformations. A few preliminary observations will prepare us for the task ahead of us.

We introduce the special improper orthogonal involution J_n :

$$x_1' = x_1, \cdots, x_{n-1}' = x_{n-1}, x_n' = -x_n. \tag{9.3}$$

A formal invariant f shall be called *even* or *odd* according as J_n changes f into f or $-f$. The transformation J_{n-1} changing the sign of the $(n - 1)^{\text{th}}$ variable only will have the same effect because the proper transformation $J_{n-1}J_n$,

$$x_{n-1}' = -x_{n-1}, \qquad x_n' = -x_n,$$

leaves f unaltered. Indeed, in denoting by $S_{n-1,n}$ the skew-symmetric matrix in which

$$s_{n-1,n} = -s_{n,n-1} = 1$$

are the only elements $\neq 0$, $J_{n-1}J_n$ is the square of the non-exceptional

$$(E - S_{n-1,n})(E + S_{n-1,n})^{-1}.$$

f being a formal invariant, the form f' into which f changes by J_n is a formal invariant also. Invariance of f' with respect to

$$A = (E - S)(E + S)^{-1}$$

is the same as invariance of f with respect to

$$A' = J_n^{-1}AJ_n = (E - S')(E + S')^{-1} \qquad (S' = J_n^{-1}SJ_n).$$

$f + f'$ is even and $f - f'$ is odd. Therefore every formal invariant is the sum of an even and an odd one. If

$$f\begin{pmatrix} x_1 & \cdots & x_{n-1} & x_n \\ y_1 & \cdots & y_{n-1} & y_n \\ \cdots & \cdots & \cdots & \cdots \end{pmatrix}$$

is a formal invariant in n-space, then

$$f\begin{pmatrix} x_1 & \cdots & x_{n-1} & 0 \\ y_1 & \cdots & y_{n-1} & 0 \\ \cdots & \cdots & \cdots & \cdots \end{pmatrix} \tag{11.1}$$

is a formal invariant in $(n - 1)$-space. The remark about J_{n-1} and J_n shows that the latter is even if the former is such. If, however, the given f be odd, then (11.1) is obviously zero because J_n is the identity in the subspace $x_n = 0$.

We are now ready for the revision of our proof in the new more formal interpretation:

THEOREM (2.11.A). *Any even formal orthogonal invariant is expressible in terms of the scalar products of its arguments. Any odd formal orthogonal invariant is a sum of terms* $[xy \cdots z]\cdot f^*$, *where* $x, y, \cdots, z$ *are any* n *of the arguments and* f^* *is an even formal invariant.*

The salient point, and the only one where a non-formal numerical argument comes in is this. Let f be a formal invariant depending on $n - 1$ arguments and let

$$x = (x_1, \cdots, x_n)$$
$$\cdots\cdots\cdots\cdots\cdots$$
$$y = (y_1, \cdots, y_n)$$

be numerically given rational values of its arguments. There exists a rational vector $\neq 0$ perpendicular to all of them. Choosing it as the nth axis, and thus performing a certain proper orthogonal transformation A according to Lemma (2.10.C), we succeed in carrying $x, \cdots, y$ into $n - 1$ vectors

$$x' = (x_1', \cdots, x_{n-1}', 0)$$
$$\cdots\cdots\cdots\cdots\cdots$$
$$y' = (y_1', \cdots, y_{n-1}', 0)$$

whose last component $= 0$. Is it true that

$$f(x, \cdots, y) = f(x', \cdots, y')? \tag{11.2}$$

Construction of A may necessitate several successive Pythagorean adjunctions, so that A lies in a real field K over κ. By Lemma (2.10.D), A is the product A_1A_2 of two non-exceptional factors

$$A_1 = (E - S_1)(E + S_1)^{-1}, \qquad A_2 = (E - S_2)(E + S_2)^{-1}$$

in K. Formal invariance of f implies invariance for A_1 and A_2 and hence for A; thus (11.2) holds good.

A form with coefficients in a field k of characteristic 0, numerically invariant under non-exceptional rational proper orthogonal transformations, is obviously formally invariant. Hence we may note as a corollary to our theorem the fact that invariance can be extended from such particular transformations to all proper orthogonal transformations in any field whatsoever over κ.

12. Arbitrary metric ground form

After attaining this degree of generality, it is easy to substitute an arbitrary non-degenerate quadratic form

$$\sum_{i,k=1}^{n} g_{ik} x_i x_k \qquad (g_{ki} = g_{ik}) \tag{12.1}$$

for the square sum hitherto used,

$$x_1^2 + \cdots + x_n^2 . \tag{12.2}$$

The coefficient matrix $G = \| g_{ik} \|$ may lie in a reference field k of characteristic 0. The linear substitutions B in k leaving (12.1) invariant,

$$B^*GB = G, \tag{12.3}$$

form a group, the orthogonal group $O_G(n)$ with the "metric ground form" (12.1).

According to Lemma (2.10.A), a non-exceptional transformation B of that kind may be written in the form

$$B = (E - T)(E + T)^{-1}, \tag{12.4}$$

where T satisfies the linear condition

$$GT + T^*G = 0.$$

Formal invariance is defined in an obvious manner in terms of this parametrization.

After a suitable adjunction of square roots extending k to a field K over k one is able to transform the new ground form (12.1) into (12.2) by means of a certain substitution H:

$$G = H^*H.$$

This is brought about by the classical inductive construction of a Cartesian coördinate system when

$$(xy) = \sum g_{ik} x_i y_k \tag{12.5}$$

is considered as the scalar product.* The substitutions B of our present group arise from the orthogonal substitutions A by transformation with H:

$$B = H^{-1}AH.$$

However, this correspondence presupposes that we operate in the larger field K.

The representation (10.2) of A leads to (12.4) with

$$T = H^{-1}SH,$$

and this indeed changes the condition

$$S^* + S = 0 \qquad \text{into} \qquad GT + T^*G = 0.$$

If $f(x, y, \cdots)$ is a formal invariant of our new group $O_G(n)$ in K, then $f(H^{-1}x, H^{-1}y, \cdots)$ is a formal orthogonal invariant in K and hence expressible in terms of bracket factors and scalar products. The same is then true for our original form $f(x, y, \cdots)$ under the proviso of course that we now define the scalar product by (12.5). This result once established, one may afterwards confine oneself to the field k again.

A case of particular interest to the analyst is the field K of all real numbers. The metric ground form is then a non-degenerate quadratic form with real coefficients, not necessarily positive definite; thus the various possibilities as described by the index of inertia present themselves. They cause no difficulties for the study of the corresponding invariants (Lorentz group).

13. The infinitesimal standpoint

We are now prepared to give our investigation a new turn by introducing the idea of infinitesimal orthogonal transformations, and we shall thus attain a still more succinct and trim interpretation of invariance. This idea has sprung from the soil of continuous variables and in order to grasp its significance we shall follow the historical development and first consider the continuum of all real numbers as our reference field again. Surprisingly enough, however, the method goes through in a formal algebraic modification for any reference field k of characteristic 0.

Take the group $O^+(3) = D$ of rotations in 3-space as an example. This group serves to describe the mobility of a top, a solid body, one point o of which, the center, is fixed in space. Let t_1, t_2 be any two moments during the motion of our top. That spot on the top which covers the space-point p at time t_1 will cover a point p' at time t_2, and the mapping $p \to p'$, i.e. the displacement achieved in the time interval t_1t_2 will be an operation $H(t_1t_2)$ of the

---

* One must be careful, however, to choose the first basic vector e such that $\alpha = (ee) \neq 0$. Then every vector x is capable of a unique decomposition

$$x = \xi e + x'$$

where x' is in the $(n - 1)$-dimensional space defined by $(ex') = 0$. The square root to be extracted at this first step is that of α.[19]

group D. "Mobility" must always be described by a *group*; for $H(tt)$ will be the identity, $H(t_2t_1)$ the inverse of $H(t_1t_2)$, and $H(t_1t_3) = H(t_2t_3)H(t_1t_2)$. A material substance distributed throughout space (or any portion of it) moves as a rigid body about o if the group of possible displacements is our group D. In this description we compare the positions of the substance at two separate times t_1, t_2, ignoring the intermediate states through which it passes. It seems more natural to describe the actual continuous motion in time as one in which the position of the top undergoes an infinitesimal rotation during each time element $(t, t + dt)$, so that the motion appears as an integral-like chain of *infinitesimal operations* of D.

We use Cartesian coördinates x_1, x_2, x_3 with o as origin; x_1, x_2, x_3 are the coördinates of p or the components of the vector arm $\mathfrak{r} = \overrightarrow{op}$. In order to avoid concepts discredited by the history of mathematics, we shall replace the infinitesimal displacement (dx_1, dx_2, dx_3) by the velocity $\dot{\mathfrak{r}}$:

$$(\dot{x}_1 = dx_1/dt, \quad \dot{x}_2 = dx_2/dt, \quad \dot{x}_3 = dx_3/dt).$$

At every moment t we have a definite velocity field in space determining the displacement of the body during the following time element dt. Each rotation being a linear transformation, the equations defining the velocity $\dot{\mathfrak{r}}$ in its dependence on the point p must be linear and homogeneous:

$$\dot{x}_i = \sum_k s_{ik}x_k \qquad (i, k = 1, 2, 3). \tag{13.1}$$

The further requirement that $x_1^2 + x_2^2 + x_3^2$ stays invariant, leads to:

$$2\sum_i x_i\dot{x}_i = 0 \quad \text{or} \quad 2\sum_{i,k} s_{ik}x_ix_k = 0.$$

If we write this equation as

$$\sum_{i,k}(s_{ik} + s_{ki})x_ix_k = 0$$

the coefficients are symmetric and hence the vanishing of the quadratic form for all x requires the vanishing of those coefficients:

$$s_{ik} + s_{ki} = 0;$$

thus $\| s_{ik} \|$ is skew-symmetric. In 3-space we put

$$s_{23} = d_1, \qquad s_{31} = d_2, \qquad s_{12} = d_3.$$

Then (13.1) changes into the well-known kinematic formula: $\dot{\mathfrak{r}}$ = vector product of the arm $\mathfrak{r} = \overrightarrow{op}$ with a constant vector $\mathfrak{d} = (d_1, d_2, d_3)$ [constant = independent of p]. If the velocity field of a substance is of this character at any moment, then the substance moves like a rigid top around o.

In euclidean n-space the infinitesimal rotations are likewise described by equations

$$dx_i = \sum_k s_{ik}x_k \qquad (i, k = 1, \cdots, n)$$

with a skew-symmetric matrix $\| s_{ik} \|$.

In a similar fashion each continuous group of transformations will contain its infinitesimal elements which are infinitely small displacements of the point field. They are conveniently replaced by their velocity fields. Composition of two infinitesimal displacements amounts to adding their velocity fields. Hence the infinitesimal elements of a transformation group will form a linear pencil; they are nothing else than the pencil of line elements issuing from the unit element I on the group manifold γ. Every element of the group, at least every element which can be reached from I along a continuous path on γ will be constructable from the infinitesimal elements by stringing these together in an integral-like chain. This theory of the reduction of a continuous group to its infinitesimal elements, to which we return in more detail in Part B of Chapter VIII, is due to the Norwegian mathematician Sophus Lie. Enriched by the fundamental idea that was his, we now turn back to our algebraic endeavors.

By applying the algebraic definition of the derivative (Chapter I, §1) to a rational function $\varphi = f/g$ of one variable, one finds

$$\varphi' = (gf' - fg')/g^2,$$

the accent denoting the derivative. Assuming the polynomials f and g to be relatively prime, one concludes from $\varphi' = 0$, or

$$gf' - fg' = 0,$$

that $f' = g' = 0$, since f' and g' are of lower degree than f and g respectively; therefore f, g, and φ are constants.

Let F be a form of preassigned degrees $\mu, \nu, \cdots$ in a number of argument vectors $x, y, \cdots$. A substitution A cogrediently performed on all arguments changes F into a new form which we now designate as $F(A)$. In introducing new arguments $dx, dy, \cdots$ besides $x, y, \cdots$ we construct the total differential dF as the polarized form

$$dF = \sum_i \frac{\partial F}{\partial x_i} dx_i + \sum_i \frac{\partial F}{\partial y_i} dy_i + \cdots. \tag{13.2}$$

By means of a given n-matrix B one substitutes

$$dx = Bx, \qquad dy = By, \cdots;$$

the resulting form $d_B F$ has the same degrees as F in all arguments. The fact that an infinitesimal orthogonal transformation is given by

$$dx = Sx \qquad (S \text{ skew-symmetric})$$

leads one to adopt the definition:

F is an *infinitesimal orthogonal invariant* if

$$d_S F = 0 \tag{13.3}$$

identically in the skew-symmetric $S = \| s_{ik} \|$. The equation (13.3) is linear with respect to the $\frac{1}{2}n(n-1)$ indeterminates s_{ik} $(i < k)$, and so this equation comprehends $\frac{1}{2}n(n-1)$ homogeneous linear equations for the coefficients of F. The link with our previous developments is provided by the following simple

THEOREM (2.13.A). *The notions of formal and infinitesimal orthogonal invariants coincide.*

We first prove that a formal invariant is an infinitesimal one. Let S be a rational skew-symmetric matrix. From the given F we form the difference

$$F\left(\frac{E - \lambda S}{E + \lambda S}\right) - F(E) = \lambda \cdot \frac{\Phi(\lambda)}{|E + \lambda S|^h} \qquad (h = \mu + \nu + \cdots), \tag{13.4}$$

using a parameter λ. The "derivative" of $(E - \lambda S)/(E + \lambda S)$ for $\lambda = 0$ equals $-2S$, as proved by the equation

$$\frac{E - \lambda S}{E + \lambda S} - E = \frac{-2\lambda S}{E + \lambda S}.$$

The numerator $\Phi(\lambda)$ in (13.4) is a polynomial whose value $\Phi(0)$ for $\lambda = 0$ is thus

$$\Phi(0) = -2 \cdot d_S F.$$

If F is a formal invariant, the left side of (13.4) vanishes identically; hence $\Phi(0) = 0$, $d_S F = 0$.

The demonstration of the converse is a little more involved; the main tool is the composition law (10.11). We have the equation

$$\frac{E - \lambda^* S}{E + \lambda^* S} = \frac{E - T}{E + T} \cdot \frac{E - \lambda S}{E + \lambda S}, \tag{13.5}$$

where λ and λ^* are two parameters and

$$T = (\lambda^* - \lambda) \cdot \frac{S}{E - \lambda\lambda^* S^2}.$$

We put

$$F\left(\frac{E - \lambda^* S}{E + \lambda^* S}\right) - F\left(\frac{E - \lambda S}{E + \lambda S}\right) = \frac{(\lambda^* - \lambda) \cdot \Phi(\lambda, \lambda^*)}{|E + \lambda S|^h \, |E + \lambda^* S|^h}, \tag{13.6}$$

where Φ is a polynomial. The derivative of

$$F\left(\frac{E - \lambda S}{E + \lambda S}\right) = \varphi(\lambda)$$

is by definition

$$\varphi'(\lambda) = \frac{\Phi(\lambda, \lambda)}{|E + \lambda S|^{2h}}. \tag{13.7}$$

Using (13.5) and applying to (13.6) the same argument as in the first part of the proof, one finds that (13.7) becomes equal to $-2dF$ when one first introduces in (13.2) the differentials $dx, dy, \cdots$ by

$$dx = \frac{S}{E - \lambda^2 S^2} x, \qquad dy = \frac{S}{E - \lambda^2 S^2} y, \cdots,$$

and then substitutes

$$\frac{E - \lambda S}{E + \lambda S}x, \qquad \frac{E - \lambda S}{E + \lambda S}y, \cdots$$

for $x, y, \cdots$. Therefore if F is an infinitesimal invariant, one obtains

$$\varphi'(\lambda) = 0, \qquad \varphi(\lambda) = \text{const.} = \varphi(0).$$

Substitution of the value $\lambda = 1$ into the identity

$$F\left(\frac{E - \lambda S}{E + \lambda S}\right) = F$$

leads to the invariance of F with respect to the transformation $(E - S)(E + S)^{-1}$ provided $|E + S| \neq 0$.

C. The Second Main Theorem

14. Statement of the proposition for the unimodular group

In n-dimensional vector space the typical basic invariants with respect to the group of all unimodular linear transformations are the Latin bracket factor $[x_1, \cdots, x_n]$ of n covariant vectors x_i, the Greek bracket factor $[\xi_1, \cdots, \xi_n]$ of n contravariant vectors ξ_i, and the mixed factor (ξx), the product of a covariant vector x by a contravariant ξ. (Lower indices are now used for distinguishing several vectors since no notation for vector components is needed.) Among these basic invariants there exist relations of the following five types:

$$\text{(I)} \qquad \sum_x \pm [x_1 x_2 \cdots x_n](\xi x_0) = 0,$$

$$\text{(II)} \qquad \sum_x \pm [x_1 x_2 \cdots x_n][x_0 y_2 \cdots y_n] = 0,$$

$$\text{(III)} \qquad \sum_\xi \pm [\xi_1 \xi_2 \cdots \xi_n](\xi_0 x) = 0,$$

$$\text{(IV)} \qquad \sum_\xi \pm [\xi_1 \xi_2 \cdots \xi_n][\xi_0 \eta_2 \cdots \eta_n] = 0,$$

$$\text{(V)} \qquad [x_1 x_2 \cdots x_n][\xi_1 \xi_2 \cdots \xi_n] - \begin{vmatrix} (\xi_1 x_1) & \cdots & (\xi_1 x_n) \\ \cdots & \cdots & \cdots \\ (\xi_n x_1) & \cdots & (\xi_n x_n) \end{vmatrix} = 0.$$

The Latin letters denote covariant vectors, the Greek ones contravariant vectors. The alternating sum $\sum_x \pm$ in (I) and (II) consisting of $n + 1$ terms refers to the sequence $x_0 \mid x_1 \cdots x_n$; the same in (III) and (IV) with respect to $\xi_0 \mid \xi_1 \cdots \xi_n$. The identities (I) and (II) follow at once from the fact that the left side is a skew-symmetric multilinear form depending on $n + 1$ vectors $x_0, x_1, \cdots, x_n$.

Theorem (2.14.A). (*Second main theorem for the unimodular group.*) *All relations holding among the basic invariants are algebraic consequences of relations of these five types.*[20]

For the sake of a precise formulation of the second main theorem, one will first have to consider quantities like

$$(14.1) \qquad [x_1 \cdots x_n], \qquad [\xi_1 \cdots \xi_n], \qquad (\xi x)$$

as independent variables ("formal standpoint"); the Latin and Greek "symbols" x and ξ are here devoid of any independent significance. Nevertheless, it is to be understood that a bracket factor containing two identical symbols is zero, and that a bracket factor like $[x_1x_2 \cdots x_n]$ changes into $\pm[x_1x_2 \cdots x_n]$ by a permutation of the x_i—with the positive sign for even, the negative for odd permutations. Let F be an integral rational function of such variables composed of certain Latin *symbols* $x_1, x_2, \cdots$ and certain Greek ones $\xi_1, \xi_2, \cdots$. All the functions J obtained by substituting into the expressions on the left side of (I) to (V) these symbols in all possible combinations for the Latin and Greek letters used there—of course Latin symbols should be substituted for Latin letters only, Greek symbols for Greek letters—form the basis of an ideal $\mathfrak{J} = \{J\}$. One returns to the old standpoint by replacing each of the Latin and Greek symbols $x_1, x_2, \cdots; \xi_1, \xi_2, \cdots$ by a variable covariant or contravariant vector respectively, and then interpreting the symbols (14.1) in their old meaning, as determinants and inner product; this procedure is what I call the substitution. The second main theorem contends: *If F goes over into* 0 *by substitution, then it belongs to the ideal* $\mathfrak{J}$.

Besides (I) to (V), I make use of the following typical expression, which vanishes by substitution:

$$(\text{VI}) \qquad \begin{vmatrix} (\xi_0 x_0) & (\xi_0 x_1) & \cdots & (\xi_0 x_n) \\ (\xi_1 x_0) & (\xi_1 x_1) & \cdots & (\xi_1 x_n) \\ \cdots & \cdots & \cdots & \cdots \\ (\xi_n x_0) & (\xi_n x_1) & \cdots & (\xi_n x_n) \end{vmatrix}$$

It belongs to the ideal $\mathfrak{J}$; for on expanding by the first column and replacing determinants like $\det(\xi_i x_k)$, $(i, k = 1, \cdots, n)$ which appear as factors, by the product

$$[x_1 \cdots x_n][\xi_1 \cdots \xi_n]$$

modulo (V), one obtains (I) (where one has to take $\xi = \xi_0$) multiplied by $[\xi_1 \cdots \xi_n]$.

Without loss of generality we may suppose F to be homogeneous in each of the Latin and Greek symbols. For F may be decomposed in such homogeneous parts according to the degrees in those symbols; and if F changes into zero by substitution, the same holds for each individual part. (When computing the degree of a monomial term of F one has of course to consider each variable of degree 1 in the symbols that occur, of degree 0 in those that do not occur.) A single term of F may contain l Latin, λ Greek bracket factors; then the total degree in the Latin symbols minus the total degree in the Greek symbols is $n(l - \lambda)$ for this term. Consequently under the assumption of a homogeneous F the difference $l - \lambda$ has the same value for all terms of F. The product of a Latin and a Greek bracket factor, $[x_1 \cdots x_n]$ and $[\xi_1 \cdots \xi_n]$, may be replaced, mod (V), by a polynomial of variables of type (ξx), namely, the determinant of the $(\xi_i x_k)$. Hence we may assume that F contains either Latin or Greek bracket factors exclusively, and each term of F the same number of them. Since our table of fundamental relations is symmetric with respect to the part played by the Latin and Greek symbols, we confine ourselves to the case where only Latin bracket factors occur in F. After these preparations, the following sharper formulation of the second main theorem will hold:

T_0. *A homogeneous F which becomes zero by substitution and contains only variables of type (ξx) is $\equiv 0$ modulo expressions of type* (VI) *alone.*

T_x. *If, however, the homogeneous F involves Latin but no Greek bracket factors, besides variables of type (ξx), then it is $\equiv 0$ modulo expressions of types* (I) *and* (II).

15. Capelli's formal congruence

Upon the homogeneous F we are going to apply that Capelli identity which involves $n + 1$ Latin arguments in n-dimensional space. However, we now regard the Latin and Greek symbols, not as vectors, but simply as ingredients of the notations (14.1). Therefore we ought first to define the polar process according to this interpretation ("formal polarization"); Capelli's relation will then hold as a congruence mod $\mathfrak{J}$ rather than as an equation.

x and y being two of the Latin symbols, the polar process $D = D_{yx}$ is assumed to satisfy the formal laws (1.1.7). We derive from them how D_{yx} affects any polynomial f as soon as we know how it affects the arguments (14.1) of f. This is defined by the following rules:

1) A variable is changed into zero by D_{yx} if its symbol does not contain the letter x;

2) $D_{yx}[xx_2 \cdots x_n] = [yx_2 \cdots x_n], \qquad D_{yx}(\xi x) = (\xi y).$

To prove Capelli's congruence we proceed exactly as before, distinguishing, however, from the beginning the two cases that our homogeneous F contains either (a) variables of type (ξx) only, or (b) bracket factors besides. One introduces symbols $x_0', x_1', \cdots, x_n'$ not occurring in F and forms the sum

$$\sum_{x'} \pm D_{x_n' x_n} \cdots D_{x_1' x_1} D_{x_0' x_0} F \tag{15.1}$$

extending alternatingly to the $(n + 1)!$ permutations of $x_0', x_1', \cdots, x_n'$. In the case (a) this expression is obviously made up of terms

$$Q \cdot \sum_{x'} \pm (\xi_0 x_0')(\xi_1 x_1') \cdots (\xi_n x_n'), \tag{15.2}$$

where Q is a monomial of the variables occurring in F. Hence (15.1) is congruent to 0 modulo determinants of type (VI). *After* forming (15.1) one substitutes $x_0, x_1, \cdots, x_n$ for $x_0', x_1', \cdots, x_n'$. That this has the same effect as in the earlier less formal interpretation, one sees by the same procedure employed there, when one now makes use of the following two facts which take the place of the usual differentiation formulae:

(α) x and x' being two different symbols, we have

$$D_{yx}\{f(x', x)\}_{x'=x} = \{D_{yx'}f(x', x) + D_{yx}f(x', x)\}_{x'=x};$$

(β) if $f(x')$ be linear in x' then $D_{yx'}f(x') = f(y)$.

According to the rules (1.1.7) it is sufficient to give the proof of (α) for the case when f is one of the variables. The only variable in which both symbols x, x' actually occur together is a bracket factor $[x'xx_3 \cdots x_n]$. The left side is 0, as $[xxx_3 \cdots x_n]$ with two equal x means 0 by definition, whereas the right side equals

$$[yxx_3 \cdots x_n] + [xyx_3 \cdots x_n],$$

which is also 0 by definition.

In this manner one finds that (15.1) goes into

$$\text{(15.3)} \qquad \sum_{x'} \pm (D_{x_n'x_n} + n\delta_{x_n'x_n}) \cdots (D_{x_2'x_2} + 2\delta_{x_2'x_2})(D_{x_1'x_1} + 1\cdot\delta_{x_1'x_1})D_{x_0'x_0}F$$

after the original symbols $x_0, x_1, \cdots, x_n$ are substituted for the new ones $x_0', x_1', \cdots, x_n'$. The alternating sum in (15.3) has now to be interpreted such that $x_0', x_1', \cdots, x_n'$ is replaced by all the permutations of $x_0, x_1, \cdots, x_n$ one after the other. $\delta_{x'x}$ means 1 or 0 according as the symbols x' and x coincide or not. *The result is that* (15.3) *is congruent to* 0 *modulo type* (VI).

In the case (b) we proceed as follows. We have to apply (15.3) on a monomial F which is a product of variables of type (ξx) and of Latin bracket factors; at least one bracket factor is present. On first extending the sum only to the permutations of $x_1', \cdots, x_n'$ we are capable of successively drawing all symbols $x_1', \cdots, x_n'$ into one bracket factor. This is done by means of the identity (8.4) for the two cases

$$L(x) = (\xi x), \qquad L(x) = [xy_2 \cdots y_n],$$

where it reduces to (I) and (II) respectively. Our transformation by which all n symbols $x_1', \cdots, x_n'$ are finally thrown into one bracket factor $[x_1', \cdots, x_n']$ is however not an identity from the "formal standpoint," as was the case in §8, but a transformation modulo expressions of type (I) and (II). Disregarding factors which do not contain the symbols $x_0', x_1', \cdots, x_n'$, the sum (15.3) is now an expression either of type (I) or of type (II). We thus arrive at the result that (15.3) is congruent to 0 modulo types (I) and (II) provided the homogeneous F involves Latin bracket factors.

16. Proof of the second main theorem for the unimodular group

If F is of degrees $r_0, r_1, \cdots, r_n$ with respect to the symbols $x_0, x_1, \cdots, x_n$ we may write Capelli's congruence just proved in the form

$$\text{(16.1)} \qquad \rho F \equiv \sum \mathcal{P}F^*.$$

Here

$$\rho = r_0(r_1 + 1) \cdots (r_n + n),$$

and hence $\rho \neq 0$ if F actually involves the symbol x_0. The polynomial F^* is of lesser rank than F and derived from F by polarization; $\mathcal{P}$ is a succession of polar processes.

We defined "formal polarization" such that it does not matter whether polarization on a given F is carried out ("formally") before or ("not formally") after substitution. Hence if F vanishes by substitution, the same will hold for any G derived from F by polarization, in particular for the forms F^* occurring in (16.1). The expressions (VI) as well as (I) and (II) change into expressions of the same structure by polarization. We have therefore, by means of the congruence (16.1), reduced the validity of the respective theorems T_0 and T_x for a given F to their validity for the lower F^*, as long as F actually contains the symbol x_0. The inductive procedure thus started will end in entirely eliminating x_0 from F. The same procedure may be repeated as long as F still contains more than n Latin symbols by assigning to $n + 1$ of these symbols the rôle played by $x_0, x_1, \cdots, x_n$ in the above. In this manner one finally comes down to F's involving not more than n Latin symbols $x_1, \cdots, x_n$. In the case (a) such an F is a function of $n\nu$ variables of the form

$$(\xi_\kappa x_k), \qquad (\kappa = 1, \cdots, \nu; k = 1, \cdots, n);$$

the number ν is not subject to any limitation. Theorem T_0 will be proved by showing: if in case (a) the number of Latin symbols is n, our F cannot become zero through the substitution unless it was zero before the substitution.

This is readily done, as one can find at once n covariant vectors $x_1, \cdots, x_n$ and ν contravariant vectors ξ_κ such that the inner products $(\xi_\kappa x_k)$ become equal to arbitrarily preassigned numbers $z_{\kappa k}$. For this purpose one has only to take

$$\begin{aligned} x_1 &= (1, 0, 0, \cdots, 0), \\ x_2 &= (0, 1, 0, \cdots, 0), \\ &\cdots\cdots\cdots\cdots\cdots\cdots \\ x_n &= (0, 0, 0, \cdots, 1), \\ \xi_\kappa &= (z_{\kappa 1}, z_{\kappa 2}, \cdots, z_{\kappa n}). \end{aligned}$$

In case (b), if the homogeneous F contains not more than n Latin symbols, it is necessarily of the form

$$[x_1 x_2 \cdots x_n]^l \cdot G\{(\xi_\kappa x_k)\},$$

where the second factor G again depends on variables of type $(\xi_\kappa x_k)$ alone. Not more than one such term can occur because the exponent l is fixed by the difference between the total degrees of F in the Latin and Greek symbols. Hence, here again it is true that F can vanish after the substitution only if it does before the substitution; indeed the vanishing of F after the substitution implies that of G.

17. The second main theorem for the orthogonal group

If Γ is the group $O(n)$ of all proper and improper orthogonal transformations, then we have only one basic type of invariant, namely the scalar product (xy). A typical relation among scalar products is the following, involving $n + 1$ vectors x and $n + 1$ vectors y:

$$J = \begin{vmatrix} (x_0 y_0) & (x_0 y_1) & \cdots & (x_0 y_n) \\ (x_1 y_0) & (x_1 y_1) & \cdots & (x_1 y_n) \\ \cdots & \cdots & \cdots & \cdots \\ (x_n y_0) & (x_n y_1) & \cdots & (x_n y_n) \end{vmatrix} = 0.$$

THEOREM (2.17.A). *(Second main theorem for the orthogonal group.) Every relation among scalar products is an algebraic consequence of relations of type J.*

m Latin "symbols" $x_1, \cdots, x_m$ being given, a "relation" is a polynomial in the $\frac{1}{2}m(m + 1)$ variables $(x_\alpha x_\beta)$ that becomes zero when one replaces the symbols $x_1, \cdots, x_m$ by arbitrary vectors, and the variable $(x_\alpha x_\beta)$ by the scalar product of the two vectors x_α, x_β ("substitution"). We agree that even from the formal standpoint $(x_\alpha x_\beta)$ and $(x_\beta x_\alpha)$ shall be considered one and the same variable. When replacing the "letters" $x_0, x_1, \cdots, x_n, y_0, y_1, \cdots, y_n$ in the expression J by any of the "symbols" x one must allow a letter y to be replaced by the same symbol x_α as a letter x; it is useless, however, to replace different letters x (or y) by the same symbol x_α since the whole expression J is skew-symmetric in the x's (and in the y's). The expressions derived from J by the described replacements of the letters x and y by the symbols x_α in all possible combinations form the basis of an ideal $\mathfrak{J}$, and the exact rendering of the second main theorem states that *every relation R is congruent to* 0 *mod* $\mathfrak{J}$.

The proof is again given by means of Capelli's congruence. The formal definition of the polar process $D = D_{yx}$ is here as follows: $D(uv) = 0$ if neither u nor v equals x; $D(xu) = (yu)$ if u is different from x; $D(xx) = 2(xy)$. The rules (α) and (β) of §15 remain valid; one need only check the case $f(x', x) = (x'x)$ for (α). In forming (15.1) we shall obtain some terms that are built quite similarly to (15.2):

$$Q \cdot \sum_{x'} \pm (x_0' y_0)(x_1' y_1) \cdots (x_n' y_n); \tag{17.1}$$

but just these terms are $\equiv 0$ modulo the ideal $\mathfrak{J}$ defined above. However, this is not the only possibility now. Suppose that a term of F, for instance, contains the factor $(x_0 x_1)$ and that the first polarization $D_{x_0' x_0}$ is performed on this factor; it then goes into $(x_0' x_1)$. The second polarization $D_{x_1' x_1}$ performed on this factor in its new form, will change it into $(x_0' x_1')$. Hence we should be prepared for the possibility that, instead of the alternating sum in (17.1), another might occur whose leading term involves, in addition to variables of the kind $(x_0' y_0)$, also variables of the kind $(x_0' x_1')$ joining two of the new symbols x'. But $\sum_{x'}$ will then certainly be $= 0$ according to the convention $(xy) = (yx)$. Capelli's congruence therefore proves to be true modulo the ideal here introduced, whose basis consists of expressions of type J alone.

Using this congruence the number of symbols may gradually be reduced to n. To finish the proof we must show: a polynomial F of the $\frac{1}{2}n(n + 1)$ variables

$$(x_\alpha x_\beta) \qquad (\alpha, \beta = 1, 2, \cdots, n) \tag{17.2}$$

will be zero before the substitution if it is changed into zero by the substitution. In §2 we alluded to a demonstration of this based on the fact that (in the field K of real numbers) vectors x_α can be determined such that the scalar products (17.2) form an arbitrarily preassigned symmetric matrix, provided the quadratic form with this matrix of coefficients is positive definite.

Here we prefer to give instead a direct algebraic proof based on induction from $n - 1$ to n, and valid in any number field. For this purpose we depend upon the following two simple lemmas concerning the vanishing of polynomials:

1) A polynomial $\varphi(t)$ of the variable t vanishes identically if $\varphi(t + a) = 0$, where a is a fixed number in the ring from which the coefficients of φ are taken.

2) A polynomial $\varphi(t_1, \cdots, t_h)$ of h variables t_i vanishes identically if $\varphi(t_1', \cdots, t_h')$ vanishes identically in t when the t_i' are connected with the variables t_i by a non-singular linear transformation

$$t_i' = \sum_{k=1}^{h} a_{ik} t_k, \qquad \det a_{ik} \neq 0.$$

We assume $x_1, \cdots, x_{n-1}$ to be numerically given and linearly independent vectors in the subspace P_{n-1} of all vectors whose last component vanishes. Then the determinant Δ of the $n - 1$ vectors

$$x_i = (a_{i1}, \cdots, a_{i,n-1}) \qquad (i = 1, \cdots, n - 1) \tag{17.3}$$

in P_{n-1} is $\neq 0$. We regard the remaining vector

$$x = x_n = (t_1, \cdots, t_{n-1}, t)$$

as variable. Then (after substitution)

$$(xx_i) = \sum_k a_{ik} t_k \qquad (i = 1, \cdots, n - 1),$$

$$(xx) = t^2 + (t_1^2 + \cdots + t_{n-1}^2).$$

We now carry out a "partial substitution," replacing the variables

$$(x_\alpha x_\beta) \qquad (\alpha, \beta = 1, \cdots, n - 1) \tag{17.4}$$

by the scalar products of the $n - 1$ constant vectors (17.3). The given polynomial F which now depends on the variables (xx) and (xx_i) may be regarded as a polynomial in (xx) with coefficients lying in the ring of polynomials of the (xx_i):

$$F = \sum_l (xx)^l \varphi_l((xx_1), \cdots, (xx_{n-1})). \tag{17.5}$$

Putting

$$t_1^2 + \cdots + t_{n-1}^2 = \alpha,$$

$$\varphi_l(\sum_k a_{ik} t_k) = \varphi_l^*(t_1, \cdots, t_{n-1}),$$

we have, since F vanishes after substitution,

$$\sum_l (t^2 + \alpha)^l \varphi_l^*(t_1, \cdots, t_{n-1}) = 0.$$

Hence

$$F^*(s + \alpha) = 0 \tag{17.6}$$

identically in the variable s, where

$$F^*(s) = \sum_l s^l \varphi_l^*(t_1, \cdots, t_{n-1}).$$

The vanishing of the polynomial $F^*(s)$ now follows from (17.6) by the first lemma, and the vanishing of all its coefficients $\varphi_l^*(t_1, \cdots, t_{n-1})$ implies that of $\varphi_l(t_1, \cdots, t_{n-1})$ identically in t according to the second lemma.

The coefficients of the polynomial F of $(x_1x_n), \cdots, (x_{n-1}x_n), (x_nx_n)$ are polynomials f in (17.4). Concerning each such coefficient f, we have learned that it vanishes if one substitutes for $(x_\alpha x_\beta)$ the scalar products of $n - 1$ vectors x_α in the space P_{n-1} whose determinant $\Delta \neq 0$. The restriction by this algebraic inequality is irrelevant. On assuming the proposition under test to hold in P_{n-1} we are able to infer from the vanishing of f after this substitution its vanishing before the substitution, and that concludes the argument leading to the formal identity $F = 0$.

In the second place, let us consider the group $O^+(n)$ of all proper orthogonal transformations. To the typical invariant (xy) one then has to add as a further fundamental invariant the bracket factor $[x_1 \cdots x_n]$, and to the relations of type $J = J_1$ the following two further types:

$$J_2 = [x_1 \cdots x_n][y_1 \cdots y_n] - \begin{vmatrix} (x_1y_1) & \cdots & (x_1y_n) \\ \cdots & \cdots & \cdots \\ (x_ny_1) & \cdots & (x_ny_n) \end{vmatrix},$$

$$J_3 = \sum_x \pm [x_1 \cdots x_n](x_0 y).$$

The second main theorem asserts that this enumeration is exhaustive for the group $O^+(n)$.

Proof: A given relation R can first be reduced modulo type J_2 to such a form that no two bracket factors ever appear multiplied together, i.e.

$$R \equiv F + G$$

where F is a function of the

$$(x_\alpha x_\beta) \qquad (\alpha, \beta = 1, \cdots, n),$$

and G a linear combination of terms of the form

$$[x_{\alpha_1} \cdots x_{\alpha_n}] \cdot F^*\{(x_\alpha x_\beta)\}.$$

By means of an improper orthogonal transformation, for instance by changing the sign of the nth component of all m vectors, one realizes at once that F and G are themselves relations. The proof we have just carried through for the full orthogonal group then shows that, before the substitution, F must be congruent to 0 modulo type J_1. The same consideration as carried through in §16 in case bracket factors occur will here show that the Capelli congruence holds for G modulo type J_3. The relation (II), appearing there in addition to (I), does not come into play here because each term of G contains only one bracket factor. By means of Capelli's congruence the general G is reduced to a G involving not more than n Latin symbols $x_1, \cdots, x_n$:

$$G = [x_1 \cdots x_n] \cdot F^*\{(x_\alpha x_\beta)\}.$$

If G be a relation, the same is true for F^* and we saw that under such circumstances F^* and hence G must vanish before the substitution.

CHAPTER III

MATRIC ALGEBRAS AND GROUP RINGS

A. Theory of Fully Reducible Matric Algebras

1. Fundamental notions concerning matric algebras. The Schur lemma

When dealing with any set $\mathfrak{A}$ of matrices A in k it is natural to introduce its *linear closure* $[\mathfrak{A}]$ in k consisting of all finite linear combinations

$$\alpha_1 A_1 + \cdots + \alpha_r A_r$$

of matrices A_i in $\mathfrak{A}$ by means of coefficients α_i in k. Abstractly speaking, the linear closure is a linear set (or vector space) of a certain order h; h indicates the maximum number of matrices in $\mathfrak{A}$ that are linearly independent in k. If a subspace of the n-dimensional vector space P (in k) on which the matrices A operate is invariant under $\mathfrak{A}$, then it is also invariant under $[\mathfrak{A}]$; in all considerations concerning invariant subspaces and reduction, it is therefore convenient to replace $\mathfrak{A}$ by the linear closure $[\mathfrak{A}]$. If $\mathfrak{A}$ be a group, then $[\mathfrak{A}]$ will be closed with respect to the following three operations: addition of two matrices, multiplication of a matrix by a number in k, and multiplication of two matrices. Such a set is called a (matric) *algebra in k* and $[\mathfrak{A}]$ the *enveloping algebra* of the group $\mathfrak{A}$. Our statement is true even when $\mathfrak{A}$ is merely a *semi-group*; by that we mean a set of matrices closed with respect to multiplication (omitting the additional assumptions characteristic for a group proper, that it contains the unit matrix, that every element A is non-singular, and that A^{-1} is a member together with A). If we start with an arbitrary set $\mathfrak{A}$ of matrices, we can first form its multiplicative closure consisting of all finite products $\prod A$, A in $\mathfrak{A}$, which is a semi-group, and then pass to its linear closure: by these two steps we ascend to the enveloping algebra.

Just as in passing from a group of (linear) transformations to the abstract group scheme, we may ignore the nature of the elements of which a matric algebra consists and fix our attention solely on the operations performed on them. An *(abstract) algebra* $\mathfrak{a}$ *in k* then appears as a set of elements a for which three operations are defined: addition $a + b$ and multiplication ab of two elements a, b, and multiplication λa of an element a by a number λ in k. But throughout this book we look upon the matric algebras as our primary object; the abstract schemes are merely a device serving to facilitate their management. We shall stick to the convention that corresponding types like A and a, $\mathfrak{A}$ and $\mathfrak{a}$ of the upper and the lower case are used to mark the transition from matrices to abstract elements. Vice versa, the matric algebra $\mathfrak{A}$ is a *faithful representation* $a \to A$ of the abstract algebra $\mathfrak{a}$. Any correspondence $a \to R(a)$ associating with the elements a of a given abstract algebra $\mathfrak{a}$ matrices

$R(a)$ of degree n is called a *k-representation of* $\mathfrak{a}$ *of degree* n provided it preserves the fundamental operations:

$$R(a + b) = R(a) + R(b), \qquad R(\lambda a) = \lambda R(a), \qquad R(ab) = R(a) \cdot R(b),$$

$$(a, b \text{ elements in } \mathfrak{a}, \lambda \text{ a number in } k).$$

The representation is *faithful* if different elements a are represented by different matrices $R(a)$. The defining operations of an algebra $\mathfrak{a}$ satisfy the following laws in which a, b, c designate arbitrary elements of $\mathfrak{a}$ and λ an arbitrary number in k:

(1) All axioms characteristic for a vector space in k (of finite dimension h);

(2) The distributive law for both factors,

$$(a + b)c = (ac) + (bc), \qquad c(a + b) = (ca) + (cb),$$

supplemented by

$$\lambda a \cdot c = \lambda(ac), \qquad c \cdot \lambda a = \lambda(ca),$$

(3) The associative law

$$(ab)c = a(bc).$$

We say that the algebra contains *the unit* e if an element e is present satisfying the relations

$$ae = ea = a$$

for all elements a. (The unit being uniquely determined, we are justified in using the definite article "the".) An algebra $\mathfrak{a}$ containing the unit is called a *division algebra* if every element $a \neq 0$ possesses an inverse a^{-1}:

$$a \cdot a^{-1} = a^{-1} \cdot a = e.$$

If, moreover, the multiplication is commutative the division algebra will be a field finite over k.

As in the case of groups, we may associate with each element a of $\mathfrak{a}$ the linear transformation

$$(a): x \to x' = ax$$

operating on a variable element x in $\mathfrak{a}$. The algebra $\mathfrak{a}$ of order h appears here in two rôles: 1) as the set of elements a to which the transformations (a) correspond, 2) as the h-dimensional vector space ρ in which the transformations operate. The correspondence $(\mathfrak{a}): a \to (a)$ defines a representation, the so-called *regular representation,* because

$$b(ax) = (ba)x;$$

the degree of the regular representation is the order of the algebra. With algebras we are in a much more fortunate position than with groups inasmuch as this process furnishes a veritable representation by *linear* transformations, not merely a realization by some transformations in a general vague functional sense. The regular representation will be faithful if the algebra $\mathfrak{a}$ contains the unit e or, more generally, if 0 is the only element a satisfying the relation $ax = 0$ for all elements x.

A matrix A commuting with each member L of a given set $\mathfrak{L}$ of matrices,

$$AL = LA,$$

is called a *commutator* of $\mathfrak{L}$. The commutators of $\mathfrak{L}$ lying in k form a k-algebra $\mathfrak{A}$ of matrices, the *commutator algebra* of $\mathfrak{L}$ in k. Indeed

$$A_1L = LA_1, \qquad A_2L = LA_2$$

imply

$$(A_1 + A_2)L = L(A_1 + A_2), \qquad (A_2A_1)L = L(A_2A_1)$$

and

$$AL = LA \quad \text{implies} \quad \lambda A \cdot L = L \cdot \lambda A$$

(λ a number in k). Only the second implication needs a proof:

$$(A_2A_1)L = A_2(A_1L) = A_2(LA_1) = (A_2L)A_1 = (LA_2)A_1 = L(A_2A_1).$$

E is always a commutator, and if A is a non-singular commutator so is A^{-1}; for $AL = LA$ may then be written as $LA^{-1} = A^{-1}L$. The following statement due to I. Schur is of paramount importance:[1]

LEMMA (3.1.A). *If $\mathfrak{L}$ is irreducible in k, then any commutator A of $\mathfrak{L}$ in k is either zero or non-singular; in other words, the commutator algebra $\mathfrak{A}$ of $\mathfrak{L}$ in k is a division algebra.*

PROOF: The linear transformation $\mathfrak{x}' = A\mathfrak{x}$ will map our vector space P upon a subspace P′, the set of all image vectors $\mathfrak{x}'$. If $\mathfrak{y} = L\mathfrak{x}$, then $\mathfrak{y}' = L\mathfrak{x}'$ for any L in $\mathfrak{L}$, on account of the assumed commutability of A with all L. Hence P′ is invariant under $\mathfrak{L}$ and, according to the further assumption that $\mathfrak{L}$ is irreducible, either zero or the full space P. In the former case $A = 0$, in the latter A is non-singular.

The notion of commutators and this lemma have a direct bearing upon the problem of *covariants* of a given type $\mathfrak{G}$ discussed at the end of Chapter I. Let us suppose $\mathfrak{G}$ to be irreducible, so that our covariants $f = (f_1, \cdots, f_n)$ are primitive quantities. The components are forms of pre-assigned degrees $\mu, \nu, \cdots$ in some argument quantities $x, y, \cdots$. All such forms make up a linear set, a vector space Φ_N whose dimensionality N is given by (1.5.4). A first remark is this: the components $f_1, \cdots, f_n$ are either all zero or they are linearly independent. Indeed the sets of coefficients $(\alpha_1, \cdots, \alpha_n)$ constituting an identical relation

$$\alpha_1 f_1 + \cdots + \alpha_n f_n = 0$$

form a vector space invariant with respect to the representation contragredient to $\mathfrak{G}$, and hence, because of the irreducibility of $\mathfrak{G}$, either the whole space

($f_i = 0$) or the zero space (f_i linearly independent). If we have several covariants of type $\mathfrak{G}$,

$$f = (f_1, \cdots, f_n), \qquad f' = (f'_1, \cdots, f'_n), \cdots,$$

the n components of each span an n-dimensional subspace P, P$'$, $\cdots$ of Φ_N. A space $\text{P}^{(j)}$ of this sequence is either completely contained in the sum of the preceding ones, $\text{P} + \text{P}' + \cdots + \text{P}^{(j-1)}$, or linearly independent of it. This is shown by a typical argument that will reoccur again and again in the future: the intersection of $\text{P}^{(j)}$ with that sum is invariant and consequently, because of the irreducibility of $\mathfrak{G}$, either zero or the whole $\text{P}^{(j)}$. We can therefore determine a complete set of covariants of type $\mathfrak{G}$,

$$f' = (f'_1, \cdots, f'_n), \cdots, \qquad f^{(\tau)} = (f_1^{(\tau)}, \cdots, f_n^{(\tau)}),$$

such that all the $n\tau$ components

$$f_i^{(\alpha)} \qquad (i = 1, \cdots, n; \alpha = 1, \cdots, \tau)$$

are linearly independent, while the components of any covariant $f = (f_1, \cdots, f_n)$ of type $\mathfrak{G}$ are linear combinations of them:

$$f = A'f' + A''f'' + \cdots.$$

The ensuing equation

$$sf = A'(sf') + A''(sf'') + \cdots$$

reads more explicitly

$$Sf = A'(Sf') + A''(Sf'') + \cdots$$

if S is the matrix corresponding to s in $\mathfrak{G}$. Thus

$$S(A'f' + A''f'' + \cdots) = A'(Sf') + A''(Sf'') + \cdots,$$

and hence, on account of the linear independence of all the $f_i^{(\alpha)}$,

$$SA' = A'S, \qquad SA'' = A''S, \cdots.$$

In other words, A', A'', $\cdots$ lie in the commutator algebra $\mathfrak{A}$ of the irreducible set $\mathfrak{G}$, and each of them is either zero or non-singular. In more symmetric shape we may state this result thus:

THEOREM (3.1.B). *Given a number of covariants $f', f'', \cdots, f^{(\tau)}$ of irreducible type $\mathfrak{G}$, either there exists no linear relation among all their $n\tau$ components or else we have a relation*

$$A'f' + A''f'' + \cdots + A^{(\tau)}f^{(\tau)} = 0$$

where the A's lie in the commutator algebra of $\mathfrak{G}$ and at least one of them is $\neq 0$.

Particularly simple conditions prevail if the reference field k is *algebraically closed*, i.e. if any k-polynomial

$$\varphi(x) = x^m + \beta_1 x^{m-1} + \cdots + \beta_m$$

of a single indeterminate and of degree $m \geqq 1$ has a root α in k and therefore splits into m linear factors $(x - \alpha_1) \cdots (x - \alpha_m)$. The so-called fundamental theorem of algebra asserts that the domain of ordinary complex numbers is algebraically closed. In such a field k Schur's lemma takes on the simpler form:

LEMMA (3.1.C). *The only commutators A of a k-irreducible matric set $\mathfrak{L}$ in an algebraically closed field k are the numerical multiples αE of the unit matrix.*

Indeed, $\alpha E - A$ will be a commutator if A is, whatever the number α. On determining α as a root of the characteristic equation,

$$\det(\alpha E - A) = 0,$$

this commutator will be singular and hence by Schur's lemma $= 0$.

Therefore, in an algebraically closed field k, the components of several covariants $f', f'', \cdots$ of the same primitive type $\mathfrak{G}$ are either all linearly independent or we have a non-trivial relation of the form

$$\alpha' f' + \alpha'' f'' + \cdots = 0.$$

In other words, there exist either n simultaneous relations

$$\alpha' f'_i + \alpha'' f''_i + \cdots = 0 \qquad (i = 1, \cdots, n)$$

or no relations at all among the components $f'_i, f''_i, \cdots$ (excluding of course, in both cases, the trivial relation with all coefficients equal to zero). These considerations are clearly an indispensible supplement to the general concept of a covariant.

There is another part to Schur's lemma dealing with two inequivalent irreducible matric sets $\mathfrak{A}_1, \mathfrak{A}_2$. We operate in an arbitrary field k again. In order to establish in the most general way the correspondence between the two sets on which the notion of equivalence depends, we assume that we are given an abstract set $\mathfrak{a}$ of elements a and that to each a there corresponds a matrix $A_1(a)$ of degree n_1 and another $A_2(a)$ of degree n_2. Equivalence prevails if the degrees are equal, $n_1 = n_2$, and if there exists a non-singular matrix B such that

$$B^{-1}A_1(a)B = A_2(a) \quad \text{or} \quad A_1(a) = BA_2(a)B^{-1}$$

for all a in $\mathfrak{a}$.

LEMMA (3.1.D). *If the two sets $A_1(a)$, $A_2(a)$ are irreducible and inequivalent, then there is no matrix B (of n_1 rows and n_2 columns) such that*

$$A_1(a)B = BA_2(a)$$

holds identically in a, except $B = 0$.

PROOF: Let P_1 and P_2 be the two vector spaces subject to the transformations $A_1(a)$ and $A_2(a)$ respectively. The matrix B can be interpreted as a linear mapping $x_1 = Bx_2$ of P_2 into P_1. The linear subspace of P_1 consisting of all vectors x_1 of the form Bx_2 is invariant, for $A_1(a)x_1 = Bx'_2$ with $x'_2 = A_2(a)x_2$. In view of the assumed irreducibility of P_1 there are only two possibilities:

either $Bx_2 = 0$ for all x_2 in P_2, i.e. $B = 0$, or the whole space P_1 is covered by the linear mapping B of P_2 onto P_1. On the other hand, the set of all vectors x_2 in P_2 such that $Bx_2 = 0$ is an invariant subspace of P_2, for $BA_2(a)x_2 = A_1(a)Bx_2 = 0$. From the irreducibility of P_2 we conclude: either $Bx_2 = 0$ for all x_2 in P_2, i.e. $B = 0$, or $x_2 = 0$ is the only vector in P_2 such that $Bx_2 = 0$, so that distinct vectors in P_2 go into distinct vectors in P_1 under the mapping B. Hence if $B \neq 0$ we conclude that B defines a one-to-one linear mapping of P_2 onto P_1. But this means that B is a non-singular square matrix ($n_1 = n_2$) and hence that $A_1(a)$ and $A_2(a)$ are equivalent, contrary to hypothesis.

Let

$$\mathfrak{G}: f, f', f'', \cdots; \qquad \mathfrak{H}: g', g'', \cdots; \cdots$$

now be several sets of covariants of the given irreducible and inequivalent types $\mathfrak{G}$, $\mathfrak{H}$, $\cdots$ written in front of them, and let us suppose again that, with the omission of f, all the components of $f', f'', \cdots; g', g'', \cdots; \cdots$ are linearly independent. Again there are only two possibilities: either every component of f is a linear combination of the components of the latter, or the table of components consists of independent members even after the addition of f. In the first case we shall have a relation of the form

$$f = (A'f' + A''f'' + \cdots) + (B'g' + B''g'' + \cdots) + \cdots.$$

If

$$s \to S(s) \text{ in } \mathfrak{G}, \qquad s \to T(s) \text{ in } \mathfrak{H}, \cdots$$

we obtain

$$S(s)A' = A'S(s), \cdots; \qquad S(s)B' = B'T(s), \cdots; \cdots.$$

Hence $A', A'', \cdots$ lie in the commutator algebra of $\mathfrak{G}$ while $B', B'', \cdots; \cdots$ according to our new lemma (3.1.D) are all zero. The result may be stated in a more symmetric form:

THEOREM (3.1.E). *If $f', f'', \cdots; g', g'', \cdots; \cdots$ are sets of covariants of given irreducible and inequivalent types $\mathfrak{G}$, $\mathfrak{H}$, $\cdots$, their components are either all linearly independent, or at least one of the sets, let us say the first one $\mathfrak{G}: f', f'', \cdots$, is bound by a relation*

$$A'f' + A''f'' + \cdots = 0,$$

where $A', A'', \cdots$ are commutators of $\mathfrak{G}$ one of which is $\neq 0$.

2. Preliminaries

We shall now embark upon a more thorough investigation of the *structure of matric algebras*. Every such algebra $\mathfrak{A}$ carries with it its commutator algebra $\mathfrak{B}$ and the simultaneous discussion of $\mathfrak{B}$ with $\mathfrak{A}$ will throw much light upon $\mathfrak{A}$.

A few preparatory observations, mostly about "degeneracy", will clear our path. We operate throughout in a given field k and all terms like matrix, vector space, algebra, irreducible, are understood "in k". A given set $\mathfrak{A}$ of matrices A may show two kinds of *degeneracy*:

1) All matrices A map the vector space P upon one and the same proper subspace P′, i.e. all images $A\mathfrak{x}$ (A in $\mathfrak{A}$, $\mathfrak{x}$ in P) lie in P′: *degeneracy of the first kind.*

2) There are vectors $\neq 0$ carried by all transformations A of $\mathfrak{A}$ into zero: *degeneracy of the second kind.*

If a matric set $\mathfrak{A} = \{A\}$ shows the first kind of degeneracy, then the set $\mathfrak{A}^*$ of the transposed matrices A^* suffers from the second disease, and vice versa; for the assumption means that all vectors $\mathfrak{y} = (y_1, \cdots, y_n)$ of form $A\mathfrak{x}$ satisfy a non-trivial condition

$$a_1y_1 + \cdots + a_ny_n = 0,$$

and hence the contravariant vector $(a_1, \cdots, a_n)$ is carried into zero by all A^*. For a single matrix, non-degenerate in either sense means non-singular. A matric set $\mathfrak{A}$ containing the unit matrix or any non-singular matrix is degenerate in neither of the two senses. The vectors $\mathfrak{x}$ satisfying the relation $A\mathfrak{x} = 0$ for all A in $\mathfrak{A}$ and the vectors of form $A\mathfrak{x}$ (A in $\mathfrak{A}$) evidently constitute invariant subspaces. Therefore if $\mathfrak{A}$ is irreducible, $\mathfrak{A}$ can not be degenerate of the first kind, unless $A\mathfrak{x} = 0$ for all A in $\mathfrak{A}$ and all vectors $\mathfrak{x}$. But then $\mathfrak{A}$ consists of the one matrix $A = 0$ and on account of the irreducibility the degree must $= 1$. The matrix algebra $\mathfrak{A}$ consisting solely of the one-rowed matrix 0 or the abstract algebra consisting of the single element 0 shall be called the *null-algebra*. Hence an irreducible matric algebra is non-degenerate of the first, and as one readily proves in the same manner, of the second kind unless it is the null-algebra.

An irreducible matric algebra $\mathfrak{A}$ when considered as an abstract algebra is called *simple*; or, twisted around the other way, *a simple algebra $\mathfrak{a}$ is one capable of a faithful irreducible representation* $\mathfrak{A}$: $a \to A$. The null-algebra shall here be explicitly excluded.

When dealing with a representation $a \to R(a)$ of an algebra $\mathfrak{a}$ containing the unit e, there will correspond to e an idempotent matrix $T(e) = J$, i.e. one satisfying $JJ = J$. In our vector space P we construct the subspaces P_0 and P_1 of those vectors $\mathfrak{x}_0$ and $\mathfrak{x}_1$ for which

$$J\mathfrak{x}_0 = 0, \qquad J\mathfrak{x}_1 = \mathfrak{x}_1$$

hold respectively. They are linearly independent, and P′ splits into $P_0 + P_1$, because

$$\mathfrak{x} = \mathfrak{x}_0 + \mathfrak{x}_1 \quad \text{implies} \quad J\mathfrak{x} = \mathfrak{x}_1$$

and thus leads to the desired unique decomposition

$$\mathfrak{x}_1 = J\mathfrak{x}, \qquad \mathfrak{x}_0 = \mathfrak{x} - J\mathfrak{x} \tag{2.1}$$

(Peirce's decomposition). In a coördinate system adapted to this decomposition, J is the unit matrix surrounded by a rim of zeros:

$$J = \begin{Vmatrix} E & 0 \\ 0 & 0 \end{Vmatrix}.$$

Every $A = R(a)$ bears a rim of zeros of the same width on account of the equations

$$AJ = JA = A$$

reflecting $ae = ea = a$. Hence we may and shall limit ourselves to the *non-degenerate representations* in which e is represented by the unit matrix E. The most general representations arise from them by adding a rim of zeros to all matrices.

The set of all k-matrices of degree n is called the *complete matric algebra* $\mathfrak{M}_n$; its order is n^2.

From a given set $\mathfrak{A} = \{A\}$ of matrices of degree n we can derive the two sets $t\mathfrak{A}$ and $\mathfrak{A}_t$ of degree nt consisting of all matrices

$$\begin{Vmatrix} A & 0 & \cdots & 0 \\ 0 & A & \cdots & 0 \\ \cdots & \cdots & \cdots & \cdots \\ 0 & 0 & \cdots & A \end{Vmatrix}, \qquad (2.2) \begin{Vmatrix} A_{11} & \cdots & A_{1t} \\ \cdots & \cdots & \cdots \\ A_{t1} & \cdots & A_{tt} \end{Vmatrix}$$

respectively (A in $\mathfrak{A}$, A_{ik} in $\mathfrak{A}$). $t\mathfrak{A}$ and $\mathfrak{A}_t$ are algebras if $\mathfrak{A}$ is so. We prove the

LEMMA (3.2.A). *$\mathfrak{A}_t$ is irreducible provided $\mathfrak{A}$ is irreducible (and not the null-algebra).*

The sub-index t thus indicates a formal process by which new simple algebras are derived from a given one.

PROOF: The vectors in our tn-dimensional space P^t may be described as sets $(\mathfrak{x}_1, \cdots, \mathfrak{x}_t)$ of arbitrary vectors $\mathfrak{x}_i$ in the underlying n-dimensional space P. Let Σ^t be a subspace of P^t, invariant under $\mathfrak{A}_t$ and containing at least one vector

$$(\mathfrak{x}_1^0, \cdots, \mathfrak{x}_t^0) \neq (0, \cdots, 0),$$

and let us suppose $\mathfrak{x}^0 = \mathfrak{x}_1^0 \neq 0$. In performing on this vector the operation (2.2) in which all A_{ik} vanish save one $A_{\tau 1} = A$ in the first column we find that

$$(0, \cdots, A\mathfrak{x}^0, \cdots, 0)$$

lies in Σ^t. Since we excluded the degenerate case we may choose A_0 in $\mathfrak{A}$ such that $A_0\mathfrak{x}^0 \neq 0$, and thus there is at least one vector of the form

$$(0, \cdots, \mathfrak{x}_\tau, \cdots, 0)$$

$\neq 0$ in Σ^t. Let $A_{\tau\tau} = A$ vary in $\mathfrak{A}$ and take all other $A_{ik} = 0$. Because of the irreducibility of $\mathfrak{A}$ we then see that *each* vector of the form

$$(0, \cdots, \mathfrak{x}_\tau, \cdots, 0)$$

is contained in Σ^t. By summing over $\tau = 1, \cdots, t$ we come to the conclusion that Σ^t coincides with P^t.

In this assemblage of disparate observations a last one may fix the typical argument already encountered in §1 and frequently reappearing later:

LEMMA (3.2.B). *Out of a given row of irreducibly invariant subspaces Σ,*

$(j = 1, \cdots, m)$ one can select a subsequence all members of which are linearly independent and make up the same sum as the total sequence.

We suppose of course that we operate in a vector space P with a given set $\mathfrak{A}$ of matrices. The intersection of any Σ_j of our row with the sum of the previous terms $\Sigma_1 + \cdots + \Sigma_{j-1}$ is an invariant subspace and hence either 0 or the whole Σ_j. We drop Σ_j in the second, we retain it in the first case.

3. Representations of a simple algebra[2]

THEOREM (3.3.A). *A division algebra $\mathfrak{a}$ is simple. Indeed its regular representation $(\mathfrak{a})$ is irreducible as well as faithful.*

PROOF: $\mathfrak{a}$ as a vector space is again denoted by ρ. A subspace ρ' of ρ, invariant with respect to all operations $x \to ax$ and containing an element $i \neq 0$, would contain every element of the form ai and hence every element c whatsoever: $a = c \cdot i^{-1}$.

THEOREM (3.3.B). *Every non-degenerate representation of a division algebra $\mathfrak{a}$ is a multiple $t(\mathfrak{a})$ of its regular representation $(\mathfrak{a})$.*

Let $a \to T(a)$ be the given non-degenerate representation in an n-dimensional vector space P whose generic vector is denoted by $\mathfrak{x}$ and which we span by a coördinate system $\mathfrak{e}_1, \cdots, \mathfrak{e}_n$. We have $T(e) = E$. The terms invariant, irreducible, when applied to subspaces of P, refer to the algebra $\mathfrak{T}$ of matrices $T(a)$. An equation $\mathfrak{x}' = a\mathfrak{x}$ is to be interpreted as meaning $\mathfrak{x}' = T(a)\mathfrak{x}$. Let P_i be the subspace consisting of all vectors $\mathfrak{x} = x\mathfrak{e}_i$, one obtains when x varies over $\mathfrak{a}$. The correspondence $x \to \mathfrak{x}$ thus established is a similitude, i.e. ax goes into $a\mathfrak{x}$; hence P_i is invariant under the transformations $T(a)$ of $\mathfrak{T}$. Either P_i is zero or this mapping of ρ on P_i is a one-to-one correspondence. Indeed, the elements x for which $x\mathfrak{e}_i = 0$ form an invariant subspace of ρ; and as $(\mathfrak{a})$ is irreducible, either every x or no x except zero satisfies $x\mathfrak{e}_i = 0$. The first case $P_i = 0$ is here excluded because $e\mathfrak{e}_i = E\mathfrak{e}_i = \mathfrak{e}_i \neq 0$. The sum $P_1 + \cdots + P_n$ contains each of the basic vectors $\mathfrak{e}_1, \cdots, \mathfrak{e}_n$ and therefore coincides with the whole space P. On applying Lemma (3.2.B) to the sequence $P_1, \cdots, P_n$ we split P into a number of linearly independent invariant subspaces $P_{\alpha_1}, \cdots, P_{\alpha_t}$ in each of which $\mathfrak{T}$ induces a representation equivalent to $(\mathfrak{a})$.

If $\mathfrak{T}$ is allowed to degenerate it will be the direct sum of a multiple of the regular representation $(\mathfrak{a})$ and a multiple of the null-representation.

THEOREM (3.3.C). *A simple algebra $\mathfrak{a}$ contains a unit element. Its regular representation is a multiple t of that faithful irreducible representation $\mathfrak{A}$: $a \to A$ by which $\mathfrak{a}$ was defined. The order h is a multiple of the degree g: $h = gt$.*

The matrices A of degree g are linear mappings in a g-dimensional vector space P. The regular representation $(\mathfrak{A})$ associates with A the linear mapping

$$(A): X \to X' = AX$$

whose argument X varies within the linear set $\mathfrak{A}$ that here appears as an h-dimensional vector space ρ. Let us pick out an irreducible invariant subspace ρ_1 of ρ. ρ_1 is similar to P under their respective transformations (A) and A.

Indeed, let A^0 be an element $\neq 0$ in ρ_1 and $\mathfrak{e}$ a vector in P such that $A^0\mathfrak{e} \neq 0$. The formula $\mathfrak{x} = X\mathfrak{e}$ (X in ρ_1) maps ρ_1 on an invariant subspace $\rho_1\mathfrak{e}$ of P by the similitude $X \to \mathfrak{x}$; for $X \to \mathfrak{x}$ entails $AX \to A\mathfrak{x}$. The subspace $\rho_1\mathfrak{e}$ is either zero or the whole space P, because of the irreducibility of $\mathfrak{A}$. The first possibility is here excluded by $A^0\mathfrak{e} \neq 0$. In the remaining case the similitude $X \to \mathfrak{x}$ is a one-to-one correspondence between ρ_1 and P due to the irreducibility of ρ_1: the X in ρ_1 for which $X\mathfrak{e} = 0$ form an invariant subspace of ρ_1, and therefore $X = 0$ is the only such element. This proves that *any irreducible part of* ($\mathfrak{A}$) *is equivalent to the representation* $\mathfrak{A}$.

Since every vector $\mathfrak{x}$ in P is representable in the form $X\mathfrak{e}$ (X in ρ_1), there exists in particular an element I_1 in ρ_1 such that $\mathfrak{e} = I_1\mathfrak{e}$. Because of the invariance of ρ_1, the matrix XI_1 lies in ρ_1 for every matrix X in ρ. Since both matrices X and XI_1 change $\mathfrak{e}$ into the same vector $\mathfrak{x} = X\mathfrak{e}$ they must coincide for an X lying in ρ_1; in particular $I_1I_1 = I_1$: the matrix I_1 is a *generating idempotent* of ρ_1 in ρ. We now apply to ρ the idea of the Peirce decomposition and introduce the invariant subspaces ρ_1 and σ_1 of those X satisfying the equations $XI_1 = X$, $XI_1 = 0$ respectively. The first subspace is indeed that one previously designated as ρ_1. The uniqueness of the decomposition of any X in ρ into a matrix X_1 in ρ_1 and Y_1 in σ_1,

$$X = X_1 + Y_1, \tag{3.1}$$

results by deriving

$$X_1 = XI_1 \qquad (Y_1 = X - XI_1) \tag{3.2}$$

from (3.1).

Repeating this process, we determine an irreducibly invariant subspace ρ_2 of σ_1, a generating idempotent I_2' of ρ_2 in σ_1, and by means of it a decomposition of σ_1 into ρ_2 and a complementary invariant subspace σ_2:

$$Y_1 = Y_1I_2' + (Y_1 - Y_1I_2') = X_2 + Y_2,$$

$$Y_1 \text{ in } \sigma_1; \quad X_2 \text{ in } \rho_2, \quad Y_2 \text{ in } \sigma_2, \text{ i.e. } Y_2I_2' = 0.$$

On using the expression (3.2) of the generic matrix Y_1 in σ_1 we obtain

$$X_2 = (X - XI_1)I_2' = XI_2, \qquad Y_2 = X - (XI_1 + XI_2), \tag{3.3}$$

where

$$I_2 = I_2' - I_1I_2'.$$

At the next step we break σ_2 into an irreducibly invariant subspace ρ_3 generated in σ_2 by the idempotent I_3' and a complementary invariant subspace σ_3. ρ_3 consists of the matrices of the form Y_2I_3' (Y_2 in σ_2) which by (3.3) reduces to

$$XI_3 \quad \{X \text{ in } \rho, I_3 = I_3' - (I_1I_3' + I_2I_3')\}.$$

The final result will be *a decomposition of ρ into irreducibly invariant subspaces* $\rho_1, \cdots, \rho_t$ according to a formula

$$X = XI_1 + XI_2 + \cdots + XI_t \qquad (X \text{ in } \rho).$$

We already know that each irreducible part of the regular representation is equivalent to the representation $\mathfrak{A}$; hence $(\mathfrak{a})$ is equivalent to the multiple t of $\mathfrak{A}$ and $h = tg$.

As XI_α is the component X_α of X lying in ρ_α we obtain in particular for $X = I_\beta$ (whose β^{th} component is I_β while all other components vanish)

$$I_\beta I_\alpha = 0 \text{ for } \beta \neq \alpha, \qquad = I_\alpha \text{ for } \beta = \alpha.$$

The sum

$$I = I_1 + I_2 + \cdots + I_t$$

satisfies the equation $AI = A$ for all A in $\mathfrak{A}$, in particular $I \cdot I = I$. A vector $\mathfrak{y}$ carried by I into zero,

$$(3.4) \qquad I\mathfrak{y} = 0,$$

satisfies the equation

$$A\mathfrak{y} = 0 \text{ for all } A \text{ in } \mathfrak{A},$$

because $A\mathfrak{y} = AI\mathfrak{y}$. Having excluded the trivial case of the null-algebra we are thus sure that only $\mathfrak{y} = 0$ has the property (3.4). The Peirce decomposition (2.1) then shows at once that $\mathfrak{x} = I\mathfrak{x}$ for every vector $\mathfrak{x}$, or that I is the unit matrix; thus $\mathfrak{A}$ *contains the unit matrix E* and hence $\mathfrak{a}$ the unit element e represented by E in $\mathfrak{A}$.

The method by which we obtained Theorem (3.3.B) may be used to prove a general statement worth mentioning although it will not figure as an indispensable part in our theoretical construction.

THEOREM (3.3.D). *If the regular representation* $(\mathfrak{a})$ *of an algebra* $\mathfrak{a}$ *splits into irreducible parts* $\mathfrak{A}_1, \mathfrak{A}_2, \cdots$ *then every representation that is non-degenerate of the first kind splits into irreducible parts each of which is equivalent to one of the* $\mathfrak{A}_i$.

PROOF: The hypothesis asserts that $\mathfrak{a}$, considered as the space ρ of the regular representation, decomposes into irreducibly invariant subspaces $\rho_1, \rho_2, \cdots, \rho_t$. Let $\mathfrak{x}$ be the generic vector and $\mathfrak{e}_1, \cdots, \mathfrak{e}_g$ a coördinate system of the space P of the given representation

$$\mathfrak{A}: a \rightarrow T(a).$$

Again, $\mathfrak{x}' = a\mathfrak{x}$ shall mean $\mathfrak{x}' = T(a)\mathfrak{x}$ and $\rho_\alpha\mathfrak{e}$ shall denote the set of all vectors $\mathfrak{x} = x\mathfrak{e}$ (x in ρ_α). We then form the table

$$\begin{array}{c} \rho_1\mathfrak{e}_1, \cdots, \rho_t\mathfrak{e}_1 \\ \cdots\cdots\cdots\cdots \\ \rho_1\mathfrak{e}_g, \cdots, \rho_t\mathfrak{e}_g \end{array}$$

to which we apply Lemma (3.2.B). The representation induced by $\mathfrak{A}$ in any of the subspaces $\rho_i\mathfrak{e}_k$ we retain is equivalent to the representation induced by the regular one in ρ_i; for such a $\rho_i\mathfrak{e}_k$ does not vanish and stands in one-to-one similarity correspondence with ρ_i by means of the mapping $x \to x\mathfrak{e}_k$ (x in ρ_i). The sum of the whole table contains every vector $a\mathfrak{x}$ (a in $\mathfrak{a}$, $\mathfrak{x}$ in P) and hence, degeneracy of the first kind being explicitly excluded, any vector whatsoever.

An immediate consequence of this and the previous proposition is

THEOREM (3.3.E). *Every non-degenerate representation of a simple algebra* $\mathfrak{a}$ *is equivalent to a multiple of its faithful irreducible representation* $\mathfrak{A}$. *In particular,* $\mathfrak{A}$ *is the only irreducible representation of* $\mathfrak{a}$.

4. Wedderburn's theorem

We now pass to the relationship of this analysis to the commutator idea. It springs from the following source:

THEOREM (3.4.A). *If an algebra* $\mathfrak{a}$ *contains a unit element* e, *the only linear transformations that commute with all transformations* (a): $x \to x' = ax$ *are of the form* $(b)'$: $x \to y = xb$ (b *an element in* $\mathfrak{a}$).

Indeed, if $y = B(x)$ is such a commutator, we must have by definition

$$B(ax) = a \cdot B(x). \tag{4.1}$$

Put $B(e) = b$ and apply (4.1) to $x = e$; one thus gets the desired formula, $B(a) = ab$ for every a.

When we designate by $\mathfrak{a}'$ the *inverse algebra* of $\mathfrak{a}$ differing from $\mathfrak{a}$ in that the product of two elements a and b is now defined as ba rather than ab, we may express our result thus: *The commutator algebra of the regular representation of* $\mathfrak{a}$ *is the regular representation of* $\mathfrak{a}'$; the relationship is therefore *mutual.*

This applies in particular to a *division algebra* $\mathfrak{a}$; then both regular representations $(\mathfrak{a})$ and $(\mathfrak{a}')$ are irreducible.

We take up again our *simple algebra* $\mathfrak{A}$ or $\mathfrak{a}$. The *commutator algebra* $\mathfrak{B}$ of $\mathfrak{A}$ is in abstracto a division algebra $\mathfrak{d}'$ (of order d), hence in concreto a multiple $t(\mathfrak{d}')$ of its regular representation $(\mathfrak{d}')$; thus the generic matrix of $\mathfrak{B}$ has the form

$$B = \begin{Vmatrix} (b)' & & \\ & \ddots & \\ & & (b)' \end{Vmatrix} \quad (t \text{ rows}), \tag{4.2}$$

where $(b)'$ varies over all the operators

$$x \to x' = xb \qquad (x \text{ variable in } \mathfrak{d})$$

belonging to the elements b of the inverse $\mathfrak{d}$. Hence $g = d \cdot t$. The commutator algebra $\bar{\mathfrak{A}}$ of $\mathfrak{B}$ consists of all matrices of the form

$$\bar{A} = \begin{Vmatrix} (a_{11}), & \cdots, & (a_{1t}) \\ \cdots & \cdots & \cdots \\ (a_{t1}), & \cdots, & (a_{tt}) \end{Vmatrix}, \tag{4.3}$$

where each (a_{ik}) is an operator

$$x \to x' = a_{ik}x \qquad (a_{ik} \text{ in } \mathfrak{d})$$

of the regular representation $(\mathfrak{d})$ of the division algebra $\mathfrak{d}$; thus

$$\bar{\mathfrak{A}} = (\mathfrak{d})_t . \tag{4.4}$$

$\bar{\mathfrak{A}}$, the commutator of the commutator of $\mathfrak{A}$, evidently contains $\mathfrak{A}$. *We wish to establish the fact that it coincides with* $\mathfrak{A}$. For that purpose we notice that the commutator of the commutator of $u\mathfrak{A}$ certainly comprises $u\bar{\mathfrak{A}}$, because the commutator algebra of $u\mathfrak{A}$ is $\mathfrak{B}_u$. Hence were $\bar{\mathfrak{A}}$ actually larger than $\mathfrak{A}$, the same would hold for $u\mathfrak{A}$, in particular for that multiple of $\mathfrak{A}$, $\tau\mathfrak{A}$, which by Theorem (3.3.C) is $\sim (\mathfrak{a})$. This, however, is contrary to our above remark that $(\mathfrak{a})$ is the commutator of the commutator $(\mathfrak{a}')$ of $(\mathfrak{a})$. Thus we are enabled to replace the equality (4.4) by Wedderburn's theorem*:

$$\mathfrak{A} = (\mathfrak{d})_t . \tag{4.5}$$

It shows that the order h of our simple algebra $\mathfrak{a}$ is $d \cdot t^2 = g \cdot t$, and hence the number τ just mentioned equals t:

$$(\mathfrak{a}) \sim t\mathfrak{A}.$$

Abstractly speaking, (4.5) states that our simple algebra is isomorphic to the algebra of t-rowed matrices

$$\left\| \begin{matrix} a_{11} & \cdots & a_{1t} \\ \cdots & \cdots & \cdots \\ a_{t1} & \cdots & a_{tt} \end{matrix} \right\| \tag{4.6}$$

whose elements a_{ik} are taken from a division algebra $\mathfrak{d}$. It goes beyond that abstract statement in telling us how to obtain the concrete matric form $\mathfrak{A}$ of $\mathfrak{a}$, namely by replacing each element a_{ik} in $\mathfrak{d}$ by the d-rowed matrix (a_{ik}). In terms of the normal form (4.6) of an arbitrary element a of $\mathfrak{a}$, the equation

$$(a)\colon x' = ax \qquad (a \text{ and } x \text{ in } \mathfrak{a})$$

reads:

$$x'_{ij} = \sum_{k=1}^{t} a_{ik}x_{kj} \qquad (a_{ik}, x_{ik} \text{ in } \mathfrak{d}).$$

Separation of the individual columns $(x_{1j}, \cdots, x_{tj})$ exhibits the way in which the regular representation $(\mathfrak{a})$ decomposes into t times $\mathfrak{A}$. The commutator algebra $\mathfrak{B}$, i.e. $\mathfrak{d}'$, and hence $\mathfrak{d}$, are *uniquely determined* by $\mathfrak{A}$.

THEOREM (3.4.B). *The relationship of an irreducible matric algebra $\mathfrak{A}$ and its commutator algebra $\mathfrak{B}$ is mutual: $\mathfrak{A}$ is the full commutator algebra of $\mathfrak{B}$. $\mathfrak{B}$ is expressed in terms of a uniquely determined division algebra $\mathfrak{d}$ of order d as $t \cdot (\mathfrak{d}')$, $\mathfrak{A}$ as $(\mathfrak{d})_t$. Besides $h = tg$ we have $g = dt$, hence $h = dt^2$.*

* This short-cut to Wedderburn's theorem was pointed out to me by R. Brauer.

We mention explicitly the following special case:

THEOREM (3.4.C). *An irreducible $\mathfrak{A}$ of degree g whose only commutators are multiples αE of the unit matrix E (case $d = 1$) is the complete matric algebra $\mathfrak{M}_g$ in k (and therefore irreducible in any field over k—"absolute irreducibility").*

If k is algebraically closed, the multiples of the unit matrix are the only commutators of $\mathfrak{A}$ and hence $\mathfrak{A} = \mathfrak{M}_g$. In this form our theorem is due to Burnside, while the general criterion, Theorem (3.4.C), holding in any field k, was given by Frobenius and I. Schur.[3]

Those elements a in $\mathfrak{a}$ that commute with all elements x,

$$ax = xa, \tag{4.7}$$

from the *centrum* $\mathfrak{z}$ of $\mathfrak{a}$. The corresponding matrix A must be at the same time a matrix B of the commutator algebra, i.e. we must have

$$(a_{ik}) = (z)\delta_{ik}, \qquad a_{ik} = z\delta_{ik} \tag{4.8}$$

in (4.3), (4.6) respectively, where z designates an element in the centrum of the division algebra $\mathfrak{d}$. Hence the centrum of $\mathfrak{a} = \mathfrak{d}_t$ is isomorphic to the centrum of the division algebra $\mathfrak{d}$. Or one might argue a little more directly as follows. On specializing $x = \| x_{ik} \|$ in the equation (4.7), i.e. in

$$\sum_k a_{ik}x_{kj} = \sum_k x_{ik}a_{kj}, \tag{4.9}$$

to $x_{ik} = \xi_{ik}e$ (ξ_{ik} numbers in k, e the unit element in $\mathfrak{d}$) one finds (4.8), namely $\| a_{ik} \| = zE$, and then (4.9) shows that z in $\mathfrak{d}$ must commute with every x in $\mathfrak{d}$.

The full reciprocity between algebra and commutator algebra is not reached until we pass from the irreducible representation $\mathfrak{A}$ of our simple algebra $\mathfrak{a}$ to a multiple $s\mathfrak{A}$. For this algebra, $s \cdot (\mathfrak{d})_t$, we readily find $t \cdot (\mathfrak{d}')_s$ to be its commutator algebra. The structure of the generic elements of our two algebras is indicated by the schemes

$$\left|\begin{array}{c|c} \begin{array}{ccc} (a_{11}) & \cdots & (a_{1t}) \\ \cdots & \cdots & \cdots \\ (a_{t1}) & \cdots & (a_{tt}) \end{array} & 0 \\ \hline 0 & \begin{array}{ccc} (a_{11}) & \cdots & (a_{1t}) \\ \cdots & \cdots & \cdots \\ (a_{t1}) & \cdots & (a_{tt}) \end{array} \end{array}\right| \qquad \left|\begin{array}{c|c} \begin{array}{ccc} (b_{11})' & \cdots & 0 \\ \cdots & \cdots & \cdots \\ 0 & \cdots & (b_{11})' \end{array} & \begin{array}{ccc} (b_{12})' & \cdots & 0 \\ \cdots & \cdots & \cdots \\ 0 & \cdots & (b_{12})' \end{array} \\ \hline \begin{array}{ccc} (b_{21})' & \cdots & 0 \\ \cdots & \cdots & \cdots \\ 0 & \cdots & (b_{21})' \end{array} & \begin{array}{ccc} (b_{22})' & \cdots & 0 \\ \cdots & \cdots & \cdots \\ 0 & \cdots & (b_{22})' \end{array} \end{array}\right| \tag{4.10}$$

where all (a_{ik}) vary independently in $(\mathfrak{d})$, all $(b_{\alpha\beta})'$ in $(\mathfrak{d}')$; $i, k = 1, \cdots, t$; $\alpha, \beta = 1, \cdots, s$. On changing our notation, *we are dealing with two algebras*

$$\mathfrak{A} \sim s(\mathfrak{d})_t, \qquad \mathfrak{B} \sim t(\mathfrak{d}'), \tag{4.11}$$

which are mutually commutators of each other; the degrees of $\mathfrak{A}$ *and* $\mathfrak{B}$ *equal* $d \cdot st$, *their respective orders are*

$$h_{\mathfrak{A}} = d \cdot t^2 \quad \text{and} \quad h_{\mathfrak{B}} = d \cdot s^2,$$

hence

$$g^2 = h_{\mathfrak{A}} \cdot h_{\mathfrak{B}}. \tag{4.12}$$

(4.11) are simultaneous equivalences in the sense that, in a coördinate system common to both, $\mathfrak{A}$ and $\mathfrak{B}$ are described by the schemes (4.10).

5. The fully reducible matric algebra and its commutator algebra

Our next concern is a natural generalization of matric algebras: the elements a may be v-uples

$$a = (A_1, A_2, \cdots, A_v) \tag{5.1}$$

of matrices in k, each component A_u being a matrix of prescribed degree g_u. Such elements may be added and multiplied among each other and multiplied by numbers in k by performing these operations on the several components separately. We want to study algebras $\mathfrak{a}$ in k consisting of such elements a. Each component like $A_1 = A_1(a)$ defines a representation $\mathfrak{A}_1$ of $\mathfrak{a} : a \to A_1$. By the simple artifice of writing our elements in the form

$$\left\| \begin{matrix} A_1 & 0 & \cdots & 0 \\ 0 & A_2 & \cdots & 0 \\ \cdots & \cdots & \cdots & \cdots \\ 0 & 0 & \cdots & A_v \end{matrix} \right\| \tag{5.2}$$

rather than (5.1) we could keep within the bounds of matric algebras. We prove:

Theorem (3.5.A). *If the component representations* $\mathfrak{A}_u$ *of a* v*-uple matric algebra* $\mathfrak{a}$ *are irreducible and inequivalent, then the* v *components* A_u *are independent of each other. The regular representation of* $\mathfrak{a}$ *is decomposable into irreducible parts each equivalent to one of the component representations.*

The asserted "independence" may be formulated in different manners. The simplest formulation is perhaps to say that if

$$a = (A_1, A_2, \cdots, A_v) \tag{5.3}$$

is contained in $\mathfrak{a}$, then the same holds for

$$\begin{aligned} a_1 &= (A_1, 0, \cdots, 0), \\ &\cdots\cdots\cdots\cdots \\ a_v &= (0, 0, \cdots, A_v). \end{aligned} \tag{5.4}$$

Or, with a varying over $\mathfrak{a}$, each component $A_u(a)$ varies independently over its whole range $\mathfrak{A}_u$. Or, $\mathfrak{a}$ is the direct sum of the algebras $\mathfrak{A}_u$.

The proof follows exactly the lines laid out in the proof of Theorem (3.3.C).

$\mathfrak{a}$ is a vector space ρ. In an irreducible invariant subspace ρ_1 of ρ we again pick out an element $a^0 \neq 0$. At least one of its v components A_u^0, let us say A_1^0, is $\neq 0$. Again we choose a vector $\mathfrak{e}$ such that $A_1^0\mathfrak{e} \neq 0$, and conclude that ρ_1 is similar to the first component space, i.e. to the representation space of $\mathfrak{A}_1$ (or that the representation induced by the regular one in ρ_1 is equivalent to $\mathfrak{A}_1$). We now add this little remark: for no element a in ρ_1 can the second component A_2 be $\neq 0$. For then, starting with such an a instead of a^0 we should find that ρ_1 is similar to the second component space, which is impossible because of the inequivalence of $\mathfrak{A}_1$ and $\mathfrak{A}_2$. After the decomposition of ρ into irreducibly invariant subspaces $\rho_1, \rho_2, \cdots$ we unite those that are similar to the first component space, those similar to the second component space, and so on, and this, according to our last observation, means that (5.3) has been broken up into terms in $\mathfrak{a}$ of the form

$$\begin{array}{cccc}
(\bar{A}_1, & 0, & \cdots, & 0), \\
(0, & \bar{A}_2, & \cdots, & 0), \\
\multicolumn{4}{c}{\dots\dots\dots\dots\dots\dots} \\
(0, & 0, & \cdots, & \bar{A}_v).
\end{array}$$

But their sum equals (5.3), hence $\bar{A}_u = A_u$, and thus we arrive at the desired result (5.4).

We finally consider a *k-algebra* $\mathfrak{a}$ *of matrices in k which is decomposable into irreducible parts.* Writing the equivalent ones among them alike, the generic element a breaks up into "blocks" of the kind

$$\begin{bmatrix} A_u(a) & & \\ & \ddots & \\ & & A_u(a) \end{bmatrix} \qquad (u = 1, \cdots, v),$$

where

$$\mathfrak{A}_u : a \to A_u(a)$$

are irreducible and mutually inequivalent representations. Lemma (3.1.D), the second part of Schur's lemma, shows that each commutator B of $\mathfrak{a}$ breaks up into blocks of the same size. The situation prevalent for the individual block is taken care of by our analysis at the end of the previous section: to the block $s_u \cdot (\mathfrak{d}_u)_{t_u}$ of the given algebra $\mathfrak{A}$ corresponds the block $t_u \cdot (\mathfrak{d}'_u)_{s_u}$ in the commutator algebra $\mathfrak{B}$; $\mathfrak{d}_u$ is a certain division algebra. Our proposition (3.5.*A*) concerning v-uple matric algebras insures the given algebra $\mathfrak{A}$ to be the *direct sum* of the blocks:

$$\mathfrak{A} = \sum_{u=1}^{v} s_u(\mathfrak{d}_u)_{t_u}.$$

In the same sense we have

$$\mathfrak{B} = \sum_{u=1}^{v} t_u(\mathfrak{d}'_u)_{s_u},$$

and by applying to $\mathfrak{B}$ the same argument, namely essentially the Schur lemma, we readily see that $\mathfrak{A}$ is the commutator algebra of $\mathfrak{B}$. Were the several blocks of $\mathfrak{A}$ not independent of each other, then the commutator algebra of $\mathfrak{B}$ would certainly have been larger than $\mathfrak{A}$! Our study thus culminates in the

THEOREM (3.5.B). *If a k-algebra $\mathfrak{A}$ of matrices in k is decomposable into irreducible parts, so is its commutator algebra $\mathfrak{B}$. $\mathfrak{A}$ is conversely the commutator algebra of $\mathfrak{B}$. Their structure is described by the simultaneous equivalences*

$$(5.5) \qquad \mathfrak{A} \sim \sum_{u=1}^{v} s_u(\mathfrak{d}_u)_{t_u}, \qquad \mathfrak{B} \sim \sum_{u=1}^{v} t_u(\mathfrak{d}'_u)_{s_u}$$

where $\mathfrak{d}_u$, $\mathfrak{d}'_u$ are inverse (abstract) division algebras.

It is hardly necessary to mention explicitly that $\mathfrak{A}$ contains the unit matrix E. The degree g of $\mathfrak{A}$ and $\mathfrak{B}$ is $= \sum_u d_u s_u t_u$ if d_u denotes the order of $\mathfrak{d}_u$; the respective orders are

$$h_{\mathfrak{A}} = \sum_u d_u t_u^2 \qquad \text{and} \qquad h_{\mathfrak{B}} = \sum_u d_u s_u^2 .$$

Therefore under all circumstances

$$g^2 \leqq h_{\mathfrak{A}} \cdot h_{\mathfrak{B}}.$$

As an abstract algebra, $\mathfrak{A}$ is identical with the algebra $\mathfrak{a}$ of the v-uples

$$a = (A_1, \cdots, A_v)$$

where each A_u varies independently within an irreducible matric set $\mathfrak{A}_u$ of degree g_u—or, still more abstractly,

$$a = (a_1, \cdots, a_v)$$

where a_u varies in the simple algebra $\mathfrak{a}_u$ (a direct sum of the $\mathfrak{a}_u$). In the same sense the centrum of $\mathfrak{a}$ is the direct sum of the centra of $\mathfrak{a}_u$, which in their turn are isomorphic to the centra of the division algebras $\mathfrak{d}_u$. According to the results gained for any irreducible set like $\mathfrak{A}_u$, the regular representation $(\mathfrak{a})$ splits into t_1 times $\mathfrak{A}_1$ plus t_2 times $\mathfrak{A}_2$ plus $\cdots$. The number t_u of times each irreducible constituent occurs is a divisor of its degree g_u:

$$(5.6) \qquad g_u = d_u t_u.$$

On making use of the general proposition (3.3.D) we infer:

THEOREM (3.5.C). *Any non-degenerate representation of the abstract scheme $\mathfrak{a}$ of the fully reducible matric algebra $\mathfrak{A}$ decomposes into irreducible constituents each of which is equivalent to one of the $\mathfrak{A}_u$ $(u = 1, \cdots, v)$. In particular, the number of times this constituent occurs in the regular representation of $\mathfrak{a}$ is the number t_u in Theorem* (3.5.B), *a divisor of g_u defined by equation* (5.6).

An immediate consequence of our final result (3.5.B) is the following criterion first put to use in a fruitful manner by R. Brauer:

THEOREM (3.5.D). *The enveloping algebra of a fully reducible matric set $\mathfrak{A}$ is the commutator algebra of the commutator algebra of $\mathfrak{A}$.*

By *fully reducible* we mean that the set $\mathfrak{A}$ of matrices decomposes into irreducible parts. Indeed, the enveloping algebra $\bar{\mathfrak{A}}$ of $\mathfrak{A}$ splits in the same fashion as $\mathfrak{A}$ itself, and its parts, as they comprehend more than those of $\mathfrak{A}$, are a fortiori irreducible. Any commutator B of $\mathfrak{A}$ is a commutator of the enveloping $\bar{\mathfrak{A}}$. Consequently Theorem (3.5.B) is applicable, to the effect that $\bar{\mathfrak{A}}$ is the commutator algebra of the set $\mathfrak{B}$ of all commutators B of $\mathfrak{A}$.

B. The Ring of a Finite Group and Its Commutator Algebra

6. Stating the problem

Part B shall deal with a situation closely kindred to that of Part A, though less general in scope. It arose from the problem of *decomposing tensor space into its irreducible components under the full linear group*, and we go at it with this application in mind. Under the influence of any non-singular transformation $A = \| a(ik) \|$ of the underlying n-dimensional vector space P, the components of an arbitrary tensor $F(i_1 \cdots i_f)$ of rank f undergo the transformation

$$\Pi_f(A) = A \times A \times \cdots \times A \text{ (f factors)} \tag{6.1}$$

that is, F is transformed into

$$F'(i_1 \cdots i_f) = \sum_{k_1, \cdots, k_f} a(i_1k_1) \cdots a(i_fk_f) \cdot F(k_1 \cdots k_f) .$$

All arguments or indices i and k range over the integers from 1 to n. The tensors of rank f form a vector space P_f of n^f dimensions. We say that A *induces* $\Pi_f(A)$ in P_f. The correspondence $A \to \Pi_f(A)$ defines a representation of the group $GL(n)$ in P_f. Examples for invariant subspaces of P_f are the sets of all *symmetric* or of all *skew-symmetric tensors*; and, in obvious generalization of these most primitive cases, *the set of all tensors satisfying certain symmetry conditions*. In order to describe what a symmetry condition is we have first to explain what it means to *apply a permutation* s of the f sub-indices $1 \cdots f$ *upon a given tensor* F. If s is the permutation

$$1 \to 1', \cdots, f \to f', \tag{6.2}$$

we shall define $F' = \mathbf{s}F$ by

$$F'(i_1i_2 \cdots i_f) = F(i_{1'}, i_{2'}, \cdots i_{f'}). \tag{6.3}$$

This is the correct way if we want the desirable relation $\mathbf{t}(\mathbf{s}F) = (\mathbf{ts})F$ to hold for any two permutations s: $\alpha \to \alpha'$ and t: $\alpha' \to \alpha''$ $\{\alpha = 1, \cdots, f\}$ whose composition ts is defined to be $\alpha \to \alpha''$. Indeed, composition in the right order is preserved if with the permutation s we associate the following (linear) transformations, first of the variables i into new variables i' and then of the function F into a new function F':

$$i'_{\alpha'} = i_\alpha, \qquad F'(i'_\alpha) = F(i_\alpha).$$

The ensuing equation

$$F'(i'_\alpha) = F(i'_{\alpha'})$$

coincides with (6.3) if we replace the letter i' by i. The permutations s of f figures form the *symmetric group* $\gamma = \pi_f$ of order $f!$.

A *linear symmetry condition* imposed upon the tensor F is a relation

$$\sum_s a(s) \cdot \mathbf{s}F = 0 \tag{6.4}$$

with arbitrary coefficients $a(s)$; it asserts that the *symmetry operator*

$$\mathbf{a} = \sum_s a(s) \cdot \mathbf{s}$$

changes F into 0: $\mathbf{a}F = 0$. Symmetry operators may be added and multiplied by numbers in the obvious manner. Moreover, successive performance of

$$\mathbf{a} = \sum_s a(s)\mathbf{s}, \qquad \mathbf{b} = \sum_s b(s)\mathbf{s}$$

(first **a**, then **b**) gives rise to a new symmetry operator **c**, the product **ba**, defined by

$$\sum_{t,t'} b(t')a(t)\mathbf{t}'\mathbf{t} = \sum_s c(s)\mathbf{s},$$

where

$$c(s) = \sum_{t't=s} b(t')a(t) = \sum_t b(st^{-1})a(t) = \sum_t b(t)a(t^{-1}s). \tag{6.5}$$

The rules by which addition and multiplication are carried out depend merely on the *structure* of the group and have nothing to do with the particular realization of the group elements s as linear transformations $F \to \mathbf{s}F$ in the space P_f. Therefore it is appropriate to take the abstract viewpoint again: any finite group γ of order h gives rise to an algebra of order h, the so-called *group ring* $\mathfrak{r}$ in k consisting of all linear combinations of the group elements s with coefficients $a(s)$ in k:

$$a = \sum_s a(s) \cdot s. \tag{6.6}$$

The sum is merely a suggestive form of writing, the "*quantity*" a of $\mathfrak{r}$ is nothing but the set of coefficients $a(s)$ or the function $a(s)$ defined on the group. What is essential is the definition of the three operations: If a has the coefficients $a(s)$ and b the coefficients $b(s)$, then $a + b$, λa, ba have the coefficients

$$a(s) + b(s), \qquad \lambda \cdot a(s) \text{ and } (6.5)$$

respectively (λ any number in k). All the axioms characteristic for an algebra are fulfilled; the associative law of multiplication is an immediate consequence of the same law for group elements. The group ring contains the unit 1. Transition from the group to the group ring facilitates its management considerably by extending the list of permissible operations from mere multiplication ba to include $a + b$, λa as well.

On returning to the symmetric group π_f of order $f!$ and the realization of its elements and of the quantities a of the group ring by linear operators $F \to \mathbf{a}F$ in tensor space one ought to observe that in general this realization is not an isomorphic or faithful one. If we put $\delta_s = +1$ or -1 according as the permutation s is even or odd, then the alternation $\sum \delta_s \cdot s$ will carry every tensor F into 0,

$$\sum_s \delta_s \cdot \mathbf{s}F = 0,$$

provided the rank f surpasses the dimensionality n.

The group of the $\Pi_f(A)$ induced in P_f by all non-singular transformations in P is suitably replaced by its enveloping algebra

$$\mathfrak{A}_f = [\Pi_f(A)]_{A \text{ in } GL(n)}.$$

It is easily seen and shall later be proved explicitly [Theorem (4.4.E)] that this algebra $\mathfrak{A}_f$ consists of all transformations

$$A_f = \|a(i_1 \cdots i_f;\, k_1 \cdots k_f)\| \tag{6.7}$$

in tensor space which are symmetric in the sense that $a(i_1 \cdots i_f;\, k_1 \cdots k_f)$ does not change if both rows of arguments $i_1 \cdots i_f$, $k_1 \cdots k_f$ are submitted to the same arbitrary permutation s, (6.2):

$$a(i_{1'} \cdots i_{f'};\, k_{1'} \cdots k_{f'}) = a(i_1 \cdots i_f;\, k_1 \cdots k_f). \tag{6.8}$$

As the term "symmetric" is used for so many other purposes, even in connection with linear transformations, I propose to substitute the word "*bisymmetric*" for the kind of symmetry here under investigation. A bisymmetric operator in tensor space can be described as one commuting with all the $f!$ symmetry operators $F \to \mathbf{s}F$. Indeed, on putting

$$a(i_{1'} \cdots i_{f'};\, k_{1'} \cdots k_{f'}) = a'(i_1 \cdots i_f;\, k_1 \cdots k_f),$$

the equation

$$G(i_1 \cdots i_f) = \sum_k a(i_1 \cdots i_f;\, k_1 \cdots k_f)F(k_1 \cdots k_f)$$

leads by (6.3) to

$$\mathbf{s}G(i_1 \cdots i_f) = \sum_k a'(i_1 \cdots i_f;\, k_1 \cdots k_f) \cdot \mathbf{s}F(k_1 \cdots k_f),$$

or, in easily understandable notation,

$$\mathbf{s}^{-1}A'_f\mathbf{s} = A_f \quad \text{or} \quad A'_f = \mathbf{s}A_f\mathbf{s}^{-1}.$$

As the permutations s enter into the definition of the algebra $\mathfrak{A}_f$, it is not surprising at all that the study of the tensor space under the influence of the group $GL(n)$ ties up with the symmetric group π_f.

The idea of replacing the group $\Pi_f(A)$, A in $GL(n)$, by the enveloping algebra $\mathfrak{A}_f$ of all bisymmetric A_f was first suggested to the author by the *application of these theories to quantum mechanics.*[4] There a tensor F depicts the *state of a physical system* consisting of f similar particles, let us say f electrons. Each observable physical quantity is represented by a linear operator A_f in tensor space, and the change $\dot{F} = dF/dt$ of the state F in time t is in particular determined by the operator H representing *energy*:

$$\dot{F}(i_1 \cdots i_f) = \sum_k h(i_1 \cdots i_f; k_1 \cdots k_f) \cdot F(k_1 \cdots k_f).$$

If all particles are alike, H will be bisymmetric in our sense, and hence no transitions are possible in time between several subspaces of the tensor space that are invariant with respect to the algebra $\mathfrak{A}_f$, *whatever the forces acting between the particles may be.* The algebra $\mathfrak{A}_f$ rather than the group of the $\Pi_f(A)$ plays the decisive part in quantum mechanics.

By now it should have become fairly obvious under what *general problem* to subsume our question concerning tensors: the group π_f is replaced by an arbitrary finite group γ, the tensor space by any vector space whose generic vector we will call f and whose dimensionality may be denoted by n, the operators $F \to \mathsf{s}F$ by any representation $\mathfrak{S}$ of the given group γ in that vector space, and finally $\mathfrak{A}_f$ by the commutator algebra of $\mathfrak{S}$. *We shall establish a complete reciprocity between the regular representation of the group ring* $\mathfrak{r}$ *of* γ *and the commutator algebra of the given representation.* Once more and in fuller detail: γ is a finite group of order h, s denotes a typical element thereof, $\mathfrak{r}$ is the corresponding group ring consisting of the quantities (6.6). Let a representation of the group be given in an n-dimensional vector space P: $s \to U(s) = \| u_{ik}(s) \|$,

$$f'_i = \sum_k u_{ik}(s) f_k \quad (i, k = 1, \cdots, n) \quad \text{or} \quad f' = U(s)f. \tag{6.9}$$

We abbreviate (6.9) as $f' = \mathsf{s}f$ and extend this representation to the group ring:

$$f \to f' = \sum_s a(s) \cdot \mathsf{s}f = \mathsf{a}f \quad \text{represents} \quad \sum a(s) \cdot s = a.$$

The linear operators a form an algebra $\mathfrak{S}$ homomorphic with the group ring. $\mathfrak{A}$ is the commutator algebra of the given representation or of the algebra $\mathfrak{S}$; its members shall now be denoted by $A = \| a_{ik} \|$. Hence

$$\tilde{f} = Af, \qquad \tilde{f}_i = \sum_k a_{ik} f_k \tag{6.10}$$

involves

$$\mathsf{s}\tilde{f} = A \cdot \mathsf{s}f \quad \text{or} \quad \mathsf{s}\tilde{f}_i = \sum_k a_{ik} \cdot \mathsf{s}f_k \tag{6.11}$$

if A is in $\mathfrak{A}$. The group ring is an h-dimensional vector space ρ that is considered as the substratum of the regular representation $(\mathfrak{r})$ of $\mathfrak{r}$ associating with the quantity a the substitution

$$(a): x \to ax$$

in ρ, while our vector space P will be regarded as the substratum of the matrices of $\mathfrak{A}$ (rather than $\mathfrak{S}$); terms like invariant, irreducible, equivalent are to be interpreted accordingly.

The first step will be to show that *the regular representation of the group ring* $\mathfrak{r}$ *is fully reducible* and this will hold in any reference field k whatsoever.* Presuming this result we could devise the following procedure: from Theorem (3.3.D) and its proof one learns how $\mathfrak{S}$ may be decomposed into irreducible parts, and then the general theorem (3.5.B) insures the full reducibility of the commutator algebra $\mathfrak{A}$. However, we shall here establish a much more complete and direct parallelism between ($\mathfrak{r}$) and $\mathfrak{A}$ without passing through $\mathfrak{S}$, and that by a much more elementary method which may also be gleaned from our special case dealing with tensors F.

How did we propose to specify an invariant subspace within the tensor space P_f? By imposing a number of symmetry conditions (6.4) on F. These conditions state that the "quantity" with the coefficients $F(s) = \mathbf{s}F$ lies in a certain linear subspace σ of ρ. If the reader is loath to operate with symmetry quantities whose coefficients are tensors rather than numbers, he will require each quantity x whose coefficients are

$$x(s) = \mathbf{s}F(i_1 \cdots i_f)$$

to lie in that subspace. Carrying over this remark to our general problem we win the following starting point.

If f is a vector, then $f_i(\cdot)$ shall denote the quantity with the coefficients

$$f_i(s) = \mathbf{s}f_i = \sum_k u_{ik}(s)f_k, \tag{6.12}$$

and one might consider $f(\cdot) = (f_1(\cdot), \cdots, f_n(\cdot))$ as a vector whose components are quantities rather than numbers. Any subspace σ in ρ determines a subspace $\Sigma = \sharp\sigma$ of the vector space P in this way: *the vector f belongs to Σ if and only if each of the n quantities $f_i(\cdot)$, $(i = 1, \cdots, n)$, lies in σ.* It is nearly trivial that a Σ constructed in this way is an invariant subspace of P (invariant with respect to the algebra $\mathfrak{A}$). Our chief aim is to make sure the converse, i.e. that each invariant Σ is of the form $\sharp\sigma$. The natural method to accomplish this is by constructing σ out of Σ, and there is no doubt how to attempt it. Let Σ be any subspace of P; *we define $\sigma = \natural\Sigma$ as the linear closure of all quantities of the form $f_i(\cdot)$ (f in Σ, $i = 1, \cdots, n$).* More explicitly, $f^{(\alpha)}$ $(\alpha = 1, \cdots, m)$ being a basis of Σ, the corresponding $\natural\Sigma$ consists of all quantities of the form

$$x = \sum_{i,\alpha} \varphi_i^{(\alpha)} f_i^{(\alpha)}(\cdot) \tag{6.13}$$

* It will hold even if k has characteristic $p \neq 0$, provided p is not a divisor of the order h of the group γ. We remark that Part A of this chapter is valid for fields of any characteristic, and Part B for fields of characteristic p not dividing h.

with arbitrary coefficients $\varphi_i^{(\alpha)}$. Again it is almost trivial that such a $\sigma = \natural\Sigma$ is invariant and moreover a subspace of

$$\rho_0 = \natural \mathrm{P} \tag{6.14}$$

(which need not be the full space ρ). *With these limitations that Σ is invariant in* P *and σ invariant in ρ_0, we want to show that the two operations $\sharp$ and $\natural$ are inverse.*

7. Full reducibility of the group ring

In carrying out the program delineated in the last section we first prove that the regular representation ($\mathfrak{r}$) of the group ring $\mathfrak{r}$ is fully reducible.[5]

THEOREM (3.7.A). *An invariant subspace σ of ρ possesses an idempotent generator e;* i.e. *xe lies in σ for every x, and $xe = x$ for every x in σ.*

The theorem implies that $e = \mathfrak{l}e$ is in σ and hence $ee = e$..

Any linear mapping $x \to y$ carrying each "vector" x into a vector y in σ and leaving the vectors x in σ unchanged will appropriately be called a *projection* of ρ into σ. If σ is any linear subspace of g dimensions one constructs such a projection easily enough: after adapting the coördinate system $e_1, \cdots, e_g; e_{g+1}, \cdots, e_h$ to the subspace σ one defines

$$e_1 \to e_1, \cdots, e_g \to e_g; \qquad e_{g+1} \to 0, \cdots, e_h \to 0.$$

We have to prove that there exists a projection of the peculiar form

$$x \to y = xe$$

provided σ is an *invariant* subspace.

We start with an arbitrary projection $x \to y$:

$$y(s) = \sum_t d(s, t)x(t), \qquad y = Dx.$$

On account of the invariance of σ,

$$ry = Drx$$

defines another projection $x \to y$ whatever element of the group we choose for r; its explicit expression is

$$y(r^{-1}s) = \sum_t d(s, t)x(r^{-1}t) \quad \text{or}$$

$$y(s) = \sum_t d(rs, rt)x(t).$$

On forming the "average" of all our projections

$$e(s, t) = \frac{1}{h}\sum_r d(rs, rt),$$

we obtain a projection again whose matrix $e(s, t)$ satisfies the relation

$$e(rs, rt) = e(s, t)$$

and is therefore of the form $e(t^{-1}s)$. Consequently our new projection is of the desired form,

$$y(s) = \sum_t x(t)e(t^{-1}s) \quad \text{or} \quad y = xe.$$

THEOREM (3.7.B). *An invariant subspace σ containing the invariant subspace $\sigma_1 \subset \sigma$ can be split according to $\sigma = \sigma_1 + \sigma'$ into σ_1 and a complementary invariant subspace σ'.*

The proof follows a scheme already used in the demonstration of Theorem (3.3.C). Let e_1 be a generating idempotent of σ_1. Those x in σ which satisfy $xe_1 = x$ form the subspace σ_1, those for which $xe_1 = 0$ another invariant subspace σ'. The Peirce decomposition yields $\sigma = \sigma_1 + \sigma'$.

If an invariant σ is decomposed into a number of linearly independent invariant subspaces,

$$\sigma = \sigma_1 + \sigma_2 + \cdots + \sigma_u,$$

we have for each element x of σ a unique decomposition

$$x = x_1 + x_2 + \cdots + x_u \qquad (x_\alpha \text{ in } \sigma_\alpha); \tag{7.1}$$

in particular, for a generating idempotent e of σ,

$$e = e_1 + e_2 + \cdots + e_u \qquad (e_\alpha \text{ in } \sigma_\alpha). \tag{7.2}$$

From (7.2) follows, for any x in σ,

$$x = xe = xe_1 + xe_n + \cdots + xe_u. \tag{7.3}$$

Since xe_α (as well as e_α) lies in σ_α because of the invariance of σ_α, (7.3) is the decomposition (7.1), which thus proves to be an immediate consequence of (7.2). Applying our observation to $x = e_\alpha$ we see that

$$e_\alpha e_\beta = e_\alpha \text{ or } 0 \text{ according as } \beta = \alpha \text{ or } \beta \neq \alpha.$$

Such quantities e_α shall be called *normal idempotents*. An idempotent e is indecomposable or *primitive* if it allows no decomposition $e = e_1 + e_2$ into two normal idempotents e_1, e_2 save the trivial ones $e = e + 0$ and $e = 0 + e$. In carrying decomposition as far as possible we end by splitting a given idempotent e into mutually normal primitive idempotents. Indeed, if e_1 can be split further,

$$e_1 = e_1' + e_1''; \qquad e_1'e_1' = e_1', \qquad e_1'e_1'' = 0,$$
$$e_1''e_1' = 0, \qquad e_1''e_1'' = e_1'',$$

one has

$$e_1'e_\alpha = e_1'e_1e_\alpha = 0, \qquad e_\alpha e_1' = e_\alpha e_1 e_1' = 0$$

for $\alpha > 1$; hence the sequence e_1', e_1''; e_2, $\cdots$ again consists of mutually normal idempotents. Our analysis shows that decomposition of the unit I into normal primitive idempotents will result in splitting up the total ρ into irreducible invariant subspaces.

At this juncture we repair to the situation on which Lemma (3.1.D) was based in order to explain the general idea of similarity or equivalence. We have two sets of matrices, $A_1(a)$ and $A_2(a)$, put in a certain correspondence by the parameter a varying in an abstract set. P_1 and P_2 are the two vector spaces of dimensionality g_1 and g_2 on whose vectors the transformations $A_1(a)$ and $A_2(a)$ operate. The equations $\mathfrak{y}_1 = a\mathfrak{x}_1$, $\mathfrak{y}_2 = a\mathfrak{x}_2$ for vectors $\mathfrak{x}_1$ in P_1 and $\mathfrak{x}_2$ in P_2 are to be interpreted as meaning $\mathfrak{y}_1 = A_1(a)\mathfrak{x}_1$, $\mathfrak{y}_2 = A_2(a)\mathfrak{x}_2$. Let Σ_1, Σ_2 be invariant subspaces of P_1, P_2 respectively. A linear mapping $\mathfrak{x}_2 \rightarrow \mathfrak{x}_1$ associating a vector $\mathfrak{x}_1$ in Σ_1 with an arbitrary vector $\mathfrak{x}_2$ in Σ_2 is called a *similarity mapping* of Σ_2 upon Σ_1 if it changes $a\mathfrak{x}_2$ into $a\mathfrak{x}_1$ (for every a). Σ_1 and Σ_2 are *similar* or *equivalent* if there exists a *one-to-one* similarity mapping $\mathfrak{x}_2 \rightleftarrows \mathfrak{x}_1$ of Σ_2 on Σ_1. The term applies in particular to subspaces Σ_1 and Σ_2 of one and the same vector space P which is the substratum of a set of transformations $A(a)$.

THEOREM (3.7.C). *A similarity mapping* $D: x \rightarrow x'$ *of an invariant subspace* σ *upon* σ' *is generated by aft multiplication with a quantity* b: $x' = xb$.

PROOF: Take as b the quantity into which a generating idempotent e of σ is sent by D. This operation being a similarity mapping, it will send xe into xb whatever x may be; but $x = xe$ if x is in σ.—Proposition and proof are a slight modification of Theorem (3.4.A).

By the general formula $x' = xb$ the image of e will be eb; hence the b we constructed here satisfies the relation $eb = b$ and, as a quantity in σ', the further relation $be' = b$ if e' is an idempotent generator of σ'. Consider the particular case where σ and σ' are equivalent and thus mapped upon each other by a one-to-one similarity correspondence $x \rightleftarrows x'$. It will be generated in the one direction by a quantity $b: x' = xb$ satisfying the relations just mentioned, in the other direction by a quantity a:

$$x = x'a \tag{7.4}$$

satisfying the analogous equations

$$ae = e'a = a; \qquad be' = eb = b. \tag{7.5}$$

e is carried by the direct mapping into b and this by the inverse mapping into ba; hence

$$ba = e \quad \text{and} \quad ab = e'. \tag{7.6}$$

Vice versa, if e and e' are given idempotents and a, b satisfy the equations (7.5), (7.6) then

$$x' = xb, \qquad x = x'a$$

establish inverse mappings $x \rightarrow x'$, $x' \rightarrow x$ of the invariant subspaces σ, σ' generated by e, e' into each other.

What we have somewhat carelessly described as invariant subspaces should perhaps more precisely be called *left invariant subspaces*, because the operations are the *left* multiplications (a): $x \rightarrow ax$ by any elements a of our ring.

A subspace invariant with respect to right multiplications $(a)': x \to xa$ deserves the name of a *right invariant subspace*. While the former has a generating idempotent e to the *right*, in the sense that it consists of all quantities of the form xe, the latter will have a generating idempotent to the *left*. The proof could be carried through in the same fashion. There is, however, a general method of passing freely from left to right. When we define $\hat{a}$ by

$$\hat{a}(s) = a(s^{-1}) \quad \text{or} \quad \hat{a}(s^{-1}) = a(s),$$

then the following lemma holds.

LEMMA (3.7.D). *$a = a_1 a_2$ implies $\hat{a} = \hat{a}_2 \hat{a}_1$.*

The roofing process therefore changes our algebra $\mathfrak{r}$ into its "inverse" $\mathfrak{r}'$. If the two invariant subspaces generated by two idempotents e, e' are equivalent then the same holds for $\hat{e}$, $\hat{e}'$; indeed, the equations (7.5), (7.6) show that $\hat{e}$, $\hat{e}'$ are tied together by $\hat{a}$, $\hat{b}$ in the same fashion as e, e' by b, a. Hence the representation induced in $\hat{\sigma}$ by the regular representation is, in the sense of equivalence, uniquely determined by that in σ, where σ, $\hat{\sigma}$ denote the subspaces generated by e and $\hat{e}$. The question as to the nature of this coupling of representations is answered by

THEOREM (3.7.E). *The representations induced in σ and $\hat{\sigma}$ by the regular representation are contragredient to each other.*

The proof rests on the notion of *trace*: the unit component $a(1) = \text{tr}\,(a)$ of a quantity a: $a(s)$ is called its trace. The trace of the product of two variable quantities,

$$\text{tr}(xy) = \sum_s x(s)y(s^{-1}) = \sum_s x(s^{-1})y(s), \tag{7.7}$$

is a symmetric bilinear non-degenerate form in ρ: a being given, $\text{tr}(xa) = 0$ cannot hold identically in x unless $a = 0$.

We compare the left and the right invariant subspaces σ and τ consisting of the quantities xe and ex respectively. We assert that $\text{tr}(xy)$ is non-degenerate if x and y vary in σ and τ respectively. Indeed, if z is any element whatever and a in σ, then

$$az = ae \cdot z = a \cdot ez = ay$$

where $y = ez$ is in τ. Hence the assumption $\text{tr}(ay) = 0$ for y in τ implies $\text{tr}(az) = 0$ for all z, whence $a = 0$. Similarly for the second factor. We now refer σ and τ each to a coördinate system a_i and b_k such that

$$x = \xi_1 a_1 + \cdots + \xi_g a_g, \qquad y = \eta_1 b_1 + \cdots + \eta_{g'} b_{g'} \tag{7.8}$$

describe σ and τ if the numbers ξ and η vary freely in k. From the non-degeneracy of $\text{tr}(xy)$ for (7.8) follows readily the coincidence $g' = g$ of the dimensions of σ and τ and the possibility of adapting the coördinate system b_k in τ to the arbitrarily chosen coördinate system a_i in σ such that

$$\text{tr}(xy) = \xi_1\eta_1 + \cdots + \xi_g\eta_g \tag{7.9}$$

for x in σ and y in τ.

With y varying in τ, $\hat{y}$ ranges over $\hat{\sigma}$. The simultaneous substitutions

$$x \to sx \ (x \text{ in } \sigma), \qquad y \to ys^{-1} \ (y \text{ in } \tau)$$

or

$$x \to sx \ (x \text{ in } \sigma), \qquad \hat{y} \to s\hat{y} \ (\hat{y} \text{ in } \hat{\sigma}) \tag{7.10}$$

leave (7.9) unaltered:

$$\text{tr}\,(sas^{-1}) = \text{tr}\,(a), \qquad \text{tr}\,(sxys^{-1}) = \text{tr}(xy). \tag{7.11}$$

In terms of the chosen coördinate systems the two substitutions (7.10) are therefore contragredient.

A final remark is concerned with the *character* $\chi(s)$ of the representation induced in σ by the regular representation of $\mathfrak{r}$. We have to compute the trace $\sum_s a(s)\chi(s)$ of the linear substitution

$$x \to y = ax \qquad (x \text{ in } \sigma). \tag{7.12}$$

When we adapt the coördinate system $e_1, \cdots, e_g; e_{g+1}, \cdots, e_h$ to the g-dimensional subspace σ we see at once that the substitution

$$x \to y = axe \qquad (x \text{ in } \rho) \tag{7.13}$$

has a matrix

*	0
*	0

where the left upper square is occupied by the matrix of (7.12). Hence we may compute the trace of (7.13) or

$$y(s) = \sum a(t)x(r)e(t') \qquad (trt' = s),$$

which equals

$$\sum_{tst'=s} a(t)e(t') = \sum_{s,t} a(t)e(s^{-1}t^{-1}s) = \sum_t a(t)\chi(t).$$

Therefore, by exchanging the letters s and t,

$$\chi(s) = \sum_t e(t^{-1}s^{-1}t). \tag{7.14}$$

A function $a(s)$ is a *class function* if it takes on equal values for elements s in the same class, i.e. for conjugate elements like s and $t^{-1}st$:

$$a(t^{-1}st) = a(s). \tag{7.15}$$

In this sense the character $\chi(s)$ of any representation $s \to U(s)$ is a class function because the trace of

$$U(t^{-1}st) = U^{-1}(t) \cdot U(s) \cdot U(t)$$

coincides with that of $U(s)$. Formula (7.14) puts this in evidence for our representations contained in the regular one. $a(s)$ is a class function if and only if the quantity a belongs to the centrum of the group ring. Indeed, $ax = xa$ or

$$\sum_t a(st)x(t^{-1}) = \sum_t x(t^{-1})a(ts) \tag{7.16}$$

leads back to (7.15),

$$a(st) = a(ts),$$

when one equates the coefficients of both sides of (7.16).

8. Formal lemmas

After indulging in all these details about finite groups and their group rings, some of which are not needed for our immediate purpose, it is time to return to the problem formulated at the end of section 6. We base the proof of the complete reciprocity between the regular representation ($\mathfrak{r}$) of the group ring $\mathfrak{r}$ and the commutator algebra $\mathfrak{A}$ on three simple lemmas.[6]

LEMMA (3.8.A).

$$a \cdot f_i(\cdot) = \sum_k \alpha_{ik} f_k(\cdot) \tag{8.1}$$

with

$$\| \alpha_{ik} \| = \sum_s a(s) U^{-1}(s). \tag{8.2}$$

PROOF: The s-coefficient of $af(\cdot)$ is the vector

$$g(s) = \sum_r a(r) \cdot \mathfrak{r}^{-1}\mathfrak{s}f = \sum_r a(r) \cdot U^{-1}(r) f(s).$$

LEMMA (3.8.B). $f_i(\cdot)a = g_i(\cdot)$, *where the vector g is defined by*

$$g = \sum_r a(r^{-1}) \cdot \mathfrak{r}f = \hat{\mathfrak{a}}f.$$

In other words, if $g = \hat{\mathfrak{a}}f$ then $g(\cdot) = f(\cdot)a$.

PROOF: $f_i(\cdot)a = x$ is indeed given by

$$x(s) = \sum_r \mathfrak{s}\mathfrak{r}f_i \cdot a(r^{-1}) = \mathfrak{s}g_i\,.$$

LEMMA (3.8.C). *An equation of the kind*

$$\hat{a}(s) = \sum_{i=1}^{n} \varphi_i \cdot \mathfrak{s}f_i = \varphi U(s) f, \tag{8.3}$$

where $\varphi = (\varphi_1, \cdots, \varphi_n)$ is a row rather than a column of numbers (contravariant vector), entails

$$a(s) = \sum_{i=1}^{n} f_i \cdot \mathfrak{s}\varphi_i \tag{8.4}$$

when $s\varphi = \varphi(s)$ *is defined as the row* $\varphi U^{-1}(s)$. *The linear transformation*

$$\| a_{ik} \| = \| \sum_s sf_i \cdot s\varphi_k \|, \text{ i.e.} \tag{8.5}$$

$$A = \sum_s U(s) f\varphi U^{-1}(s) \tag{8.6}$$

commutes with the operators s.

PROOF: (8.4) or

$$a(s) = \varphi U^{-1}(s) f$$

follows at once from (8.3) because $U^{-1}(s) = U(s^{-1})$. Moreover

$$U(t) A U^{-1}(t) = \sum_s U(ts) f\varphi U^{-1}(ts) = A.$$

9. Reciprocity between group ring and commutator algebra

With $\acute{A} = \| a_{ik} \|$ in the commutator algebra, the relation

$$\bar{f}_i = \sum_k a_{ik} f_k \tag{9.1}$$

according to (6.11) entails

$$\bar{f}_i(\cdot) = \sum_k a_{ik} f_k(\cdot). \tag{9.2}$$

This remark proves $\sharp\sigma$ to be invariant: $\bar{f} = Af$ lies in $\sharp\sigma$ together with f because the quantities $\bar{f}_i(\cdot)$, (9.2), lie in σ if the $f_k(\cdot)$ do. On the other hand, by Lemma (3.8.A), $\natural\Sigma$ is invariant for any linear subspace Σ of P: $a \cdot f_i(\cdot)$ is a linear combination of the $f_k(\cdot)$ for any vector f in Σ. Moreover, one has by definition

$$\natural\sharp\sigma \subset \sigma, \qquad \Sigma \subset \sharp\natural\Sigma.$$

Our aim is to replace the inclusion $\subset$ by equality $=$ for invariant $\Sigma \subset$ P and $\sigma \subset \rho_0$, and we therefore exhibit the complete reciprocity between ρ_0 and P as established by $\sharp$ and $\natural$ in the following two theorems:

THEOREM (3.9.A). *If* σ (σ', σ_1, σ_2) *are any invariant subspaces of* ρ_0 *and* $\Sigma = \sharp\sigma$, *then*

$$\sigma' \subset \sigma, \qquad \sigma = \sigma_1 + \sigma_2, \qquad \sigma_1 \sim \sigma_2$$

imply

$$\Sigma' \subset \Sigma, \qquad \Sigma = \Sigma_1 + \Sigma_2, \qquad \Sigma_1 \sim \Sigma_2$$

respectively, while conversely,

$$\sigma = \natural\Sigma.$$

THEOREM (3.9.B). *If* Σ (Σ', Σ_1, Σ_2) *are any invariant subspaces of* P *and* $\sigma = \natural\Sigma$, *then*

$$\Sigma = \sharp\sigma$$

and

$$\Sigma' \subset \Sigma, \qquad \Sigma = \Sigma_1 + \Sigma_2, \qquad \Sigma_1 \sim \Sigma_2$$

imply

$$\sigma' \subset \sigma, \qquad \sigma = \sigma_1 + \sigma_2, \qquad \sigma_1 \sim \sigma_2$$

respectively.

Before proceeding to the proofs, we make this remark. If $\hat{e}$ is the idempotent generator of an invariant σ, then the corresponding $\Sigma = \sharp\sigma$ consists of all vectors of the form $\mathbf{e}f$. Indeed, $g = \mathbf{e}f$ lies in Σ because $g_i(\cdot) = f_i(\cdot)\cdot\hat{e}$ by Lemma (3.8.B), and for each f in Σ one has $\mathbf{e}f = f$.

We prove the first part of Theorem (3.9.A) by observing that when the decomposition $\sigma = \sigma_1 + \sigma_2$ is applied to an idempotent generator $\hat{e}$ of σ, the resulting equation $\hat{e} = \hat{e}_1 + \hat{e}_2$ leads to the following decomposition $\Sigma = \Sigma_1 + \Sigma_2$ of $\Sigma = \sharp\sigma$:

$$f = \mathbf{e}f = \mathbf{e}_1 f + \mathbf{e}_2 f = f^{(1)} + f^{(2)} \qquad (f \text{ in } \Sigma).$$

Lemma (3.7.D) allows us to shear all the roofs off the relations

$$\hat{e}_1\hat{e}_2 = \hat{e}_2\hat{e}_1 = 0 \qquad (\hat{e}_1\hat{e}_1 = \hat{e}_1,\ \hat{e}_2\hat{e}_2 = \hat{e}_2),$$

thus warranting the independence of the parts Σ_1, Σ_2:

$$\mathbf{e}_1 f^{(2)} = 0, \qquad \mathbf{e}_2 f^{(1)} = 0.$$

The similarity correspondence $x_1 \rightleftarrows x_2$ between σ_1 and σ_2 described by

$$x_2 = x_1\hat{b}, \qquad x_1 = x_2\hat{a}$$

gives rise to the mutually inverse transformations

$$f^{(2)} = \mathbf{b}f^{(1)}, \qquad f^{(1)} = \mathbf{a}f^{(2)}$$

between the vectors $f^{(1)}, f^{(2)}$ of $\Sigma_1 = \sharp\sigma_1$ and $\Sigma_2 = \sharp\sigma_2$. By (6.11) these formulas establish a similarity correspondence:

$$f_*^{(1)} = Af^{(1)} \quad \text{entails} \quad \mathbf{b}f_*^{(1)} = A(\mathbf{b}f^{(1)}) \qquad \{A \text{ in } \mathfrak{A}\}.$$

To secure the last part of Theorem (3.9.A), $\sigma \subset \natural\Sigma$, we construct $\natural\Sigma$ by means of an idempotent generator $\hat{e}$ of σ as follows: if $g^{(\alpha)}$ $(\alpha = 1, \cdots, n)$ ranges over a basis of the complete vector space P, all $f^{(\alpha)} = \mathbf{e}g^{(\alpha)}$ lie in $\Sigma = \sharp\sigma$ and hence

$$y = \sum_{\alpha,i} \varphi_i^{(\alpha)}\cdot f_i^{(\alpha)}(\cdot)$$

lies in $\natural\Sigma$. On introducing

$$x = \sum_{\alpha,i} \varphi_i^{(\alpha)} g_i^{(\alpha)}(\cdot),$$

we have $y = x\hat{e}$. So $x\hat{e}$ lies in $\natural\Sigma$ if x lies in $\rho_0 = \natural$P. But each x in σ satisfies both conditions: x in ρ_0 and $x\hat{e} = x$.

The converse theorem (3.9.B) exhibits the really important facts. Its assertion that

$$\sigma = \natural\Sigma \quad \text{implies} \quad \Sigma = \#\sigma \tag{9.3}$$

for any subspace Σ invariant under $\mathfrak{A}$ is the backbone of the whole theory. Let $\hat{e}$ be an idempotent generator of $\sigma = \natural\Sigma$. Like all elements of σ it is of the form

$$\hat{e}(s) = \sum_{\alpha,i} \varphi_i^{(\alpha)} \cdot \mathbf{s} f_i^{(\alpha)},$$

where

$$f^{(\alpha)} = (f_1^{(\alpha)}, \cdots, f_n^{(\alpha)})$$

ranges over a basis of Σ. Hence by Lemma (3.8.C)

$$e(s) = \sum_{\alpha,k} \mathbf{s}\varphi_k^{(\alpha)} \cdot f_k^{(\alpha)},$$

and any vector $g = \mathbf{e}f$ of $\#\sigma$ is given by

$$g_i = \sum_\alpha \left\{ \sum_k a_{ik}^{(\alpha)} f_k^{(\alpha)} \right\} = \sum_\alpha g_i^{(\alpha)}, \tag{9.4}$$

where

$$a_{ik}^{(\alpha)} = \sum_s \mathbf{s} f_i \cdot \mathbf{s}\varphi_k^{(\alpha)}.$$

Each term $g^{(\alpha)}$ of the sum (9.4), $g = \sum_\alpha g^{(\alpha)}$, arises from $f^{(\alpha)}$ by a linear transformation $A^{(\alpha)} = \| a_{ik}^{(\alpha)} \|$ which commutes with all $\mathbf{s}$ according to the same lemma. Hence Σ, being invariant with respect to the transformations A of the commutator algebra $\mathfrak{A}$, contains $g^{(\alpha)}$ as well as $f^{(\alpha)}$. This proves our statement: g in Σ or $\#\sigma \subset \Sigma$.

The *decomposition* $\Sigma = \Sigma_1 + \Sigma_2$ implies by definition that each quantity x in $\sigma = \natural\Sigma$ can be written as a sum $x_1 + x_2$, x_1 in $\sigma_1 = \natural\Sigma_1$, x_2 in $\sigma_2 = \natural\Sigma_2$. It remains to prove that σ_1 and σ_2 are linearly independent, or that the intersection $\sigma^* = \sigma_1 \cap \sigma_2$ is empty provided $\Sigma^* = \Sigma_1 \cap \Sigma_2$ be empty. But according to the part of Theorem (3.9.B) already proved,

$$\#\sigma^* \subset \#\sigma_1 = \Sigma_1, \qquad \#\sigma^* \subset \Sigma_2;$$

hence $\#\sigma^* \subset \Sigma^*$ and, by the last part of Theorem (3.9.A), $\sigma^* \subset \natural\Sigma^*$.

The *transition from* $\Sigma_1 \sim \Sigma_2$ *to* $\sigma_1 \sim \sigma_2$ for $\sigma_1 = \natural\Sigma_1$, $\sigma_2 = \natural\Sigma_2$ is to be based on the following statement, the proof of which is contained in Lemma (3.8.B);

LEMMA (3.9.C). *ρ_0 is right as well as left invariant.*

Therefore ρ_0 has an idempotent generator $\dot{i}$ to the left: i in ρ_0, $ix = x$ for every x in ρ_0.

Let $f^{(\alpha)}$ be a basis for Σ_1 and let the given similarity mapping of Σ_1 on Σ_2 send $f^{(\alpha)}$ into $g^{(\alpha)}$. When we put

$$x = \sum_{\alpha,i} \varphi_i^{(\alpha)} f_i^{(\alpha)}(\cdot), \qquad y = \sum_{\alpha,i} \varphi_i^{(\alpha)} g_i^{(\alpha)}(\cdot) \qquad \{\varphi_i^{(\alpha)} \text{ arbitrary numbers}\}$$

the correspondence $x \to y$ between an x and a y with the same coefficients $\varphi_i^{(\alpha)}$ will establish a similarity mapping of $\sigma_1 = \natural\Sigma_1$ on $\sigma_2 = \natural\Sigma_2$ because, by Lemma (3.8.A), we obtain

$$ax = \sum_{\alpha,k} \psi_k^{(\alpha)} f_k^{(\alpha)}(\cdot), \qquad ay = \sum_{\alpha,k} \psi_k^{(\alpha)} g_k^{(\alpha)}(\cdot), \qquad \text{with} \qquad \psi_k^{(\alpha)} = \sum_i \varphi_i^{(\alpha)} \cdot \alpha_{ik}.$$

This definition of $x \to y$ however goes through only *if* $x = 0$ *implies* $y = 0$. We first prove that

$$\hat{x}f = 0 \quad \text{implies} \quad \hat{y}f = 0$$

(for any vector f). By Lemma (3.8.C) the vector $F = \hat{x}f$ is a sum of terms $F^{(\alpha)}$ the αth of which arises from $f^{(\alpha)}$ by the transformation

$$A^{(\alpha)} = \| a_{ik}^{(\alpha)} \| = \| \sum_s \mathsf{s}f_i \cdot \mathsf{s}\varphi_k^{(\alpha)} \|.$$

Since $A^{(\alpha)}$ is in $\mathfrak{A}$ the given similarity mapping of Σ_1 on Σ_2 sends $F^{(\alpha)}$ into the corresponding part $G^{(\alpha)}$ of $G = \hat{y}f$ and hence F into G. Therefore $F = 0$ implies $G = 0$ and more especially: when the numbers $\varphi_i^{(\alpha)}$ satisfy the equation $x = 0$ we must have $\hat{y}f = 0$ for every vector f, or by Lemma (3.8.B), $f_i(\cdot)y = 0$. Hence the given quantity y satisfies $zy = 0$ for every z in ρ_0, in particular for $z = i$. But as y itself lies in ρ_0 the ensuing equation $iy = 0$ yields the desired result, $y = 0$.

The complete reciprocity established by Theorems (3.9.A & B) involves the fact that the process $\#$ not only changes $\sigma = 0$ into $\#\sigma = 0$ and a *part* $\sigma' \subset \sigma$ into a *part* $\#\sigma' \subset \#\sigma$, but also a $\sigma \neq 0$ into a $\#\sigma \neq 0$ and a *proper* part into a *proper* part, provided the σ's are invariant subspaces of ρ_0. *The decomposition of ρ_0 into irreducibly invariant subspaces σ_α leads to a decomposition of* P *into subspaces $\Sigma_\alpha = \mathbf{e}_\alpha$P irreducibly invariant under the algebra $\mathfrak{A}$; and both decompositions run absolutely parallel even as to the pertaining equivalences.*

We make the additional remark that any invariant subspace Σ of P breaks up into irreducible invariant subspaces each of which is similar to one of these parts Σ_α of P. This is true for any vector space which is fully reducible under a set of linear transformations, but the proof is especially simple in the present case. Indeed, e being a generating idempotent of Σ, we deduce from the decomposition P $= \Sigma_1 + \Sigma_2 + \cdots$, or

$$f = \sum_\alpha f_\alpha \qquad (f_\alpha \text{ in } \Sigma_\alpha),$$

the equation

$$\mathbf{e}f = \sum_\alpha \mathbf{e}f_\alpha .$$

As Σ_α is irreducible, the mapping $f_\alpha \to \mathbf{e}f_\alpha$ either carries every f_α in Σ_α into 0 or it is a one-to-one similarity mapping of Σ_α upon a subspace of Σ.

Theorem (3.9.D). P *breaks up into irreducible parts* $\Sigma_\alpha = \mathbf{e}_\alpha$P. *Any invariant subspace Σ of* P *is decomposable into irreducible parts each of which is similar to one of the* $\mathbf{e}_\alpha$P.

For a full appreciation of this method (II) we compare it with the method (I) alluded to in §6, which passes from $\mathfrak{r}$ to $\mathfrak{S}$ by means of Theorem (3.3.D) and from $\mathfrak{S}$ to $\mathfrak{A}$ by Theorem (3.5.B) and by the constructions given in the proofs of both propositions.[7] They depend on the choice of a coördinate system of the space P, and neither of the two steps displays such a thoroughgoing parallelism as revealed here. Coincidence between the numbers of the equivalent parts is not to be expected. Contrary to the restriction of ρ to ρ_0 in (II), the procedure (I) replaces ρ by ρ modulo that two-sided invariant subspace ρ^0 whose elements a satisfy the equation $\mathbf{a}f = 0$ identically in the vector f. Running over the whole gamut of demonstrations again, one might condense the essence of the procedure (I) as follows: the unit I of $\mathfrak{r}$ (or rather of $\mathfrak{r} \bmod \rho^0$) is decomposed into mutually normal primitive idempotents,

$$\mathsf{I} = e_1 + e_2 + \cdots ; \tag{9.5}$$

then

$$x = xe_1 + xe_2 + \cdots \quad \text{and} \quad f = \mathbf{e}_1 f + \mathbf{e}_2 f + \cdots \tag{9.6}$$

result in corresponding decompositions of ρ and P into irreducibly invariant subspaces (with respect to ($\mathfrak{r}$) and $\mathfrak{A}$ respectively). Indeed, if $\mathfrak{S}$, which is now to be identified with $\mathfrak{B}$, is written as the direct sum of v simple algebras $\mathfrak{b}_u$ as in the proof of Theorem (3.5.B), and the elements of each simple constituent $\mathfrak{b}_u$ as s-rowed matrices

$$\begin{Vmatrix} b_{11} & \cdots & b_{1s} \\ \cdots & \cdots & \cdots \\ b_{s1} & \cdots & b_{ss} \end{Vmatrix} \tag{9.7}$$

($s = s_u$) whose elements lie in a division algebra $\mathfrak{d} = \mathfrak{d}_u$, then one obtains the primitive idempotents $e_\alpha^{(u)}$ by equating in (9.7) one of the diagonal elements $b_{\alpha\alpha}$ to the unit e of $\mathfrak{d}$ and all other elements to 0, including those in the other simple constituents $\mathfrak{b}_{u'}$ ($u' \neq u$). The number of terms in (9.5) is $s_1 + \cdots + s_v$. The coördinate system in our vector space breaks up into the parts discerned by the triple index

$$\overset{(u)}{i\alpha} \; [u = 1, \cdots, v; i = 1, \cdots, t_u ; \alpha = 1, \cdots, s_u].$$

The vectors of the form $\mathbf{e}_1^{(1)} f$ are those of which all components vanish except the parts

$$\overset{(1)}{11}, \overset{(1)}{21}, \cdots, \overset{(1)}{t1} \qquad (t = t_1),$$

and these are submitted by $\mathfrak{A}$ to the transformations of the irreducible algebra $\mathfrak{A}_1$:

$$A = \begin{Vmatrix} (a_{11}), & \cdots, & (a_{1t}) \\ \cdots & \cdots & \cdots \\ (a_{t1}), & \cdots, & (a_{tt}) \end{Vmatrix} \qquad [a_{ik} \text{ in } \mathfrak{d}]$$

$\{\mathfrak{d} = \mathfrak{d}_1, t = t_1\}$.

On putting it this way, the essential difference between methods (I) and (II) is that in (I) we match

$$xe \text{ against } \mathbf{e}f,$$

while in (II)

$$x\hat{e} \text{ is matched against } \mathbf{e}f.$$

Only the second procedure establishes a correspondence independent of the choice of the generating idempotent e. So one may say that the more complete results attained by (II) are due to the presence of the operation ^ in a group ring, while missing in an arbitrary (semi-simple) algebra.[8]

10. A generalization

The simplest axioms of projective geometry (excluding those of order and continuity) show that the projective $(n-1)$-space is isomorphic to the following algebraic model: any ratio of n numbers in a given field k,

$$f_1:f_2: \cdots :f_n$$

(with the exclusion of the n-uple $0:0: \cdots :0$), represents a *point*; the *straight line* joining two distinct points a_i and b_i is given by the parametric representation

$$f_i = ua_i + vb_i$$

(with $u:v$ ranging over all pairs in k except $0:0$). If one wants the field k to be commutative one has to admit some special case of Pascal's theorem among the axioms. A non-singular linear transformation

$$f'_i = \sum_{k=1}^{n} u_{ik} f_k, \qquad f' = Uf \tag{10.1}$$

represents a *collineation*, i.e. a mapping $f \to f'$ by which points on a straight line pass into points of a straight line. The question arises as to whether this is the most general collineation. The answer was believed by all geometers to be in the affirmative, and this was established as the so-called *fundamental theorem of projective geometry*: any collineation leaving fixed the projective coördinate system is the identity. The projective coördinate system consists of the n vertices

$$(1:0:0: \cdots :0), \qquad (0:1:0: \cdots :0), \cdots, (0:0:0: \cdots :1)$$

and the "unit point"

$$(1:1:1: \cdots :1).$$

The statement is true indeed if k is the real field but false for an unspecified k (false even when k is the continuum of all complex numbers in which the classical algebraic geometry operates). For any *automorphism* $\alpha \to \alpha^*$ of k

gives rise to a collineation $f_i \to f_i^s$ leaving the coördinate system fixed. (An automorphism of a field is a one-to-one mapping $\alpha \to \alpha^s$ of the field upon itself which preserves the fundamental operations $+$ and $\times$:

$$(\alpha + \beta)^s = \alpha^s + \beta^s, \qquad (\alpha\beta)^s = \alpha^s\beta^s,$$

and hence carries 0 into 0, 1 into 1.) *The fundamental theorem of projective geometry in its universally valid form simply asserts that each collineation preserving the coördinate system is in this sense an automorphism of k.* The proof rests on the fact that the concepts of a number in k, and of addition and multiplication in k, are geometrically defined in terms of the fundamental notions entering into the projective geometric axioms. The most general collineation is a combination of an automorphism $\alpha \to \alpha^s$ with a linear substitution (10.1):

$$f_i' = \sum_k u_{ik} f_k^s, \qquad f' = Uf^s. \tag{10.2}$$

As compared with these *"semi-linear" substitutions*, the linear substitutions proper, though not forced into abdication altogether, play merely a secondary rôle: they represent *projectivities*. Two $(n - 1)$-dimensional hyperplanes immersed in a projective n-space can be mapped upon each other by a perspectivity (central projection). A projectivity is brought about by any chain of perspectivities which ultimately returns to the initial hyperplane and thus effects a certain collinear mapping of that plane on itself; and it is this type of collineation, realizable by a chain of perspectivities in a higher dimensional space, which corresponds to the linear substitutions without automorphisms.

In view of these conditions it seems natural to generalize the theory of representations of a finite group γ so as to include semi-linear transformations. We therefore assume that with every element s of the given finite group γ there is associated an automorphism of the field k, denoted by $\alpha \to \alpha^s$:

$$(\alpha^s)^t = \alpha^{ts}.$$

All representations we are going to study are based on that correspondence given once for all; an individual representation of degree n represents s by a semi-linear operator of the following type:

$$f_i' = \sum_k u_{ik}(s) f_k^s, \qquad f' = U(s)f^s. \tag{10.3}$$

The law of composition now reads

$$U(st) = U(s) \cdot U^s(t).$$

On abbreviating (10.3) again by $f' = \mathbf{s}f$, the operator associated with the quantity $a = \sum a(s) \cdot s$ of the group ring will be

$$f \to f' = \mathbf{a}f = \sum_s a(s) \cdot \mathbf{s}f. \tag{10.4}$$

If multiplication of the ring elements a is to follow the mode in which the corresponding operators $\mathbf{a}$ are composed, $c = ba$ must be given by

$$c(s) = \sum_{tt'=s} b(t)a^t(t').$$

This law defines *a modified group ring* $\mathfrak{r}$ depending on what automorphisms are associated with the group elements s.

The operators (10.4) corresponding to all quantities a of that group ring will form an algebra $\mathfrak{S}$, although one will hesitate to call it a matric algebra. All *ordinary* linear transformations

$$\bar{f}_i = \sum_k a_{ik} f_k, \qquad \bar{f} = Af,$$

which commute with the h operators $f \to \mathbf{s}f$ form a set $\mathfrak{A}$ of matrices which we shall call the commutator algebra again although it is not an algebra in the strict sense, at least not in k. $\mathfrak{A}$ is closed with respect to matrix addition and multiplication; of the membership of αA (α a number, A in $\mathfrak{A}$), however, one can be sure only if α is a self-conjugate number in k, satisfying $\alpha^s = \alpha$ for all occurring automorphisms s. The self-conjugate numbers form a subfield k_0 over which k stands as a relative field of finite order w; the automorphisms s form the Galois group of k over k_0. $\mathfrak{A}$ is an algebra in k_0, and for many purposes it might be convenient to consider k_0 rather than k the ground field. By making use of a basis of k over k_0, the operators $f \to \mathbf{a}f$ of $\mathfrak{S}$ will then turn into ordinary linear operators in k_0 of degree nw.

Whatever the merits of this standpoint may be, there is no need for accepting it when one sets out to carry the theory expounded in section B of this chapter over to the semi-linear case. All results will stand unimpaired, and only trivial modifications are necessary with the proofs. The roof operation should now be defined by

$$\hat{a}(s) = a^s(s^{-1}).$$

As to the trace, the form $\operatorname{tr}(xy)$ will not be symmetric and (7.11) is to be replaced by

$$\operatorname{tr}(sxs^{-1}) = \operatorname{tr}^s(x).$$

In all that is to follow we shall never have occasion to fall back on semi-linear transformations. None the less we were prompted by three reasons to mention this generalization: first, because it costs us almost no effort; second, it seems to indicate the natural scope of the method (II); and, finally, because the semi-linear transformations have come to the fore in a number of recent investigations in different contexts.[9]

CHAPTER IV

THE SYMMETRIC GROUP AND THE FULL LINEAR GROUP

1. Representation of a finite group in an algebraically closed field

The next step in carrying out the program laid down in §6 of the last chapter is the explicit decomposition of the regular representation of the *symmetric group* π_f of f figures into its irreducible constituents. This construction, while yielding a complete set of inequivalent irreducible representations of π_f, will go through wholly within the rational ground field κ. Nevertheless such results will be obtained as the general theory provides only for algebraically closed reference fields. The reason for this behavior lies in the fact, peculiar to the symmetric group, that the irreducible parts in κ happen to be absolutely irreducible. For a full appreciation of our explicit and elementary construction we deem it advisable to remind the reader briefly of the essential facts concerning representations of a finite group in an algebraically closed field k. They are: *orthogonality* and *completeness*.

1. *Orthogonality.* Let

$$s \longrightarrow U(s) = \| u_{ik}(s) \| \quad \text{and} \quad V(s) = \| v_{pq}(s) \|$$

be two irreducible representations of degrees g and g' of the finite group γ of order h, $(i, k = 1, \cdots, g;\ p, q = 1, \cdots, g')$, and let $\| \hat{v}_{pq}(s) \| = \| v_{qp}(s^{-1}) \|$ be the contragredient matrix $\hat{V}$ of V. $\mathfrak{M}_s$ designating the mean value $\frac{1}{h}\sum_s$, one then has

$$\mathfrak{M}_s\{u_{ik}(s)\hat{v}_{pq}(s)\} = 0, \tag{1.1}$$

provided the two representations are inequivalent, while for $V(s) = U(s)$ the averages are given by

$$\mathfrak{M}_s\{u_{ik}(s)\hat{u}_{pq}(s)\} = \begin{cases} 1/g & \text{for} \quad p = i,\ q = k, \\ 0 & \text{otherwise.} \end{cases} \tag{1.2}$$

The following simple proof is due to I. Schur.[1] By means of an arbitrary matrix $B = \| b_{ip} \|$ of g rows and g' columns one forms

$$\sum_t U(t)BV^{-1}(t) = C,$$

the sum extending to all group elements t. One verifies at once

$$U(s)CV^{-1}(s) = C \quad \text{or} \quad U(s)C = CV(s);$$

hence by Lemma (3.1.D), in case of inequivalence, $C = 0$: these are the relations (1.1). In the other case, $V(s) = U(s)$, Lemma (3.1.C) permits us in an algebraically closed field k to draw the conclusion $C = \mu E$, where the number μ must depend linearly on the arbitrary B:

$$\mu = \sum_{k,q} \mu_{kq} b_{kq}.$$

Thus

$$\sum_s \sum_{k,q} u_{ik}(s) b_{kq} \hat{u}_{pq}(s) = \delta_{ip} \sum_{k,q} \mu_{kq} b_{kq}, \quad \text{or}$$

$$\sum_s u_{ik}(s) \hat{u}_{pq}(s) = \delta_{ip} \mu_{kq}. \tag{1.3}$$

A matrix $\| u_{ik} \|$ and its contragredient $\| \hat{u}_{ik} \|$ are related by the equations

$$\sum_i u_{ik} \hat{u}_{iq} = \delta_{kq}.$$

Hence by equating p with i and summing over i one gets

$$h\delta_{kq} = g\mu_{kq},$$

and so (1.3) reduces to (1.2).

For the characters

$$\chi_u(s) = \sum_i u_{ii}(s), \qquad \chi_v(s) = \sum_p v_{pp}(s)$$

one obtains

$$\mathfrak{M}_s\{\chi_u(s)\chi_v(s^{-1})\} = \begin{cases} 0 \text{ if the two representations are inequivalent,} \\ 1 \text{ if they are equivalent.} \end{cases} \tag{1.4}$$

An immediate consequence of this orthogonality is the fact that all the components $u_{ik}(s)$, $v_{pq}(s)$, $\cdots$ of all the inequivalent irreducible representations of γ are linearly independent, and likewise their characters $\chi_u(s)$, $\chi_v(s)$, $\cdots$.

2. *Completeness* states that they form a complete linear basis for all functions or for all class functions respectively, or in other words: the number of inequivalent irreducible representations is equal to the number of classes of conjugate elements in the group γ, and the square sum $g^2 + g'^2 + \cdots$ of their degrees is equal to the order h of γ. While orthogonality assigns a property to any two given representations the proof of completeness obviously depends on the *construction* of a complete set of inequivalent irreducible representations. But the common source whence all these representations spring has been disclosed by Chapter III and found in the regular representation $(\mathfrak{r})$ of the group ring $\mathfrak{r}$ associating with each element

$$a = \sum_s a(s)s$$

of $\mathfrak{r}$ the substitution

$$(a): x \rightarrow y = ax \qquad (x \text{ variable in } \mathfrak{r}). \tag{1.5}$$

We know by §7 of that chapter that $(\mathfrak{r})$ is fully reducible, and hence according to Theorem (3.5.B) is a direct sum of simple algebras each of which is a complete matric algebra in a certain division algebra over k. *However in an algebraically closed field k there is no other division algebra than k itself.* This follows from the same argument by which we specialized Schur's lemma for an algebraically closed field and found that the only commutators are multiples of the unit matrix. Hence in terms of an appropriate basis e_{ik}, e'_{pq}, $\cdots$ the generic element a of our group ring appears as a number of matrices

$$\| \alpha_{ik} \|, \| \alpha'_{pq} \|, \cdots$$

whose elements α_{ik}, α'_{pq}, $\cdots$ vary freely in k:

$$a = \sum \alpha_{ik} e_{ik} + \sum \alpha'_{pq} e'_{pq} + \cdots . \tag{1.6}$$

$$a \rightarrow \| \alpha_{ik} \|, \qquad a \rightarrow \| \alpha'_{pq} \|, \cdots \tag{1.7}$$

are the inequivalent irreducible representations contained in the regular one, and if one writes (1.5) in the form

$$\eta_{ij} = \sum_k \alpha_{ik} \xi_{kj}, \qquad \eta'_{pr} = \sum_q \alpha'_{pq} \xi'_{qr}, \cdots$$

one realizes anew that each of them occurs as often (g, g', $\cdots$ times) as its degree indicates (the number t in (3.5.6) related to g by the equation $g = d \cdot t$ now coincides with g, as the order d of the division algebra equals 1). (1.6) implies the desired completeness:

$$h = g^2 + g'^2 + \cdots .$$

$a(s)$ is a class function if a belongs to the centrum, and according to Chapter III, §4, these are obtained by specializing the matrices (1.7) to multiples αE_g, $\alpha' E_{g'}$, $\cdots$ of the unit matrices, or

$$a = \alpha \sum_i e_{ii} + \alpha' \sum_p e'_{pp} + \cdots .$$

Thus

$$\epsilon = \sum e_{ii}, \qquad \epsilon' = \sum e'_{pp}, \cdots \tag{1.8}$$

form a basis of the centrum. Their number is equal to that of all the irreducible representations of γ. The e_{ii}, e'_{pp}, $\cdots$ furnish that full decomposition of I into normal idempotent primitive parts we dealt with in §9, Chapter III.

$$\mathsf{I} = \epsilon + \epsilon' + \cdots \tag{1.9}$$

is another normal idempotent decomposition of I which stops short with the irreducible *two-sided* invariant subspaces or the simple algebras of which $(\mathfrak{r})$ is the direct sum:

$$x = x\epsilon + x\epsilon' + \cdots ;$$

$$x\epsilon = \sum_{i,k} \xi_{ik} e_{ik}, \cdots .$$

Here we could conclude. It will, however, help to clarify the situation when we add the following considerations. The rule for multiplying matrices is equivalent to the following multiplication table of our basis:

$$(1.10)\quad \begin{cases} e_{ik}e_{kj} = e_{ij}, \quad e'_{pq}e'_{qr} = e'_{pr}, \cdots; \\ \text{all other products} = 0. \end{cases}$$

On computing the trace of the substitution (1.5) first in terms of the "natural basis" s, and then in terms of our new basis e_{ik}, e'_{pq}, $\cdots$, we find

$$h \cdot a(1) = h \cdot \mathrm{tr}(a) = g \sum_i \alpha_{ii} + g' \sum_p \alpha'_{pp} + \cdots,$$

or

$$\mathrm{tr}(e_{ik}) = 0 \quad (i \neq k), \qquad \mathrm{tr}(e_{ii}) = g/h;$$

$$\cdot \quad \cdot \quad \cdot \quad \cdot \quad \cdot \quad \cdot \quad \cdot \quad \cdot \quad \cdot \quad \cdot \quad \cdot \quad \cdot$$

The multiplication table (1.10) then shows that only the following products of basic elements have a trace $\neq 0$:

$$(1.11)\quad \mathrm{tr}(e_{ik}e_{ki}) = g/h, \qquad \mathrm{tr}(e'_{pq}e'_{qp}) = g'/h, \cdots.$$

Therefore (1.6) leads to

$$(1.12)\quad \mathrm{tr}(ae_{ki}) = \frac{g}{h}\alpha_{ik}.$$

But

$$\| \alpha_{ik} \| = \sum_s a(s)U(s), \qquad \alpha_{ik} = \sum_s a(s)u_{ik}(s)$$

yields the representation $U(s) = \| u_{ik}(s) \|$ of the group. Considering that

$$\mathrm{tr}(ae_{ki}) = \sum_s a(s)e_{ki}(s^{-1})$$

we thus infer from (1.12) the equations

$$(1.13)\quad u_{ik}(s) = \frac{h}{g} e_{ki}(s^{-1}), \qquad \chi_u(s) = \frac{h}{g} \epsilon(s^{-1}),$$

in the light of which the relations (1.11) contain a new constructive proof of the orthogonality relations.

Combining the formula (1.13) for the character $\chi_u(s)$ with the one (3.7.14) expressing it in terms of an idempotent e generating the corresponding invariant subspace, we see that

$$(1.14)\quad \epsilon(s) = \frac{g}{h} \sum_t e(t^{-1}st).$$

We shall avail ourselves of the hint contained in this equation in the constructive decomposition of the regular representation of the symmetric group, to which we now proceed.

2. The Young symmetrizers. A combinatorial lemma

We shall depict permutations s of f figures 1, 2, $\cdots$, f in the following way. We have a chess board consisting of f "fields" labeled with the ciphers from 1 to f, and f men which can be put on the f fields. There are $f!$ different positions or ways to do this. The men may be made discernible by different colors or, if one is not afraid of possible confusions, again by the labels 1, 2, $\cdots$, f. A move s is a transition from one such position to another one; it is described by the permutation

$$s: 1 \to 1', 2 \to 2', \cdots \tag{2.1}$$

if the man in field 1 is moved by s into field $1'$, the man in field 2 into field $2'$, etc. What we wish to emphasize is that the permutation (2.1) should be read as moving the man in field 1 to field $1'$ and so forth, and not as replacing the man No. 1 in his field by man No. $1'$, $\cdots$. Hence the move shown in our

board	1 | 2 | 3 | 4 | 5 | 6	1 | 2 | 3 | 4 | 5 | 6
men	①②③④⑤⑥	⑤①⑥④②③

figure is represented by the permutation

$$1 \to 2,\ 2 \to 5,\ 3 \to 6,\ 4 \to 4,\ 5 \to 1,\ 6 \to 3$$

and not by the inverse

$$1 \to 5,\ 2 \to 1,\ 3 \to 6,\ 4 \to 4,\ 5 \to 2,\ 6 \to 3.$$

A move s, (2.1), followed by the move

$$t: 1' \to 1'', 2' \to 2'', \cdots$$

results in the move called

$$ts: 1 \to 1'', 2 \to 2'', \cdots.$$

Sometimes it is convenient to have both initial and final position of the men simultaneously before our eyes. Then we need two chess boards and two identical sets of men. All this seemingly superfluous pedantry amounts to an earnest effort to keep clear the order of composition of permutations.

The problem we are to face is how to generate from an arbitrary tensor $F(i_1 \cdots i_f)$ of rank f a tensor $\mathbf{e}F$ of "highest possible symmetry" by means of an idempotent symmetry operator $\mathbf{e}$. Off hand we know two simple procedures of this kind, symmetrization and alternation:

$$\frac{1}{f!}\sum_s \mathbf{s} \qquad \text{and} \qquad \frac{1}{f!}\sum_s \delta_s \cdot \mathbf{s},$$

generating the class of symmetric and of anti-symmetric tensors; δ_s, as may be remembered, is $= +1$ or -1 according as the permutation s be even or odd. One can break the line of subindices 1, $\cdots$, f into several sections of lengths

$f_1, f_2, \cdots$ as indicated by the diagram $T = T(f_1, f_2, \cdots)$ and perform symmetrization with respect to the arguments i_α in each row separately. The rows shall be arranged in order of decreasing lengths f_k:

$$f_1 \geqq f_2 \geqq f_3 \geqq \cdots; \qquad f_1 + f_2 + f_3 + \cdots = f. \tag{2.2}$$

1	2	3	4	5	6	7	8
9	10	11	12	13			
14	15	16	17	18			
19	20	21	22				
23	24						

Diagram T corresponding to the partition

$24 = 8 + 5 + 5 + 4 + 2$

With respect to a given diagram T of f fields we denote by p any permutation which does not exchange men standing in different *rows* of T, and by q any permutation which in the same sense preserves the columns. Our partial symmetrization is then expressed by the sum

$$a = \sum_p p; \tag{2.3}$$

the result of its application on an arbitrary tensor is a tensor symmetric in the first f_1 arguments, in the next f_2 arguments, etc. We here suppose the fields of our chess board to be labeled by the figures $1, 2, \cdots, f$ in their natural order. This process will not result in a *primitive* symmetry class, i.e. in a class of tensors of highest possible symmetry. In order to increase the symmetry conditions we could follow symmetrization by alternation. If we perform alternation with respect to certain of the arguments or fields which we pick out at random on our chess board the result will certainly be zero if two of these arguments lie in the same row; for if $F(ik)$ is symmetric then $F(ik) - F(ki)$ is zero. Hence the best we can do is to select a field in each row, and without any essential loss of generality we assume them to be the fields in the first column. These considerations suggest the idea of following the symmetrization (2.3) with alternation with respect to the columns:

$$b = \sum_q \delta_q \cdot q. \tag{2.4}$$

The final process we associate with the diagram T will then be

$$c = ba = \sum_{p,q} \delta_q \cdot qp. \tag{2.5}$$

These "symmetrizers" c were first invented by A. Young. In studying their properties we shall follow G. Frobenius or rather the simplified arrangement of Frobenius' proofs due to J. v. Neumann.[2] We want to show three things:

1) that c is idempotent but for a numerical factor $\mu \neq 0$, i.e. that we have an equation

$$cc = \mu c; \tag{2.6}$$

2) that the idempotent $c/\mu = e$ is primitive, so that the set of all tensors $\mathbf{c}F$ is an irreducibly invariant subspace of tensor space with respect to the algebra $\mathfrak{A}$ of bisymmetric transformations, or that the set of quantities of the form $x\hat{c}$ is an irreducibly invariant subspace of the group ring considered as a $f!$-dimensional vector space; and 3) that invariant subspaces generated in this way by symmetrizers c and c' corresponding to distinct schemes T and T' are inequivalent.

Is there any chance that we shall then obtain all irreducible representations of the symmetric group π_f? The number of different schemes T equals the number of "partitions" of f into summands $f_1, f_2, \cdots$:

$$f = f_1 + f_2 + \cdots, \qquad f_1 \geqq f_2 \geqq \cdots > 0. \tag{2.2}$$

The distribution of permutations in classes of conjugate elements is ascertained by writing an arbitrary permutation s as the product of a number of disjoint cycles. A cycle like (1234) is the permutation carrying $1 \to 2$, $2 \to 3$, $3 \to 4$, $4 \to 1$. The number and lengths of its several cycles describe the class of s: a conjugate element is obtained by entering the figures in a different order into the scheme of its cycles. For instance

$$s = (1234)(56)(7)(8) \quad \text{and} \quad s' = (7251)(38)(4)(6)$$

are conjugate: $s' = rsr^{-1}$ where r is the permutation

$$12345678 \quad \text{into} \quad 72513846.$$

If there are α_1 cycles of length 1, α_2 cycles of length 2, $\cdots$, then these numbers $\alpha_1, \alpha_2, \cdots$ satisfying the conditions $\alpha_1 \geqq 0$, $\alpha_2 \geqq 0$, $\cdots$ and

$$f = 1\alpha_1 + 2\alpha_2 + 3\alpha_3 + \cdots, \tag{2.7}$$

describe the class. The number of classes is therefore equal to the number of solutions of (2.7) by non-negative integers. By putting

$$f_\nu = \alpha_\nu + \alpha_{\nu+1} + \cdots$$

the inequalities $\alpha_\nu \geqq 0$ are changed into $f_1 \geqq f_2 \geqq \cdots$ and the equation (2.7) into $f = f_1 + f_2 + \cdots$. Hence the number of classes is equal to the number of different diagrams. This fact holds out a fair promise that Young's construction will yield the full harvest of possible irreducible representations.

The p's form a group (of order $f_1!f_2! \cdots$) and so do the q's. If a permutation s can be put into the form qp then this decomposition into a p and a q is unique. Indeed

$$qp = q'p' \quad \text{implies} \quad q'^{-1}q = p'p^{-1}.$$

But a p can not equal a q unless it is the identical permutation I; for this is the only permutation exchanging men between neither rows nor columns of T. Hence

$$p'p^{-1} = q'^{-1}q = \mathsf{I} \quad \text{or} \quad p = p', q = q'.$$

Consequently the definition of c may be given in the form:

$$c(s) = \begin{cases} \delta_q \text{ if } s = qp, \\ 0 \text{ otherwise;} \end{cases}$$

all its coefficients are either 0 or ± 1.

We might have labeled the fields of our diagram T by the figures $1, 2, \cdots, f$ in a different order $r_1, r_2, \cdots, r_f$. The same move as described before by the permutation s would then be described by rsr^{-1} where r is the permutation

$$1 \to r_1, 2 \to r_2, \cdots,$$

and c is thus replaced by $c_r = rcr^{-1}$ with the coefficients

$$c_r(s) = c(r^{-1}sr).$$

The interpretation of the quantities x of our group ring $\mathfrak{r}$ as operators on tensors was here used merely for heuristic purposes; we shall now forget about it and rather focus our attention on the $f!$-dimensional space $\mathfrak{r}$ or ρ. The invariant subspace of all quantities $x\hat{c}$ shall be designated by $\rho_T = \rho(f_1, f_2, \cdots)$ and the corresponding representation of the ring $\mathfrak{r}$ or the group π_f (induced in ρ_T by the regular representation) by $\langle\rho(f_1, f_2 \cdots)\rangle$. It should be trivial that the labeling of the fields is without influence at least as far as equivalence is concerned. Indeed the subspace of the quantities $y = x\hat{c}$ is equivalent to the subspace of the quantities $y' = x'\hat{c}_r = (x'r^{-1})\hat{c}r$ as is readily shown by the one-to-one similarity mapping

$$y' = yr \qquad (x' = xr).$$

The investigation of the Young symmetrizers will rest on a single combinatorial lemma dealing with two diagrams T, T'. We arrange all diagrams in alphabetic order such that $T = (f_1, f_2, \cdots)$ precedes or stands higher than $T' = (f_1', f_2', \cdots)$ if the first non-vanishing difference

$$f_1 - f_1', \quad f_2 - f_2', \cdots$$

is positive. Any distribution of our f men over the f fields of T shall be denoted by $\{T\}$. The move changing a given $\{T\}$ into a given $\{T'\}$ is a permutation

$$s: \{T\} \to \{T'\}.$$

The fields in each diagram T are supposed to be labeled by $1, 2, \cdots, f$ in their natural order.

LEMMA (4.2.A). *Let $\{T\}$ and $\{T'\}$ be any distributions in the schemes T and T', and let T be not lower than T'. Then there are two possibilities:*

1) *either there are two men standing in the same row of* $\{T\}$ *and in the same column of* $\{T'\}$;

2) *or* $T = T'$ *and the permutation* $\{T\} \to \{T'\}$ *is of the form* qp.

PROOF. p', q' shall have the same significance with respect to T' as p, q have for T. The f_1 men occupying the first row of T in $\{T\}$ must be distributed in some way over the f_1' columns of T' in $\{T'\}$. If $f_1 > f_1'$, at least two of those men must lie in the same column of T': alternative 1). If however $f_1 = f_1'$, and alternative 1) is false, then each of them must lie in a different column of T'; by a permutation q' preserving the columns of $\{T'\}$ we can bring them to head their respective columns. We then omit these f_1 men which occupy the first row in $\{T\}$ as well as in the modified $q'\{T'\}$ and cancel the first row in both schemes. We then argue in the same way for the second row which now has become the first row in the decapitated schemes T, T', and continue this procedure. Either we shall encounter at some stage ν a longer row in T than in T':

$$f_1 = f_1', \cdots, f_{\nu-1} = f_{\nu-1}', f_\nu > f_\nu',$$

or $T = T'$. In the first case the same argument shows that the first alternative must hold, while in the second case this alternative can be avoided only if a certain q' causes the modified $q'\{T'\}$ to coincide with $\{T\}$ as to the men occupying each row, with however no regard to their order therein. Hence a certain shuffling p of the men in the rows of $\{T\}$ together with the rearrangement q' of the men in the several columns of $\{T'\}$ will bring the two distributions into full coincidence:

$$T = T', \qquad p\{T\} = q'\{T'\}.$$

This equation, or $\{T'\} = q'^{-1}p\{T\}$, means that the permutation $\{T\} \to \{T'\}$ has the form qp. Indeed, since the numbers are attached to the fields rather than to the men, a q' is a q if $T' = T$.

We restate our lemma in a second form now separating the two cases: T higher than T', and $T = T'$. In the latter case one should replace the notation $\{T'\}$, there meaning another position on the chess board T, by $\{T\}'$. If the first alternative 1) of Lemma (4.2.A) prevails we denote by u the transposition of the two men in their initial position in $\{T\}$ and by v' the same for their final position in $\{T'\}$. s again being the permutation $\{T\} \to \{T'\}$ we then have

$$su = v's;$$

u is a p and v' a q'.

LEMMA (4.2.B). *Let* T *be a given diagram. If* s *is not of the form* qp *then there are two transpositions* u, v *such that* $su = vs$ *and such that* u *is a* p, *while* v *is a* q.

LEMMA (4.2.C). *If* T *is higher than* T' *and* s *is an arbitrary permutation, we can find a transposition* u *of type* p *and another one* v' *of type* q' *such that* $su = v's$.

Indeed we take any position $\{T\}$ and introduce that position $\{T'\}$ on the second chess board T' into which $\{T\}$ passes by s.

3. The irreducible representations of the symmetric group

The coefficients $c(s)$ of c evidently satisfy the following relations:

(3.1) $$c(sp) = c(s), \quad c(qs) = \delta_q \cdot c(s).$$

LEMMA (4.3.A). *Any quantity d satisfying the same conditions*

(3.2) $$d(sp) = d(s), \quad d(qs) = \delta_q \cdot d(s),$$

is a numerical multiple of c:

(3.3) $$d = \lambda c, \quad d(s) = \lambda c(s); \quad \lambda = d(\mathsf{I}).$$

PROOF. From (3.2) we derive $d(p) = d(\mathsf{I})$, then $d(qp) = \delta_q d(p) = \delta_q d(\mathsf{I})$, and so conclude the validity of (3.3) for permutations s of the form qp.

If s is not of this form we make use of the two transpositions u, v supplied by Lemma (4.2.B) and thus obtain as a consequence of (3.2)

$$d(su) = d(s), \quad d(vs) = -d(s),$$

which together with $su = vs$ results in $d(s) = -d(s)$ or $d(s) = 0$.

LEMMA (4.3.B). *If T is higher than T′ any quantity d satisfying the relations*

$$d(sp) = d(s), \quad d(q's) = \delta_{q'} \cdot d(s)$$

is necessarily zero.

PROOF by Lemma (4.2.C) following the same argument as in the second part of the previous proof.

THEOREM (4.3.C). *cxc is a multiple of c; in particular* $cc = \mu c$.

PROOF. The quantity $d = cxc$,

$$d(s) = \sum c(t)x(r)c(t'), \quad trt' = s,$$

has the properties (3.2).

THEOREM (4.3.D). *T being higher than T′, any quantity of the form c′xc is zero; in particular* $c'c = 0$.

PROOF analogous.

In Chapter VII incidental use will be made of the slightly sharper relation

(3.4) $$c'xa = 0$$

involving a, (2. 3), instead of c.

The μ in Theorem (4.3.C) is the integer

(3.5) $$\mu = \sum_s c(s)c(s^{-1}).$$

It is important to know that it is $\neq 0$. We prove that it is positive. According to (3.5) the number μ tells us how much oftener it occurs that a solution of the equation

$$q_1 p_1 q_2 p_2 = \mathsf{I}$$

has two q's, q_1 and q_2, of equal parity $\delta_{q_1} = \delta_{q_2}$ than of opposite parity.

THEOREM (4.3.E). *If g is the dimensionality of the subspace ρ_T, then*

$$\mu g = f!. \tag{3.6}$$

PROOF. ρ_T consists of all quantities of the form $x\hat{c}$. The substitution $x \to x' = x\hat{c}$ changes every quantity x into a quantity x' of ρ_T and within ρ_T it is multiplication by μ:

$$(x\hat{c})\hat{c} = x(\hat{c}\hat{c}) = \mu \cdot (x\hat{c}).$$

If one adapts the coördinate system to the subspace ρ_T one sees at once that the trace of that substitution equals $g\mu$. But by use of the natural basis, the permutations s, we find that the trace of

$$x'(s) = \sum_t x(t)\hat{c}(t^{-1}s)$$

is equal to $f!\hat{c}(I) = f!$

Incidentally our theorem shows that $g = g(f_1 f_2 \cdots)$, the degree of the representation $\langle \rho(f_1 f_2 \cdots) \rangle$, is a divisor of the order $f!$ of our group. The corresponding statement in the general theory, that the degrees of absolutely irreducible representations of a finite group are divisors of its order, is true, as a matter of fact, but lies relatively deep; we therefore declined to prove it in §1.

We now introduce the idempotent

$$e = \frac{1}{\mu} c = \frac{g}{f!} c \tag{3.7}$$

and show:

THEOREM (4.3.F). *e is a primitive idempotent.*

Indeed let

$$e = e_1 + e_2$$

be a decomposition of e:

$$e_1 e_1 = e_1, \qquad e_1 e_2 = 0,$$

$$e_2 e_1 = 0, \qquad e_2 e_2 = \epsilon_2 .$$

Then $ee_1 = e_1$, $ee_1 e = e_1 e = e_1$, and hence by Theorem (4.3.C), $e_1 = \lambda e$. Because e_1 is idempotent the numerical factor λ satisfies $\lambda^2 = \lambda$. Therefore λ is either 1 or 0, and $e = e + 0$, $e = 0 + e$ are the only possible decompositions of the idempotent e.

THEOREM (4.3.G). *If $T \neq T'$, the two irreducibly invariant subspaces ρ_T, $\rho_{T'}$ are inequivalent.*

PROOF. Suppose T higher than T'. By Theorem (4.3.D) every quantity $x = x\hat{c}'$ in $\rho_{T'}$ then satisfies the equation $\hat{c}x = 0$ while in ρ_T there is at least the one quantity $x = \hat{c}$ for which $\hat{c}x \neq 0$. This precludes the possibility of a one-to-one similarity mapping of ρ_T upon $\rho_{T'}$.

(1.14) leads us to introduce the quantity ϵ by

$$\epsilon(s) = \frac{1}{\mu} \sum_t e(t^{-1}st) = \frac{1}{\mu^2} \sum_t c(t^{-1}st); \tag{3.8}$$

$\epsilon(s)$ is evidently a class function, and hence ϵ lies in the centrum of the group ring.

Lemma (4.3.H). *$\epsilon'\epsilon = 0$ or ϵ according as ϵ and ϵ' belong to distinct diagrams T, T' or to the same diagram T.*

Proof. We may write

$$\epsilon = \frac{1}{\mu^2} \sum_t tct^{-1}. \tag{3.9}$$

Hence the equation $\epsilon'\epsilon = 0$ is an immediate consequence of Theorem (4.3.D) if T stands higher than T'. But as ϵ lies in the centrum, $\epsilon'\epsilon = \epsilon\epsilon'$; hence $\epsilon\epsilon' = 0$ under the same assumption, and finally $\epsilon'\epsilon = 0$ whether T stands higher or lower than T'.

The product $\epsilon c (= c\epsilon)$ equals

$$\sum_t tct^{-1}c \tag{3.10}$$

but for the factor μ^2. By Theorem (4.3.C), $ct^{-1}c = \lambda \cdot c$ where

$$\lambda = \sum c(s)c(s'), \qquad st^{-1}s' = 1,$$

or

$$\lambda = \sum_s c(ts^{-1})c(s) = cc(t) = \mu c(t).$$

Hence (3.10) equals

$$\mu c \sum c(t)t = \mu cc = \mu^2 c,$$

or

$$\epsilon c = c.$$

As ϵ is in the centrum this implies

$$\epsilon tct^{-1} = tct^{-1}$$

and thus by summing over t

$$\epsilon\epsilon = \epsilon.$$

Our lemma implies the orthogonality relations for the characters

$$\chi(s) = \frac{f!}{g} \epsilon(s^{-1}) = \mu \cdot \epsilon(s^{-1}). \tag{3.11}$$

It shows that the class functions $\epsilon(s)$, $\epsilon'(s)$, $\cdots$ corresponding to the different diagrams T, T', $\cdots$ are linearly independent. As the number of different diagrams is equal to the number of classes in π_f, every class function must be a linear combination of those basic functions. In particular an equation like

$$1 = \lambda \cdot \epsilon + \lambda' \cdot \epsilon' + \cdots$$

must hold with numerical coefficients $\lambda, \lambda', \cdots$. On multiplying by ϵ the lemma yields

$$\epsilon = \lambda \cdot \epsilon \quad \text{or} \quad \lambda = 1.$$

Hence we have the decomposition of I,

$$\mathsf{I} = \epsilon + \epsilon' + \cdots, \tag{3.12}$$

which in its way, i.e. in the domain of class functions, is the utmost one can get.

THEOREM (4.3.J). *The unit element* I *of the group ring splits into the several central* ϵ *which correspond to Young's symmetry diagrams* T.

(3.12) may be written as

$$x = \frac{1}{\mu^2} \sum_r x\hat{c}_r + \frac{1}{\mu'^2} \sum_r x\hat{c}'_r + \cdots,$$

and this shows that the sum of the invariant subspaces generated by all the $\hat{c}_r, \hat{c}'_r, \cdots$ makes up the total $f!$-dimensional space. Because these subspaces are irreducible we can according to Lemma (3.2.B) pick out a certain number among them that are linearly independent and whose sum is the total space. This is the decomposition of the regular representation into irreducible constituents. We know a priori, and can deduce it anew from (3.12) when written in the form

$$f!\,\mathsf{I}(s) = g\chi(s) + g'\chi'(s) + \cdots, \tag{3.13}$$

how often each irreducible component will appear: as often (g times) as its degree indicates. [In (3.13) the left side is the character of the regular representation.] However we shall not attempt to carry through this decomposition in the same explicit constructive manner as the decomposition (3.12) into the two-sided invariant subspaces.[3] The latter suffices for all important purposes and, involving no arbitrariness, has the advantage of uniqueness.

4. Decomposition of tensor space

As in Chapter III §6, we now denote by $\mathfrak{A}_f$ the algebra of all bisymmetric linear transformations

$$A_f = \| a(i_1, \cdots, i_f; k_1, \cdots, k_f) \|$$

in the tensor space P_f of rank f.

THEOREM (4.4.A). P_f *splits into a number of irreducible invariant subspaces* Σ_α *with respect to the algebra* $\mathfrak{A}_f$ *in which* $\mathfrak{A}_f$ *induces the representations* $\langle \Sigma_\alpha \rangle$. *Any invariant subspace of* P_f *decomposes into irreducible parts each of which is similar to one of the subspaces* Σ_α.

More generally: any representation of $\mathfrak{A}_f$ *is fully reducible; if irreducible it is equivalent to one of the* $\langle \Sigma_\alpha \rangle$.

Not only did we prove the first part of this theorem [Theorem (3.9.D)], but we have now given a quite elementary construction of the irreducible subspaces into which P_f splits. The tensors of each of them can be generated by a Young symmetrizer working on a perfectly arbitrary tensor. The second part follows from the general Theorem (3.5.C).

In applying the method of part B in Chapter III one should know that two-sided invariant subspace $\rho_0 = \natural P_f$ of ρ which is the linear closure of all quantities $F(\cdot\,; i_1 \cdots i_f)$ with the coefficients $F(s; i_1 \cdots i_f) = \mathbf{s}F(i_1 \cdots i_f)$, ($F$ any tensor, the i_α having any values between 1 and n). For this purpose one ought to observe:

LEMMA (4.4.B). $\mathbf{c}F = 0$ *for any tensor* F *if the diagram* T *of the Young symmeterizer* $\mathbf{c}$ *contains more than* n *rows. In the opposite case there exists a tensor* F_0 *such that* $\mathbf{c}F_0 \neq 0$.

PROOF. $\mathbf{c}F$ is anti-symmetric in the arguments of the first column of T. Hence if this column is longer than n our tensor must vanish.

If on the contrary T consists of $m \leqq n$ rows we introduce the tensor Φ_0 all of whose components are zero except the one

$$\Phi_0 \begin{pmatrix} 1 & 1 & \cdots\cdots & 1 \\ 2 & 2 & \cdots\cdot & 2 \\ \cdots & \cdots & \cdots & \\ m & \cdots & m & \end{pmatrix} = 1. \tag{4.1}$$

The arguments $i_1, \cdots, i_f$ are here arranged according to the diagram T. This Φ_0 is symmetric in the arguments of each row; the process $\mathbf{c}$ when applied to Φ_0 effects only the alternation with respect to the columns. Hence $F_0 = \mathbf{c}\Phi_0$ is, but for a simple numerical factor $\neq 0$, that tensor which has the component δ_q for any argument values arising from those in (4.1) by a permutation q of the columns, and all other components zero.

The lemma shows that the partitions that matter in tensor space are those

$$f_1 + f_2 + \cdots + f_n = f, \qquad f_1 \geqq f_2 \geqq \cdots \geqq f_n \geqq 0, \tag{4.2}$$

into n summands, while in the analysis of the symmetric group all partitions whatsoever come in. The difference disappears if and only if $n \geqq f$. Allowance for the possibility $f_n = 0$ in (4.2) enables us to formulate the limitation as one to n rather than $\leqq n$ summands.

In view of the definition (3.8), $\boldsymbol{\varepsilon}F = 0$ if $\boldsymbol{\varepsilon}$ corresponds to one of the diagrams excluded by that restriction. Hence

$$F = \boldsymbol{\varepsilon}F + \boldsymbol{\varepsilon}'F + \cdots, \tag{4.3}$$

where the sum ranges only over the diagrams of not more than n rows. For any x in ρ_0 we have in the same sense

$$x = x\hat{\varepsilon} + x\hat{\varepsilon}' + \cdots,$$

or

$$x = xi = ix$$

where the sum

$$i = \hat{\epsilon} + \hat{\epsilon}' + \cdots \tag{4.4}$$

extends only over those diagrams. i lies in the centrum of ρ. Incidentally, $\hat{\epsilon} = \epsilon$ and hence $\hat{\imath} = i$, since s^{-1} is conjugate to s in the symmetric group (its cycles have the same length as those of s). We assert that i itself lies in ρ_0 and is therefore the unit element in ρ_0. To prove this we proceed as follows. The $\hat{c}$ belonging to a diagram of not more than n rows lies in ρ_0. Indeed the set ρ_T^0 of all quantities of the form $x\hat{c}$ with x in ρ_0 is a (left-) invariant subspace contained in the space ρ_T of all quantities $x\hat{c}$ where x ranges freely over the whole ρ, since ρ_0 is left-invariant. ρ_T being irreducible, the part ρ_T^0 is either 0 or the whole ρ_T. The first possibility is excluded by the previous lemma, and hence $\rho_T^0 = \rho_T$. Since ρ_0 is right- as well as left-invariant, ρ_T^0 is contained in ρ_0. As $\hat{c} = 1\cdot\hat{c}$ lies in ρ_T, this shows that $\hat{c}$ lies in ρ_0. Again making use of the two-sided invariance of ρ_0 we infer that $\hat{\epsilon}$, (3.9), is in ρ_0, and as this applies to any term in the sum (4.4), so does i.

THEOREM (4.4.C). *The sum*

$$i = \epsilon + \epsilon' + \cdots$$

extending to all diagrams of not more than n rows is the unit in ρ_0.

Thus we can specify the statement of Theorem (4.4.A) to this effect:

THEOREM (4.4.D). *If* **c** *is a Young symmetrizer corresponding to a partition* (4.2) *into n summands, then the tensors* **c***F form a non-vanishing irreducibly invariant subspace* $\mathrm{P}(f_1f_2 \cdots f_n)$ *of the tensor space* P_f. *In the decomposition of* P_f *this part will occur as often as the dimensionality g of the subspace in* ρ *of all quantities of the form* $x\hat{c}$ *indicates, a number which is given by*

$$g = \frac{f!}{\mu}, \qquad \mu = \sum_s c(s)c(s^{-1}). \tag{4.5}$$

Different diagrams give rise to inequivalent subspaces. Each irreducible invariant subspace is similar to one of the spaces $\mathrm{P}(f_1 \cdots f_n)$.

Long since we should have observed that the symmetric tensors as well as the anti-symmetric tensors form an irreducibly invariant subspace, though the latter is zero if $f > n$. They correspond to the partitions

$$f = f \quad \text{and} \quad f = 1 + 1 + \cdots$$

respectively.

We still have to insert the cornerstone into this whole edifice by proving that $\mathfrak{A}_f$ is the enveloping algebra of the group of all transformations

$$\Pi_f(A) = A \times A \times \cdots \times A \qquad (f \text{ factors})$$

induced in tensor space by the non-singular linear transformations $A = \| a(ik) \|$ in vector space. This was the observation that primarily set loose our whole flight of investigations from Chapter III, B, on. But with the same effort we can accomplish more by taking into simultaneous consideration all the tensor spaces $P_0, P_1, \cdots, P_f$ of any rank $v \leqq f$. We introduce the algebra $\mathfrak{A}^{(f)}$ whose elements

$$A^{(f)} = (A_f, A_{f-1}, \cdots, A_0)$$

are composed of arbitrary bisymmetric matrices

$$A_v = \| a(i_1 \cdots i_v ; k_1 \cdots k_v) \|$$

in P_v, $(v = f, f-1, \cdots, 0)$. The $A^{(f)}$ induced by the non-singular A in vector space is defined as

$$\Pi^{(f)}(A) = (\Pi_f(A), \Pi_{f-1}(A), \cdots, \Pi_0(A)). \tag{4.6}$$

THEOREM (4.4.E). *$\mathfrak{A}^{(f)}$ is the enveloping algebra of the group of the $\Pi^{(f)}(A)$ induced by all non-singular linear transformations A in vector space.*

PROOF. Were the enveloping algebra $\bar{\mathfrak{A}}^{(f)}$ actually smaller than $\mathfrak{A}^{(f)}$ there would be a linear relation

$$\sum_{v=0}^{f} \sum_{i,k} \gamma(i_1 \cdots i_v; k_1 \cdots k_v) a(i_1 \cdots i_v; k_1 \cdots k_v) = 0 \tag{4.7}$$

with fixed "bisymmetric" coefficients γ holding for all elements of $\bar{\mathfrak{A}}^{(f)}$ or for all elements (4.6), i.e. the following polynomial of the n^2 variables $a(ik)$,

$$\sum_{v=0}^{f} \sum_{i,k} \gamma(i_1 \cdots i_v; k_1 \cdots k_v) a(i_1k_1) \cdots a(i_vk_v),$$

vanishes for all values of the variables satisfying the algebraic inequality

$$\det(a(ik)) \neq 0.$$

But then it vanishes identically. When we look upon the pair (ik) as a single index j we realize that the coefficients γ of this polynomial are written in symmetric form and hence all the coefficients γ must vanish—contrary to our assumption of a non-trivial relation (4.7). In consequence of our last proposition we can assert:[4]

THEOREM (4.4.F). *The statements* (4.4.A) *and* (4.4.D) *are true even when the terms invariant, irreducible, etc. are interpreted relatively to the group of transformations induced in tensor space by the group $GL(n)$ in vector space.*

Any representation $A \to T(A)$ of $GL(n)$ where the elements of $T(A)$ are forms of degree f with respect to the elements $a(ik)$ of A is fully reducible, and its irreducible constituents are equivalent to those $\langle P(f_1 \cdots f_n) \rangle$ that proceed from the decomposition of P_f. (If the elements of $T(A)$ are *polynomials* of degree f, full reduction prevails likewise, and the irreducible constituents are equivalent to the $\langle P(f_1, \cdots, f_n) \rangle$ with $f_1 + \cdots + f_n \leqq f$.)

No two irreducible representations of $GL(n)$ coming from tensor spaces of different rank are equivalent.

$\langle P(f_1 \cdots f_n) \rangle$, of course, denotes that representation of the full linear group whose substratum is the tensor set $P(f_1 \cdots f_n)$.

We have to add a proof of the last remark of our theorem. Let the two ranks be f and $v < f$. According to Theorem (4.4.E), we may consider $(P_f, P_{f-1}, \cdots, P_0)$ as the substratum of all the

$$A^{(f)} = (A_f, A_{f-1}, \cdots, A_0)$$

in $\mathfrak{A}^{(f)}$. Let the tensors F_f and F_v of rank f and v range over irreducibly invariant subspaces of P_f and P_v respectively, and suppose we are given a one-to-one similarity mapping $F_f \rightleftarrows F_v$. The same mapping must match $A_f F_f$ against $A_v F_v$. But there is a particular $A^{(f)}$ for which A_f is the unit matrix and $A_v = 0$; we then find that every F_f is matched against zero. Hence both subspaces must be the zero space.

5. Quantities. Expansion

Under $GL(n)$ the manifold of all tensors in $P(f_1 \cdots f_n)$ is the range of a quantity of type $\langle P(f_1 \cdots f_n)\rangle$ or, as we like to say, of *signature* $(f_1, \cdots, f_n)$. We $\times$-multiply two such quantities of signatures $(f_1, \cdots, f_n)$ and $(g_1, \cdots, g_n)$ respectively by considering all tensors

$$F(i_1 \cdots i_f\, k_1 \cdots k_g)$$

of rank $h = f + g$ which, as functions of the first $f = f_1 + \cdots + f_n$ arguments i, lie in $P(f_1 \cdots f_n)$, and as functions of the last $g = g_1 + \cdots + g_n$ arguments k lie in $P(g_1 \cdots g_n)$. A basis for these tensors would be obtained from the product

$$F(i_1 \cdots i_f)G(k_1 \cdots k_g)$$

by letting F range over a basis of $P(f_1 \cdots f_n)$ and G over a basis of $P(g_1 \cdots g_n)$. Like every other invariant subspace of P_h it breaks up into irreducible invariant parts each of which is similar to a $P(h_1 \cdots h_n)$ with $h_1 + \cdots + h_n = h$.

The class of primitive types with which we are concerned is therefore closed with respect to the operation of multiplication (followed by decomposition into primitive parts). If we admit the covariant vector as one type our class is obviously the smallest one fulfilling this requirement, because the tensor of rank f is the $\times$-product of f vectors. The contravariant vector ξ however, is a simple example of a type not falling under our scheme. Should we limit ourselves to unimodular substitutions A, then ξ would behave like the bracket product

$$[x^{(1)} \cdots x^{(n-1)}] \tag{5.1}$$

of $n - 1$ convariant vectors which we used in Chapter II, §8, or as a skew-symmetric tensor of rank $n - 1$, or as a quantity of signature $(1, \cdots, 1, 0)$. But with respect to arbitrary elements A the law of transformation for ξ differs from that for (5.1) by a factor

$$\Delta^{-1}, \qquad \Delta = \det A.$$

This suggests the following general remark. From a given representation $R(A)$ we can derive the representation

$$A \to |A|^e R(A) \tag{5.2}$$

of the same degree; e denotes any integral exponent. If $R(A)$ is the representation of signature $(f_1 \cdots f_n)$ and $e \geqq 0$, then (5.2) has the signature $(f_1 + e, \cdots, f_n + e)$. Indeed, put e columns of lengths n in front of the diagram $T(f_1 \cdots f_n)$ so as to obtain $T(f_1 + e, \cdots, f_n + e)$. Consider the tensors F skew-symmetric in the arguments of each of these columns and of the symmetry $T(f_1 \cdots f_n)$ in the rest of the arguments; they obviously form an irreducible invariant subspace which could be described as

$$\mathrm{P}(e, \cdots, e) \times \mathrm{P}(f_1, \cdots, f_n). \tag{5.3}$$

$\mathrm{P}(e + f_1, \cdots, e + f_n)$ is part of (5.3) and hence, considering the latter's irreducibility, must coincide with it. The fact thus proved enables us to assign to the representation (5.2) the signature $(f_1 + e, \cdots, f_n + e)$, if $R(A)$ is the representation $\langle f_1, \cdots, f_n \rangle$, irrespective of whether e is $\geqq 0$ or < 0; for it depends on the sums $f_1 + e, \cdots, f_n + e$ only. The effect of this slight generalization is that the condition $f_n \geqq 0$ is removed: any integers $f_1, \cdots, f_n$ in decreasing order,

$$f_1 \geqq f_2 \geqq \cdots \geqq f_n, \tag{5.4}$$

constitute a signature for a certain irreducible representation. We shall use the word *quantic* coined by Cayley for the corresponding primitive quantities. The contravariant vector is now included as the quantic of signature $(0, 0, \cdots, -1)$. The $\times$-product of two quantics of signatures $(f_1 \cdots f_n)$, $(g_1 \cdots g_n)$ breaks up into a certain number of independent quantics of certain signatures $(h_1 \cdots h_n)$: the closure with respect to multiplication is not lost by our generalization. All types are obtained from the symmetry diagrams $T(f_1 \cdots f_{n-1}\, 0)$ with $n - 1$ rows only by adding the factor Δ^e with any integral exponent e.

Beside the covariant tensors heretofore considered, one ought to consider the contravariant tensors $\Phi(i_1 \cdots i_f)$ whose transformation law differs from (3.6.1) by substituting the contragredient matrix $\| \check{a}(ik) \|$ for $\| a(ik) \|$. What is the relationship between the covariant tensors of a given symmetry $T(f_1 \cdots f_n)$ and the contravariant tensors of the same symmetry? I maintain: while the former constitute a primitive quantity of signature $(f_1 \cdots f_n)$, the latter constitute one of signature $(-f_n, \cdots, -f_1)$. In particular, a contravariant form of degree r,

$$\varphi(x) = \sum \frac{r!}{r_1! \cdots r_n!} \varphi_{r_1 \cdots r_n} x_1^{r_1} \cdots x_n^{r_n}, \qquad (r_1 + \cdots + r_n = r)$$

is a primitive quantity of signature $(0, \cdots, 0, -r)$. Let us consider the invariant multilinear forms

$$\sum \Phi(i_1 i_2 \cdots i_f)\, x_{i_1} y_{i_2} \cdots z_{i_f}$$

depending on f covariant vectors $x, y, \cdots, z$, where Φ ranges over the irreducible set $P^*(f_1 \cdots f_n)$ of all contravariant tensors of symmetry $T(f_1 \cdots f_n)$. By making the substitution similar to (5.1),

$$x = [\xi^{(1)} \cdots \xi^{(n-1)}], \cdots, \qquad z = [\zeta^{(1)} \cdots \zeta^{(n-1)}],$$

we obtain an invariant set Σ of forms which depend linearly on $(n - 1)f$ contravariant vectors, or of covariant tensors of rank $(n - 1)f$. The corresponding representations differ by the factor Δ^f; hence Σ is irreducible just as $P^*(f_1 \cdots f_n)$, and is therefore similar to some $P(f_1^* \cdots f_n^*)$. This shows that the contravariant tensors of symmetry $T(f_1 \cdots f_n)$ constitute a primitive quantity of some signature

$$(f_1', \cdots, f_n'); \qquad f_i' = f_i^* - f.$$

In order to prove that

$$f_i' = -f_{n+1-i} \tag{5.5}$$

I make formal use of *characters*. I shall be careful to arrange the argument so as to allow immediate transference to the orthogonal and other groups. Merely the "diagonal" transformations

$$x_i' = \epsilon_i x_i \tag{5.6}$$

in $GL(n)$ will be considered. Each tensor component $F_\alpha = F(i_1 \cdots i_f)$ takes on a factor, its "weight",

$$\eta_\alpha = \epsilon_{i_1} \cdots \epsilon_{i_f},$$

under the influence of (5.6). Any invariant subspace Σ of P_f is characterized by a certain number of linear equations among the n^f tensor components F_α. According to these equations a certain number of tensor components F_β (of weights η_β) are linearly independent in Σ whereas each tensor component F_α is linearly expressible by this basis F_β within Σ. Incidentally F_α can be a combination of only such F_β as are of the same weight as F_α. This is a special case of the general Theorem (3.1.E). We repeat the argument. Let

$$F_\alpha = \sum_\beta b_\beta F_\beta \tag{5.7}$$

be the relation holding for all tensors F in Σ. As Σ is invariant under the substitution (5.6), the tensor with the components $\eta_\alpha F_\alpha$ also lies in Σ:

$$\eta_\alpha F_\alpha = \sum_\beta b_\beta \eta_\beta F_\beta . \tag{5.8}$$

On multiplying (5.7) by η_α and subtracting from (5.8) one finds

$$\sum_\beta b_\beta(\eta_\beta - \eta_\alpha)F_\beta = 0 .$$

Consequently

$$b_\beta(\eta_\beta - \eta_\alpha) = 0$$

and hence $b_\beta = 0$ whenever $\eta_\beta \neq \eta_\alpha$. We therefore may determine the basis F_β by picking out a basis among the tensor components *of each weight separately*; the terms corresponding to distinct weights will be linearly independent by themselves. Using the basic F_β as coördinates in Σ, the transformation induced by (5.6) in Σ is also in diagonal form, and if $k_{m_1 \cdots m_n}$ is the number of basic components of weight

$$\eta = \epsilon_1^{m_1} \cdots \epsilon_n^{m_n},$$

then its trace, *the character, is the polynomial*

$$\sum k_{m_1 \cdots m_n} \epsilon_1^{m_1} \cdots \epsilon_n^{m_n}$$

with non-negative integral coefficients k.

We arrange the terms of this polynomial in the familar lexicographic order. For $\Sigma = \text{P}(f_1 \cdots f_n)$ the highest possible weight is $\epsilon_1^{f_1} \cdots \epsilon_n^{f_n}$ and according to Lemma (4.4.B) and its proof, there is exactly one linearly independent component of that weight, namely

$$F\begin{pmatrix} 1 & 1 & 1 & \cdots \\ 2 & 2 & \cdots & \\ \cdots & \cdots & \cdots & \\ n & \cdots & & \end{pmatrix}.$$

Consequently the character begins with the term

$$1 \cdot \epsilon_1^{f_1} \cdots \epsilon_n^{f_n}.$$

In the same manner one readily sees that the lowest term is $\epsilon_1^{f_n} \cdots \epsilon_n^{f_1}$. The result holds for the representation of signature $(f_1, \cdots, f_n)$ whether f_n is $\geqq 0$ or < 0.

(5.6) is accompanied by the transformation

$$\xi_i' = \frac{1}{\epsilon_i} \xi_i$$

for contravariant vectors ξ. Hence the character of $\text{P}^*(f_1 \cdots f_n)$ arises from that of $\text{P}(f_1 \cdots f_n)$ by replacing each ϵ_i by $1/\epsilon_i$. This reverses the lexicographic order, and the highest term in the character of $\text{P}^*(f_1 \cdots f_n)$ is therefore

$$\epsilon_1^{-f_n} \cdots \epsilon_n^{-f_1},$$

which proves our statement (5.5).

$$\sum F(i_1 \cdots i_f)\Phi(i_1 \cdots i_f)$$

being invariant, the representations $\langle \text{P}(f_1 \cdots f_n) \rangle$ and $\langle \text{P}^*(f_1 \cdots f_n) \rangle$ are contragredient to each other. Thus

$$\langle f_1, \cdots, f_n \rangle \text{ and } \langle -f_n, \cdots, -f_1 \rangle$$

are contragredient types, and this remains true even if $f_n < 0$.

THEOREM (4.5.A). *The contravariant tensors of symmetry*

$$T(f_1 \cdots f_n) \quad (f_n \geqq 0)$$

constitute a primitive quantity of signature $(-f_n, \cdots, -f_1)$. *The irreducible representations of signatures*

$$(f_1, \cdots, f_n) \text{ and } (-f_n, \cdots, -f_1)$$

are contragredient to each other (even if $f_n < 0$).

The generalization introduced by the factor Δ^e with negative exponent e has the effect that the types of quantics are closed not only with respect to multiplication, but also under "conversion", the process of changing a type to its contragredient.

At the very beginning of our investigation we established the Capelli identities and showed how to reduce inductively by means of them forms depending on any number of vector arguments to forms in not more than n or even $n - 1$ arguments. We are now enabled to carry out that induction in a somewhat more explicit way. The "expansion" (4.3) may be looked upon as its result[5]. With a tensor $F(i_1 \cdots i_f)$ we associate the multilinear form

$$L(\xi^{(1)}, \cdots, \xi^{(f)}) = \sum_{i_1 \cdots i_f} F(i_1 \cdots i_f)\xi^{(1)}_{i_1} \cdots \xi^{(f)}_{i_f}$$

of f (contravariant) vectors $\xi^{(1)}, \cdots, \xi^{(f)}$. We carry out the operation $\mathbf{c}$ on F in two steps: first the symmetrization $\mathbf{a} = \sum \mathbf{p}$ and then the alternation $\mathbf{b} = \sum \delta_q \cdot \mathbf{q}$. The first step may be accomplished in two stages: we identify the first f_1 vectors $\xi^{(1)} = \cdots = \xi^{(f_1)} = \xi$, then the next f_2, $\xi^{(f_1+1)} = \cdots = \xi^{(f_1+f_2)} = \eta$, etc., and then completely polarize the form $L_0(\xi, \eta, \cdots)$ thus obtained which does not depend on more than n arguments $\xi, \eta, \cdots$. In view of (3.9) the equation (4.3) then states that L is a linear combination of forms which arise from forms $L_0(\xi, \eta, \cdots)$ of not more than n arguments by complete polarization; in the latter process the arguments $\xi^{(1)}, \cdots, \xi^{(f)}$ are used in all of their $f!$ possible arrangements, while $L_0(\xi, \eta, \cdots)$ originates from L by identifying its arguments $\xi^{(1)}, \cdots, \xi^{(f)}$ in some way (among each other and) with the arguments $\xi, \eta, \cdots$ of L_0. If L is invariant with respect to a given group of linear transformations so will be each $L_0(\xi, \eta, \cdots)$ and the forms into which L_0 changes by polarization.

We may go one step further in agreement with what we called Capelli's special identity. If a diagram T contains $e > 0$ columns of length n, then T belongs to a partition

$$(f_1 + e, \cdots, f_{n-1} + e, e)$$

and the tensors F of the symmetry T or the associated forms L are then of the type

$$[\xi'\eta' \cdots \zeta'] \cdots [\xi^{(e)}\eta^{(e)} \cdots \zeta^{(e)}] \cdot L'$$

where L' belongs to the diagram $T(f_1, \cdots, f_{n-1}, 0)$ of $n - 1$ rows only. Hence L' arises by polarization and alternation from a form of $n - 1$ arguments and will at least be a relative invariant under the given linear group provided L is such (the weight having changed by a factor Δ^{-e}).

The importance of the full linear group $GL(n)$ lies in the fact that any group Γ of linear transformations is a subgroup of $GL(n)$ and hence decomposition of tensor space with respect to $GL(n)$ must precede decomposition relative to Γ. One should, however, not overemphasize this relationship; for after all each group stands in its own right and does not deserve to be looked upon merely as a subgroup of something else, be it even Her All-embracing Majesty $GL(n)$.

CHAPTER V

THE ORTHOGONAL GROUP

A. THE ENVELOPING ALGEBRA AND THE ORTHOGONAL IDEAL

1. Vector invariants of the unimodular group again

For each of the classical groups in n-space P like $GL(n)$ and $O(n)$ we shall find an appertaining algebra $\mathfrak{A}_f$ of bisymmetric transformations in tensor space P_f. They are contained in the algebra of *all* bisymmetric transformations, and in this rôle of a common container the latter shall from now on be designated by $\mathfrak{K}_f$. When we want to study all tensor spaces P_v of rank $v \leqq f$ simultaneously, we string together the bisymmetric transformations A_v in P_v to form the single matrix

$$A^{(f)} = \left\| \begin{matrix} A_f & & & \\ & A_{f-1} & & \\ & & \ddots & \\ & & & A_0 \end{matrix} \right\|, \tag{1.1}$$

$$A_v = \|a(i_1 \cdots i_v; k_1 \cdots k_v)\|.$$

All the matrices $A^{(f)}$ with arbitrary bisymmetric components A_v form the algebra $\mathfrak{K}^{(f)}$.

Given any group Γ of linear transformations $A = \|a(ik)\|$ in the n-dimensional vector space P, the corresponding product transformations

$$\Pi_f(A) = A \times A \times \cdots \times A \qquad (A \text{ in } \Gamma)$$

in P_f form a group $\Pi_f(\Gamma)$ homomorphic with Γ. $\Pi^{(f)}(A)$ is the string (1.1) of matrices $A_v = \Pi_v(A)$, and $\Pi^{(f)}(\Gamma)$ is the group over which $\Pi^{(f)}(A)$ varies with A ranging over Γ. We set out to determine the enveloping algebra[1] of $\Pi_f(\Gamma)$ by making use of our general criterion, Theorem (3.5.D). For that purpose we are to consider the commutator algebra $\mathfrak{B}_f$ of $\Pi_f(\Gamma)$. The matrix

$$B = \|b(i_1 \cdots i_f; k_1 \cdots k_f)\| \tag{1.2}$$

commutes with all $\Pi_f(A)$ if

$$(A^{-1} \times \cdots \times A^{-1})B(A \times \cdots \times A) = B. \tag{1.3}$$

(1.3) states that the form

$$\sum_{i;k} b(i_1 \cdots i_f; k_1 \cdots k_f)\xi_{i_1}^{(1)} \cdots \xi_{i_f}^{(f)} y_{k_1}^{(1)} \cdots y_{k_f}^{(f)} \tag{1.4}$$

depending on f covariant and f contravariant vectors

$$y^{(1)}, \cdots, y^{(f)}; \qquad \xi^{(1)}, \cdots, \xi^{(f)}$$

is an invariant with respect to the group Γ. Hence the determination of the commutator algebra $\mathfrak{B}_f$ is equivalent to the problem of vector invariants as solved by the first main theorem. When we turn to $\Pi^{(f)}(\Gamma)$ we are to ask under what conditions a matrix

$$B = \begin{Vmatrix} B_{ff} & \cdots & B_{f0} \\ \cdots & \cdots & \cdots \\ B_{0f} & \cdots & B_{00} \end{Vmatrix} \tag{1.5}$$

commutes with all $\Pi^{(f)}(A)$, A in Γ. In the same manner we find that

$$B_{uv} = ||b(i_1 \cdots i_u; k_1 \cdots k_v)|| \qquad (0 \leqq u, v \leqq f)$$

is to be the coefficient matrix of an invariant

$$\sum_{i;k} b(i_1 \cdots i_u; k_1 \cdots k_v)\xi^{(1)}_{i_1} \cdots \xi^{(u)}_{i_u} y^{(1)}_{k_1} \cdots y^{(v)}_{k_v} \tag{1.6}$$

depending on u contravariant and v covariant vectors ξ and y.

In the case of the full linear group $GL(n)$ we were able to find the enveloping algebras in the simplest way; they turned out to be $\mathfrak{K}_f$, $\mathfrak{K}^{(f)}$ respectively. Here we may therefore use our principle for proving anew the first main theorem concerning its vector invariants, as shall be done in the present section. In more intricate cases, however, and in particular for the orthogonal group, the same principle will allow us by means of Theorem (2.9.A) to deduce the enveloping algebra from the integrity basis for vector invariants.

Within the linear set P_f of all tensors $F(i_1 \cdots i_f)$ of rank f in n-space, the symmetry operators $\mathbf{a} = \sum a(s) \cdot \mathbf{s}$ form a fully reducible matric algebra $\mathfrak{S}$; the permutation

$$s: 1 \to 1', 2 \to 2', \cdots, f \to f' \tag{1.7}$$

appears as the matrix

$$||\delta(i_1k_{1'})\, \delta(i_2k_{2'}) \cdots \delta(i_fk_{f'})||. \tag{1.8}$$

The commutator algebra of $\mathfrak{S}$ is the algebra $\mathfrak{K}_f$ of all bisymmetric matrices A_f [Theorem (4.4.E)]. Hence, according to Theorem (3.5.B), $\mathfrak{S}$ is the commutator algebra of $\mathfrak{K}_f$. Consequently the coefficient matrix of any invariant (1.4) must be a linear combination of matrices of the form (1.8), to which the invariant

$$(\xi^{(1)}y^{(1')})(\xi^{(2)}y^{(2')}) \cdots (\xi^{(f)}y^{(f')}) \tag{1.9}$$

corresponds. In other words: any invariant depending linearly on f covariant and f contravariant vectors y and ξ is expressible in terms of the products of type (ξy). In (1.9) the permutation (1.7) appears as matching f "male" symbols y with f "females" ξ.

From this one succeeds fairly easily in establishing again our old table

$$[xy \cdots z], \qquad (\xi x), \qquad [\xi\eta \cdots \zeta]$$

as a full set of typical basic invariants for the unimodular group. One has to consider invariants J depending on a number of Latin and Greek arguments. By complete polarization one may assume J to be linear in each argument. One then shows that the difference h between the number of Latin and Greek arguments must be a multiple of n, $h = ng$, and that J changes into $a^g \cdot J$ by a linear transformation of determinant a. Indeed, consider the effect on J of the following two substitutions:

$$aE = \left\| \begin{matrix} a & & & \\ & a & & \\ & & \ddots & \\ & & & a \end{matrix} \right\| \quad \text{and} \quad (a) = \left\| \begin{matrix} a & & & \\ & 1 & & \\ & & \ddots & \\ & & & 1 \end{matrix} \right\|$$

The first changes J into $a^h \cdot J$, the second into a sum $\sum_i a^i \cdot F_i$; the exponents h and i are integers, though not necessarily positive. Replacing a by a^n in the second procedure, and considering that (a^n) and aE differ by a unimodular transformation A_0,

$$(a^n) = aA_0,$$

one obtains the identity in a:

$$a^h \cdot J = \sum_i F_i \cdot a^{ni}.$$

This shows that on the right side but one term $i = g$ is present and $h = ng$, $F_g = J$. Therefore the substitution (a), and consequently every substitution of determinant a, changes J into $a^g \cdot J$.

In the case $g = 0$, where one has the same number f of Latin and of Greek arguments, J is an absolute invariant and thus expressible in terms of the (ξx). If, however, the number of Latin arguments surpasses that of the Greek by n ($g = 1$) we introduce n auxiliary Greek arguments $\xi', \eta', \cdots, \zeta'$, and express the absolute invariant

$$J \cdot [\xi'\eta' \cdots \zeta'] \tag{1.10}$$

in terms of the (ξx). (1.10) is skew-symmetric in $\xi', \eta', \cdots, \zeta'$; so if we perform all the $n!$ permutations of $\xi', \eta', \cdots, \zeta'$ and form the alternating sum, we obtain an expression for

$$n! J \cdot [\xi'\eta' \cdots \zeta']$$

where on the right side a factor like

$$(\xi' x) \cdots (\zeta' z)$$

in each term has been replaced by the determinant

$$\left| \begin{matrix} (\xi' x), & \cdots, & (\zeta' x) \\ \cdots & \cdots & \cdots \\ (\xi' z), & \cdots, & (\zeta' z) \end{matrix} \right|.$$

Since this determinant

$$= [xy \cdots z]\cdot[\xi' \, \eta' \cdots \zeta']$$

one is able to cancel the auxiliary factor $[\xi' \, \eta' \cdots \zeta']$. This process is to be repeated if the number of Latin arguments surpasses that of the Greek by $2n$ or $3n$ or $\cdots$, and it is clear what to do when, conversely, the latter number surpasses the former.

2. The enveloping algebra of the orthogonal group

The following lemma is important in many instances:

Lemma (5.2.A). *Any set of orthogonal transformations in a real field k is fully reducible.*

Proof: Let P be the vector space on which the orthogonal transformations C of the given set $\{C\}$ operate. P′ being any subspace of P invariant under $\{C\}$ we construct the subspace P″ perpendicular to P′ in P: a vector $\mathfrak{x}''$ lies in P″ if it is perpendicular to all vectors $\mathfrak{x}'$ in P′. It is clear that P″, as well as P′, is invariant under all the orthogonal substitutions C. If n' be the dimensionality and $\mathfrak{e}_1, \cdots, \mathfrak{e}_{n'}$ a basis of P′, then the generic vector $\mathfrak{x}''$ of P″ is submitted to n' independent homogeneous linear equations

$$(\mathfrak{e}_1 \mathfrak{x}'') = 0, \cdots, (\mathfrak{e}_{n'} \mathfrak{x}'') = 0.$$

Hence the dimensionality of P″ is $n'' = n - n'$, and the decomposition P = P′ + P″ is secured as soon as one knows that $\mathfrak{x}' + \mathfrak{x}''$ ($\mathfrak{x}'$ in P′, $\mathfrak{x}''$ in P″) can not be zero unless both summands vanish. But by scalar multiplication of

$$\mathfrak{x}' + \mathfrak{x}'' = 0$$

with $\mathfrak{x}'$ one finds $(\mathfrak{x}'\mathfrak{x}') = 0$, and hence, by making use of the real nature of the underlying field, $\mathfrak{x}' = 0$.

Let now $A = \|a(ik)\|$ vary within the orthogonal group $O(n)$. $\Pi_f(A)$ will then vary in *the group* $\Pi_f(O)$ *induced by* $O(n)$ *in the tensor space* P_f *of rank f.* In that tensor space we may introduce a *metric* by defining the scalar product of two tensors F and G as

$$(F, G) = \sum F(i_1 \cdots i_f)G(i_1 \cdots i_f) \qquad (i_1, \cdots, i_f = 1, \cdots, n). \tag{2.1}$$

We then make the observation that the matrix $\Pi_f(A)$ in P_f is orthogonal as well as A itself in P. Therefore our lemma proves that *in a real field* $\Pi_f(O)$ *is fully reducible.*

The matrices $A^{(f)} = \Pi^{(f)}(A)$ induced by *orthogonal* A evidently satisfy the linear equations

$$\sum_{\kappa=1}^{n} a(i_1 i_2 i_3 \cdots i_v; kkk_3 \cdots k_v) = \delta(i_1 i_2)\cdot a(i_3 \cdots i_v; k_3 \cdots k_v), \tag{2.2$_i$}$$

$$\sum_{i=1}^{n} a(iii_3 \cdots i_v; k_1 k_2 k_3 \cdots k_v) = \delta(k_1 k_2)\cdot a(i_3 \cdots i_v; k_3 \cdots k_v) \tag{2.2$_k$}$$

for $v = 2, \cdots, f$. All matrices $A^{(f)}$, (1.1), in $\mathfrak{R}^{(f)}$ which fulfill these conditions constitute a matric algebra $\mathfrak{A}^{(f)}$, as one readily verifies, and its leading terms A_f an algebra* $\mathfrak{A}_f$. We maintain:

THEOREM (5.2.B). *In a real Pythagorean field,* $\mathfrak{A}^{(f)}$ *is the enveloping algebra of the group* $\Pi^{(f)}(O)$.

This proposition was first proved by the author (for the symplectic rather than the orthogonal group).[3] The simpler method followed here is due to R. Brauer, l.c.[1] As we observed at the beginning, the set $\Pi^{(f)}(O)$ is fully reducible and so the criterion of Theorem (3.5.D) can be brought into play. A matrix (1.5) commuting with the $\Pi^{(f)}(A)$ consists of coefficient matrices B_{uv} of invariants

$$\sum_{i,k} b(i_1 \cdots i_u; k_1 \cdots k_v) x_{i_1}^{(1)} \cdots x_{i_u}^{(u)} y_{k_1}^{(1)} \cdots y_{k_v}^{(v)} \qquad (u, v = 0, \cdots, f) \tag{2.4}$$

depending linearly on $u + v$ vectors

$$x^{(1)}, \cdots, x^{(u)}; \qquad y^{(1)}, \cdots, y^{(v)}.$$

The distinction between covariant and contravariant is absent in the case of the orthogonal group.

Our proof of the main theorem for orthogonal invariants in its first form went through in a real Pythagorean field. Therefore, under the assumption in our theorem, we know that (2.4) is to be a linear combination of terms each of which matches our $u + v$ arguments $x^{(1)}, \cdots, x^{(u)}; y^{(1)}, \cdots, y^{(v)}$ in pairs to form scalar products. This time, however, there is no discrimination of two "sexes". After a suitable arrangement of the arguments, the term under consideration will be a product of α consecutive factors

$$(x^{(1)}x^{(2)})(x^{(3)}x^{(4)}) \cdots,$$

of γ consecutive factors

$$(y^{(1)}y^{(2)})(y^{(3)}y^{(4)}) \cdots,$$

and of β consecutive factors

$$\cdots (x^{(u-1)}y^{(v-1)})(x^{(u)}y^{(v)})$$

$(u = 2\alpha + \beta, v = 2\gamma + \beta)$. The corresponding B_{uv} is described by

$$b(i_1 \cdots i_u; k_1 \cdots k_v) = \delta(i_1 i_2) \cdots \bullet \delta(k_1 k_2) \cdots \bullet \cdots \delta(i_u k_v).$$

For an $A^{(f)}$ satisfying the equations (2.2), the matrix $B_{uv}A_v$ as well as $A_u B_{uv}$ turns out to be

$$\| \delta(i_1 i_2) \cdots \delta(k_1 k_2) \cdots \bullet a(i_{2\alpha+1} \cdots i_u; k_{2\gamma+1} \cdots k_v) \|.$$

* $\mathfrak{A}_f$ is readily shown to be defined by the equations[2]

$$\delta(k_1 k_2) \sum_k a(i_1 i_2 i_3 \cdots i_f; kkk_3 \cdots k_f) = \delta(i_1 i_2) \sum_i a(iii_3 \cdots i_f; k_1 k_2 k_3 \cdots k_f). \tag{2.3$_f$}$$

This would be the appropriate formulation if one does not care to combine P_f with the tensor spaces of lower rank.

In computing the product $B_{uv}A_v$ one makes use of the first set of those equations, while in evaluating A_uB_{uv} one uses the second set. The postulate of bisymmetry takes care of the freedom of rearranging the order of indices i or k. Therefore each matrix $A^{(f)}$ of our algebra $\mathfrak{A}^{(f)}$ commutes with all the commutators B and hence the criterion (3.5.D) carries us to the goal.

From a somewhat different aspect, what we have here accomplished may be described as the determination of all relations of degree f,

$$\sum_{v=0}^{f}\sum_{i,k}\gamma(i_1\cdots i_v;k_1\cdots k_v)a(i_1k_1)\cdots a(i_vk_v)=0, \tag{2.5}$$

holding among orthogonal matrices $||a(ik)||$. In such a relation we suppose the coefficients γ to be written in "bisymmetric" form. Then the corresponding relation

$$\sum_{v=0}^{f}\sum_{i,k}\gamma(i_1\cdots i_v;k_1\cdots k_v)a(i_1\cdots i_v;k_1\cdots k_v)=0 \tag{2.6}$$

must hold for all matrices $A^{(f)}$ of our algebra $\mathfrak{A}^{(f)}$. When we look upon all the quantities

$$a(i_1\cdots i_v;k_1\cdots k_v)$$

as independent variables, taking, however, the bisymmetry into account from the beginning, (2.6) must be a linear combination of the left sides of the equations (2.2). This is an application of the well-known principle that if a linear form

$$L(x)=l_1x_1+\cdots+l_Nx_N$$

vanishes for all values x_p that simultaneously annual linear forms $L_1, L_2, \cdots$, then L is of necessity a linear combination of these forms. Considering now $a(ik)$ as independent variables and specializing $A^{(f)}=\Pi^{(f)}(A)$, we find that the left side of (2.5) is a linear combination of the polynomials

$$\left\{\sum_k a(i_1k)a(i_2k)-\delta(i_1i_2)\right\}a(i_3k_3)\cdots a(i_vk_v),$$
$$\left\{\sum_i a(ik_1)a(ik_2)-\delta(k_1k_2)\right\}a(i_3k_3)\cdots a(i_vk_v).$$

In other words:

THEOREM (5.2.C). *Any polynomial $\Phi(a(ik))$ of formal degree f which vanishes for all orthogonal matrices $||a(ik)||$ is a combination*

$$\sum L_{i_1i_2}D_{i_1i_2}+\sum L^*_{k_1k_2}D^*_{k_1k_2}$$

of the $n(n+1)$ particular such polynomials of degree 2:

$$D_{i_1i_2}=\sum_k a(i_1k)a(i_2k)-\delta(i_1i_2),$$
$$D^*_{k_1k_2}=\sum_i a(ik_1)a(ik_2)-\delta(k_1k_2)$$

by means of coefficients $L_{i_1i_2}$, $L^*_{k_1k_2}$ that are polynomials of formal degree $f-2$.

3. Giving the result its formal setting

Again we try to relax the conditions imposed upon the reference field and to formalize the set of "all" orthogonal transformations. Let us first operate in the rational ground-field κ. Because of its reality, the $\Pi_f(A)$ induced by any set of rational orthogonal transformations A form a set fully reducible in κ. The second point where the nature of the reference field plays a rôle is the invariant-theoretic main theorem. Here we take recourse to its formalized interpretation. We consider the set $\mathfrak{C}^{(f)}$ of all $\Pi^{(f)}(A)$ induced by rational non-exceptional proper orthogonal matrices

$$(3.1) \qquad A = \frac{E - S}{E + S}$$

and by the one improper J_n, (2.9.3), and deduce from the formalized main theorem that its enveloping algebra in κ is the same $\mathfrak{A}^{(f)}$ as defined by our relations (2.2). But as these are homogeneous linear relations with *rational* coefficients, an element $A^{(f)}$ of $\mathfrak{A}^{(f)}$ *lying in an arbitrary field* k *over* κ is a linear combination of a finite number of (rational) elements in $\mathfrak{C}^{(f)}$ We find it convenient to enunciate our result in two parts:

Theorem (5.3.A). *Both propositions* (5.2.B *and* C) *hold in any field* k *of characteristic* 0.

Supplement. *Even the more restricted set* $\mathfrak{C}^{(f)}$ *of those* $\Pi^{(f)}(A)$ *induced by* J_n *and the rational non-exceptional proper orthogonal* A *is sufficiently large to engender the whole* $\mathfrak{A}^{(f)}$ *as its enveloping algebra.*

This supplement takes on a neater form in the language of Theorem (5.2.C). We consider polynomials $\Phi(A) = \Phi(a(ik))$ of all the n^2 elements $a(ik)$ of an arbitrary matrix $A = ||a(ik)||$ with coefficients taken from a field k of characteristic 0. Those Φ that vanish after the substitution

$$(3.2) \qquad A = \frac{E - S}{E + S}, \qquad S = ||\, s_{ik} \,|| \text{ skew-symmetric,}$$

identically in the $\frac{1}{2}\, n(n - 1)$ variables s_{ik} $(i < k)$ form an *ideal* $\mathfrak{o}$, the *orthogonal ideal* (in k). $\mathfrak{o}$ is a *prime* ideal. Indeed, if a product of two polynomials vanishes after the substitution (3.2) identically in s_{ik} $(i < k)$, so does one of the factors.

Theorem (5.3.B). *For that part of the orthogonal prime ideal* $\mathfrak{o}$ *in* k *whose elements* Φ *satisfy the one further condition* $\Phi(J_n) = 0$, *the polynomials* $D_{i_1 i_2}$, $D^*_{k_1 k_2}$, *whose vanishing defines the orthogonal group, form a basis in the sense of Theorem* (5.2.C).

Let us point out the following implication:

Corollary. *A* k*-polynomial* $\Phi(A)$ *annulled by* $A = J_n$ *and the substitution* (3.2) *identically in the skew-symmetric* S *vanishes for every orthogonal matrix* A *in* k.

With all restrictions as to the nature of the reference field out of the way, there is no obstacle that prevents us from carrying over all results to an arbitrary non-degenerate metric ground-form

$$\sum_{i,k} \gamma_{ik} x_i x_k .$$

4. The orthogonal prime ideal

The beauty of our last results is somewhat marred by the supplementary condition $\Phi(J_n) = 0$ whose part it is to bring the improper orthogonal transformations into the fold. Our wish to include them came from the higher simplicity of the invariant-theoretic main theorem for the full orthogonal group. If we now begin to feel encumbered by it, there is no other way out than to examine the *proper* orthogonal group.

A proper orthogonal matrix $A = \| a(ik) \|$ satisfies the relations

$$D_{i_1 i_2} \equiv \sum_k a(i_1k)a(i_2k) - \delta(i_1i_2) = 0, \tag{4.1}$$

$$D^*_{k_1 k_2} \equiv \sum_i a(ik_1)a(ik_2) - \delta(k_1k_2) = 0, \tag{4.2}$$

and

$$\Delta \equiv \begin{vmatrix} a(11), & \cdots, & a(1n) \\ \cdots & \cdots & \cdots \\ a(n1), & \cdots, & a(nn) \end{vmatrix} - 1 = 0. \tag{4.3}$$

As a consequence we obtained the following generalization (1.3.6) of (4.3):

$$\Delta\begin{pmatrix} i_1 & \cdots & i_\rho \\ k_1 & \cdots & k_\rho \end{pmatrix} \equiv \begin{vmatrix} a(i_1k_1), & \cdots, & a(i_1k_\rho) \\ \cdots & \cdots & \cdots \\ a(i_\rho k_1), & \cdots, & a(i_\rho k_\rho) \end{vmatrix} - \begin{vmatrix} a(\iota_1\kappa_1), & \cdots, & a(\iota_1\kappa_\sigma) \\ \cdots & \cdots & \cdots \\ a(\iota_\sigma\kappa_1), & \cdots, & a(\iota_\sigma\kappa_\sigma) \end{vmatrix} = 0. \tag{4.4}$$

Here $i_1 \cdots i_\rho \iota_1 \cdots \iota_\sigma$ and $k_1 \cdots k_\rho \kappa_1 \cdots \kappa_\sigma$ are even permutations of the row $1, 2, \cdots, n$. Hence $\rho + \sigma = n$ and we may suppose without loss of generality

$$\rho \geqq \sigma \quad \text{and} \quad i_1 < \cdots < i_\rho, \qquad k_1 < \cdots < k_\rho . \tag{4.5}$$

For the $A^{(f)}$ induced by a proper orthogonal A, (4.4) leads to relations of the type

$$\begin{aligned} &\sum_{1' \cdots \rho'} \pm a(i_1 \cdots i_\rho i_{\rho+1} \cdots i_v; k_{1'} \cdots k_{\rho'} k_{\rho+1} \cdots k_v) \\ &\quad = \sum_{1' \cdots \sigma'} \pm a(\iota_1 \cdots \iota_\sigma i_{\rho+1} \cdots i_v; \kappa_{1'} \cdots \kappa_{\sigma'} k_{\rho+1} \cdots k_v). \end{aligned} \tag{4.6}$$

Again $\rho \geqq \sigma$ and $v \leqq f$, while $i_1 \cdots i_\rho \iota_1 \cdots \iota_\sigma$ and $k_1 \cdots k_\rho \kappa_1 \cdots \kappa_\sigma$ are even permutations of $1, \cdots, n$. On the left side we have the alternating sum ranging over all permutations $1' \cdots \rho'$ of $1 \cdots \rho$, and similarly on the right side. Owing to the bisymmetry of A_v, the left side is skew-symmetric in $i_1 \cdots i_\rho$, and so is the right side with respect to $\iota_1 \cdots \iota_\sigma$. The left side is identical with what one obtains by performing the alternation on the row indices i:

$$\sum_{1' \cdots \rho'} \pm a(i_{1'} \cdots i_{\rho'} i_{\rho+1} \cdots i_v; k_1 \cdots k_\rho k_{\rho+1} \cdots k_v).$$

The equations (4.6) together with (2.2) define an algebra $\mathfrak{A}_+^{(f)}$ within $\mathfrak{R}^{(f)}$.

THEOREM (5.4.A). *$\mathfrak{A}_+^{(f)}$ as defined by (4.6) and (2.2) is the embedding algebra of the group $\Pi^{(f)}(O^+(n))$.*

The proof follows the same lines as before. (2.4) being an invariant for *proper* orthogonal transformations, it is the sum of terms each of which may contain, besides the factors mentioned before, *one* factor of the type

$$[x^{(1)} \cdots x^{(\rho)} y^{(1)} \cdots y^{(\sigma)}].$$

To show that every element $A^{(f)}$ of $\mathfrak{A}_+^{(f)}$ commutes with such a B, one has to make use of the equations (4.6). It is a piece of straight-forward calculation; I leave it to the reader because in print the burden of indices would make things look more complicated than they really are.*

Again one is allowed to limit oneself within the group $\Pi^{(f)}(O^+(n))$ to the non-exceptional rational A; and thus results the

THEOREM (5.4.B). *Every polynomial of formal degree f in the orthogonal ideal $\mathfrak{o}$ is a combination*

$$\sum L_{i_1 i_2} D_{i_1 i_2} + \sum L^*_{k_1 k_2} D^*_{k_1 k_2} + \sum_{\rho \geqq n-\rho} \Lambda\begin{pmatrix} i_1 & \cdots & i_\rho \\ k_1 & \cdots & k_\rho \end{pmatrix} \Delta\begin{pmatrix} i_1 & \cdots & i_\rho \\ k_1 & \cdots & k_\rho \end{pmatrix}$$

of the basic polynomials (4.1), (4.2), (4.4) *by means of coefficients*

$$L_{i_1 i_2}, \qquad L^*_{k_1 k_2}, \qquad \Lambda\begin{pmatrix} i_1 & \cdots & i_\rho \\ k_1 & \cdots & k_\rho \end{pmatrix}$$

which are polynomials of the formal degrees $f - 2, f - 2, f - \rho$ respectively.

One knows that the equations

$$|A|^2 = 1, \qquad D_{i_1 i_2} = 0$$

are a consequence of the equations $D^*_{k_1 k_2} = 0$ which define orthogonality. Is this true not only in the numerical sense, but also in the formal sense of ideals? In other words, is it true that, for independent variables $a(ik)$,

$$|A|^2 - 1 \equiv 0 \quad \text{and} \quad D_{i_1 i_2} \equiv 0 \qquad (\text{modd } D^*_{k_1 k_2})?$$

The answer is affirmative. As to the first equation one merely has to repeat the numerical proof by considering $|A|^2$ the determinant of A^*A. Moreover $D_{i_1 i_2}$ and $D^*_{k_1 k_2}$ are the elements of the matrices $AA^* - E$ and $A^*A - E$ respectively, and with regard to them one argues as follows:

$$A^*(AA^* - E) = (A^*A - E)A^* \equiv 0 \qquad (\text{modd } D^*_{k_1 k_2}).$$

---

* The algebra of the first terms[2] would be defined by (2.3$_f$) for odd dimensionality n; if n be even, $n = 2\nu$, one has to add those of the equations (4.6) that correspond to $v = f$, $\rho = \sigma = \nu$.

Using the minors of A^* one infers from this congruence the further relations

$$\begin{aligned} |A|\cdot(AA^* - E) &\equiv 0, \\ |A|^2\cdot(AA^* - E) &\equiv 0 \end{aligned} \qquad (\text{modd } D^*_{k_1k_2}),$$

and finally, by means of $|A|^2 \equiv 1$,

$$AA^* - E \equiv 0 \quad \text{or} \quad D_{i_1i_2} \equiv 0 \qquad (\text{modd } D^*_{k_1k_2}).$$

So each $D = D_{i_1i_2}$ is expressible as a sum

$$\sum L^*_{k_1k_2}D^*_{k_1k_2}.$$

However, this expression does not satisfy the requirement concerning formal degrees which was a characteristic feature of our theorems (5.2.C) and (5.4.B), and here would demand the $L^*_{k_1k_2}$ to be of degree 0 (the degree will rise to $2n$, according to our proof). Therefore the $D_{i_1i_2}$ are superfluous besides the $D^*_{k_1k_2}$ when we aim only at determining a basis of the ideal $\mathfrak{o}$; they are not redundant, though, if the condition about the formal degree is to be observed.

It is true in the same formal ideal sense that all the relations (4.4) are a consequence of $D^*_{k_1k_2} = 0$ together with $|A| = 1$. Let $M = \| A(ik) \|$ be the matrix of the minors of degree $n - 1$ of $A = \| a(ik) \|$. Then

$$M^*A = |A|E$$

is an identity in the $a(ik)$ which leads to

$$M^*A \equiv A^*A \qquad (\text{modd } D^*_{k_1k_2}, \Delta);$$

hence with respect to the same module

$$(M^* - A^*)\,|A| \equiv 0$$

and, taking into account $|A| \equiv 1$,

$$M^* \equiv A^* \quad \text{or} \quad M \equiv A. \tag{4.7}$$

This is the relation

$$\Delta\begin{pmatrix} i_1 & \cdots & i_{n-1} \\ k_1 & \cdots & k_{n-1} \end{pmatrix} \equiv 0. \tag{4.8}$$

The deduction of the lower Δ's from (4.7) briefly referred to in Chapter I, §3, holds good if equations are replaced by congruences (modd $D^*_{k_1k_2}$, Δ). Hence by throwing away the subtler postulate about the formal degree, our theorem simplifies to the statement that

$$\sum_i a(ik_1)a(ik_2) - \delta(k_1k_2), \quad \det(a(ik)) - 1 \tag{4.9}$$

is a basis of $\mathfrak{o}$. (Of course the $D_{i_1i_2}$ may here take the place of the $D^*_{k_1k_2}$.)

We have the choice between the following three definitions of the orthogonal ideal $\mathfrak{o}$ in the field k:

(1) Φ is in $\mathfrak{o}$ if $\Phi(A)$ vanishes after the substitution (3.2) identically in s_{ik}.
(2) Φ is in $\mathfrak{o}$ if $\Phi(A)$ vanishes for every proper orthogonal matrix A in k.
(3) $\mathfrak{o}$ is the ideal with the basis (4.9).

When we first distinguish the three definitions by the symbols $\mathfrak{o}_1$, $\mathfrak{o}_2$, $\mathfrak{o}_3$ we obviously have

$$\mathfrak{o}_3 \subseteqq \mathfrak{o}_2 \subseteqq \mathfrak{o}_1 .$$

But since we have now proved $\mathfrak{o}_1 = \mathfrak{o}_3$, all three agree. The most convenient arrangement is to start with the natural definition (2); (1) then shows that $\mathfrak{o}$ is a prime ideal with (3.2) as its "generic zero" ("allgemeine Nullstelle"—see van der Waerden, *Moderne Algebra*, vol. ii, p. 60) whereas (3) gives the finite ideal basis of $\mathfrak{o}$.

THEOREM (5.4.C). *The ideal $\mathfrak{o}$ consisting of the k-polynomials $\Phi(A)$ which vanish for all proper orthogonal matrices A in k is a prime ideal with the generic zero* (3.2) *and with the finite ideal basis* (4.9).

The first part, when expressed in "geometric" language means: *the proper orthogonal group is an irreducible algebraic manifold in the n^2-dimensional space of all matrices; and this holds good in any number field k of characteristic* 0.

The finer point comes out best when we pass to homogeneous variables by substituting $a(ik)/a$ for $a(ik)$. We then operate in the domain of $n^2 + 1$ variables $a(ik)$, a and consider homogeneous forms of them. In particular we now write

$$D_{i_1 i_2} = \sum_k a(i_1 k)a(i_2 k) - a^2\delta(i_1 i_2), \tag{4.10}$$

$$D^*_{k_1 k_2} = \sum_i a(ik_1)a(ik_2) - a^2\delta(k_1 k_2), \tag{4.11}$$

$$\Delta\begin{pmatrix} i_1 & \cdots & i_\rho \\ k_1 & \cdots & k_\rho\end{pmatrix} = \begin{vmatrix} a(i_1k_1), & \cdots, & a(i_1k_\rho) \\ \cdots & \cdots & \cdots \\ a(i_\rho k_1), & \cdots, & a(i_\rho k_\rho)\end{vmatrix} - a^{\rho-\sigma}\begin{vmatrix} a(\iota_1\kappa_1), & \cdots, & a(\iota_1\kappa_\sigma) \\ \cdots & \cdots & \cdots \\ a(\iota_\sigma\kappa_1), & \cdots, & a(\iota_\sigma\kappa_\sigma)\end{vmatrix}, \tag{4.12}$$

with the conventions used in (4.4), including $\rho \geqq \sigma$. The orthogonal group has now been extended by the dilations; the extended group is an irreducible algebraic manifold in the n^2-dimensional projective space with the $n^2 + 1$ homogeneous coördinates $a(ik)$, a, and our Theorem (5.4.B) determines its natural basis:

THEOREM (5.4.D). *The ideal of forms of the $n^2 + 1$ variables $a(ik)$, a whose basis is made up by the quantities* (4.10), (4.11), (4.12) *is a prime ideal. For it contains all and only such forms $\Phi(a(ik), a)$ as are annulled by the substitution*

$$a = 1, \qquad A = \frac{E - S}{E + S}, \quad S \text{ skew-symmetric.}$$

5. An abstract algebra related to the orthogonal group

It is of some interest to define explicitly that abstract algebra which for the orthogonal group plays the same part as the group of permutations s for the full linear group. The permutations appeared here in this disguise: we have f "male" symbols $y_1, \cdots, y_f$ and f "female" symbols $\xi_1, \cdots, \xi_f$, and each "unit" (basic element) s of the group ring is a matching of them in heterosexual pairs (ξy), as in (1.9). The composition st of two such units s and t written in the letters ξy, ηz respectively, is accomplished by contracting a product like $(\xi y_i)(\eta_i z)$ into (ξz). When one interprets s as an operator in the domain of tensors of rank f, the "males" y have to be interpreted as covariant and the "females" ξ as contravariant vectors, while the pairing (ξy) signifies formation of their product. This representation, however, is faithful only if the dimensionality n of our space is $\geqq f$; otherwise there will arise linear dependences among our units as, for example,

$$\begin{vmatrix} (\xi_1 y_1), & \cdots, & (\xi_1 y_f) \\ \cdots & \cdots & \cdots \\ (\xi_f y_1), & \cdots, & (\xi_f y_f) \end{vmatrix} = 0, \tag{5.1}$$

which are absent in the abstract domain.

The analogy is obvious. Let us, in the case of the orthogonal group, limit ourselves to the one matrix A_f rather than consider the whole row $A_f, A_{f-1}, \cdots, A_0$, and accordingly pick out the part B_{ff} in the commuting matrices B, (1.5), as formerly defined. The B_{ff} form an algebra ω_f^n every unit of which may again be described as a matching in pairs of f symbols x and f symbols y, but without any discrimination of "sex." Composition of two units with the symbols xy and yz respectively is accomplished by a rule similar to the previous: a product like

$$(uy_i)(y_iv)$$

is replaced by (uv), a product (y_iy_i) when it arises in the course of the contracting process, by the number n. Here u is either x or y and v either z or y. The result of contraction is independent of the order in which the contractions are performed; for the only two possibilities are "chains" of type

$$(uy_1)(y_1y_2) \cdots (y_{l-1}v) \qquad (u, v \text{ either } x \text{ or } z)$$

and "rings" of type

$$(y_1y_2)(y_2y_3) \cdots (y_ly_1)$$

which are to be contracted into

$$(uv) \quad \text{and} \quad n$$

respectively. (Incidentally a chain of odd length l links an x with an x or a z with a z, whereas a chain of even length joins an x with a z; rings are necessarily of even length l.) What we obtain is almost a group but not quite so: the product of two units instead of being a unit may be a unit multiplied by a power of n; and some of the units, as a matter of fact all those containing "homosexual" pairs, have no inverse. By an easy counting one finds

$1 \cdot 3 \cdots (2f - 1)$ to be the order of ω_f^n. The number n plays a rôle in the definition of our algebra ω_f^n from the beginning, whereas this was not the case for the permutation group π_f. There n did not come in until we passed from the abstract algebra to its representation by operators in the tensor space P_f of rank f corresponding to the vector space P of n dimensions. The analogous representation of our present algebra ω_f^n interprets the symbols x and y as vectors in an orthogonal n-space, and the coupling (xx), (xy) or (yy) as scalar products. ω_f^n appeared heretofore only in this concrete form. The representation is sure to be faithful when $n \geqq 2f$; for the scalar products of $2f$ vectors $x_1, \cdots, x_f, y_1, \cdots, y_f$ are algebraically independent in a space of $2f$ or more dimensions. It fails to be faithful if $n < f$, for then one has the relation (5.1) with x for ξ. I shall not take the trouble to fill the gap left between these two limits.[4]

More important is the question of full reducibility. $\mathfrak{A}_f$ is the commutator algebra of ω_f^n in its concrete form or representation $\mathfrak{B}_f = \{B_{ff}\}$. Suppose that we had at our disposal a theory of the algebra ω_f^n of the same type as the theory of the symmetric group π_f expounded in Chapter IV, in particular telling us that ω_f^n is fully reducible in the ground-field κ and that the irreducible parts stay irreducible in any field k over κ. By the general theorems of Chapter III, Part A, we could then derive from it the decomposition of $\mathfrak{A}_f$ into its (absolutely) irreducible constituents. We might even hope to obtain a more elementary and complete correspondence similar to that established in Chapter III, Part B. We shall actually embark upon a construction of the latter type, but by some simple prestidigitation we shall succeed in making contact with the familiar symmetric group π_f rather than with this somewhat enigmatic algebra ω_f^n. In all this the reader is asked to forget about $\mathfrak{A}^{(f)}$ being the enveloping algebra of $\Pi^{(f)}(O)$. $\mathfrak{A}^{(f)}$ is studied for its own sake, and only after this has been accomplished shall we return by means of that forgotten fact to the group $O(n)$.

When we replace A_f by the whole row $(A_f, A_{f-1}, \cdots, A_0)$ we are led to consider the algebra of matrices

$$\| b_{uv} \| \qquad (u, v = f, f-1, \cdots, 0)$$

where b_{uv} is a linear combination of units defined by the pairwise matching of u symbols x and v symbols y and where the multiplication of such a unit e_{uv} with a unit e'_{vw} is defined accordingly.

B. The Irreducible Representations

6. Decomposition by the trace operation

The object of our investigation will now be the tensor space P_f under the influence of the algebra $\mathfrak{A}_f$, or, should one prefer to consider all tensor spaces $\mathbf{P}_v$ of rank $v \leqq f$ simultaneously, the sum space $P^{(f)} = (P_f, P_{f-1}, \cdots, P_0)$ under the influence of $\mathfrak{A}^{(f)}$. Decomposition into irreducibly invariant subspaces once having been accomplished under this algebra, the results of Part A will show it to be at the same time a full decomposition into irreducible subspaces under the orthogonal group. The trick mentioned at the end of the last section enabling us to shun the algebra ω_f^n consists in the following simple idea.

From any tensor $F(i_1 i_2 \cdots i_f)$ of rank f we can form the 12-*trace*

$$F_{12}(i_3 \cdots i_f) = \sum_{i=1}^{n} F(i i i_3 \cdots i_f),$$

which is a tensor of rank $f - 2$. This process, usually called contraction (Verjüngung) in Tensor Calculus, is invariant under the algebra $\mathfrak{A}^{(f)}$:

$$A^{(f)} = (A_f, A_{f-1}, A_{f-2}, \cdots, A_0)$$

being any element of $\mathfrak{A}^{(f)}$, the trace $\bar{F}_{12}$ of $\bar{F}$ arises by A_{f-2} from the trace F_{12} of F if A_f sends the tensor F of rank f into $\bar{F}$; one has simply to apply the equation $v = f$ of the second set (2.2). There are $f(f-1)/2$ traces $F_{\alpha\beta}$ ($\alpha < \beta$; $\alpha, \beta = 1, \cdots, f$) of a tensor F of rank f. In P_f we consider *the invariant subspace* P_f^0 *of those tensors whose traces are all zero.* Decomposition of P_f into P_f^0 and a complementary invariant subspace is provided by

THEOREM (5.6.A). *Every tensor $F(i_1 \cdots i_f)$ can be uniquely decomposed into two summands the first of which, F^0, has all its traces = 0, while the second is of the form*

$$\Phi(i_1 \cdots i_f) = \delta(i_1 i_2) \cdot F^{12}(i_3 \cdots i_f) + \cdots \qquad (f(f-1)/2 \text{ summands}). \tag{6.1}$$

Let $\mathrm{P}_f^\dagger$ be the manifold of all tensors of form (6.1). The requirement that a tensor F be perpendicular to $\mathrm{P}_f^\dagger$,

$$(F, \Phi) = 0 \qquad \text{for all } \Phi\text{'s in } \mathrm{P}_f^\dagger,$$

obviously means that all $f(f-1)/2$ traces of F vanish. Hence the manifold P_f^0 of all tensors F^0 with vanishing traces is the subspace perpendicular to $\mathrm{P}_f^\dagger$, and our proposition is an immediate consequence of Lemma (5.2.A). This argument assumes the reference field k to be *real*.

Closer examination, however, reveals the fact that it goes through in any field whatsoever of characteristic 0, because it actually operates within the ground field κ of rational numbers. We determine a basis of the subspace $\mathrm{P}_f^\dagger$ by assigning to one of the components of one of the $f(f-1)/2$ tensors $F^{12}, \cdots$ in (6.1) the value 1, to all others the value 0. The $\frac{1}{2}f(f-1) \cdot n^{f-2}$ basic Φ's which we thus obtain by varying our choice in all possible manners may be arranged in a simple row and then those be dropped that are linearly dependent on the preceding ones. We thus find a basis $\Phi_1, \cdots, \Phi_M$ of the M-dimensional subspace $\mathrm{P}_f^\dagger$ consisting of linearly independent tensors whose components are rational integers. Construction of the part

$$\Phi = x_1\Phi_1 + \cdots + x_M\Phi_M \tag{6.2}$$

of a given tensor F then requires the solution of the set of linear equations $(F, \Phi_\alpha) = (\Phi, \Phi_\alpha)$, i.e.

$$(F, \Phi_\alpha) = \sum_\beta e_{\alpha\beta} x_\beta \qquad (\alpha, \beta = 1, \cdots, M), \tag{6.3}$$

where the coefficients are the rational integers

$$e_{\alpha\beta} = (\Phi_\alpha, \Phi_\beta).$$

For *rational* x_α the quadratic form

$$\sum_{\alpha,\beta} e_{\alpha\beta} x_\alpha x_\beta,$$

being the square (Φ, Φ) of the rational tensor (6.2), is > 0 unless all the x_α vanish. Consequently the det $(e_{\alpha\beta})$, a rational integer, is $\neq 0$ and *remains so in any field k over κ.* The solvability of the set of equations (6.3) is thereby ensured (in k).

The decomposition $\mathrm{P}_f = \mathrm{P}_f^0 + \mathrm{P}_f^\dagger$ is so important to us because $\mathrm{P}_f^\dagger$ is invariant with respect to $\mathfrak{A}_f$ just as P_f^0. According to the equation $v = f$ of the first set (2.2) the substitution A_f carries

$$\delta(i_1 i_2) \cdot F^{12}(i_3 \cdots i_f) \quad \text{into} \quad \delta(i_1 i_2) \cdot \bar{F}^{12}(i_3 \cdots i_f)$$

when A_{f-2} sends F^{12} into $\bar{F}^{12}$.

By repeating our process on the tensors $F^{12}, \cdots$ of rank $f - 2$ in (6.1), we finally split the arbitrary tensor $F(i_1 \cdots i_f)$ into summands of the form

$$\delta(i_{\alpha_1} i_{\alpha_1'}) \cdots \delta(i_{\alpha_r} i_{\alpha_r'}) \cdot \varphi(i_{\beta_1} \cdots i_{\beta_v}) \qquad (2r + v = f), \tag{6.4}$$

where the tensor $\varphi(i_1 \cdots i_v)$ of rank v is in P_v^0, i.e. has all its traces $= 0$. $\alpha_1 \alpha_{1'} \mid \cdots \mid \alpha_r \alpha_{r'} \mid \beta_1 \cdots \beta_v$ is any dissection of the row of indices $1, 2, \cdots, f$ into r portions of length 2 and one of length v; the arrangement of the r portions of length 2 and the order of the individual members within each portion is immaterial. F is decomposed into parts

$$F = F^0 + F^1 + F^2 + \cdots \tag{6.5}$$

each of a definite "valence" $v = f, f - 2, f - 4, \cdots$. This decomposition is unique; the respective subspaces $\mathrm{P}_f^0, \mathrm{P}_f^1, \mathrm{P}_f^2, \cdots$ of P_f are linearly independent. Indeed, in Theorem (5.6.A) we proved that F^0 is unique. In the decomposition

$$F = G^{(1)} + \Phi^{(1)}; \qquad G^{(1)} = F^0 + F^1, \qquad \Phi^{(1)} = F^2 + \cdots,$$

$G^{(1)}$ has all its "double traces" of the type

$$G_{12,34}(i_5 \cdots i_f) = \sum_{i,k} G(iikki_5 \cdots i_f)$$

$= 0$ while $\Phi^{(1)}$ is a sum of terms of the type

$$\delta(i_1 i_2) \delta(i_3 i_4) F^{12,34}(i_5 \cdots i_f).$$

By an argument similar to that used in Theorem (5.6.A) we then recognize that this decomposition, or $G^{(1)}$, is uniquely determined; and so on. Much ambiguity, however, is involved in splitting F^r into the individual summands (6.4) of valence $f - 2r$.

In a certain way we have thus succeeded in replacing the tensor space P_f by the spaces P_v^0 of trace 0 with $v = f, f - 2, \cdots$. The substitutions A_v of our algebra $\mathfrak{A}^{(f)}$ are much more easily characterized within these subspaces P_r^0. We first observe that any permutation s: $F \rightarrow \mathbf{s}F$ carries a tensor F with vanishing traces into a tensor of the same kind; hence $\mathbf{s}$ is a substitution within P_f^0. A_f is a bisymmetric substitution within P_f^0, i.e. one commuting with all $f!$ permutations $\mathbf{s}$. The following theorem states the complete converse thereof:

THEOREM (5.6.B). *A given bisymmetric substitution A^0 in P_f^0 and a given element*

$$A^{(f-1)} = (A_{f-1}, \cdots, A_0) \tag{6.6}$$

of the algebra $\mathfrak{A}^{(f-1)}$ uniquely determine a substitution A_f in P_f such that A_f coincides with A^0 within P_f^0 and

$$A^{(f)} = (A_f, A_{f-1}, \cdots, A_0) \tag{6.7}$$

is an element of $\mathfrak{A}^{(f)}$.

Or: if A_v^0 is a given series of bisymmetric substitutions in P_v^0 ($v = f, f - 1, \cdots, 0$) then there is a uniquely determined element (6.7) *of $\mathfrak{A}^{(f)}$ such that A_v coincides with the given A_v^0 within P_v^0.*

It is a matter of taste whether one prefers to prove the first formulation by means of the decomposition dealt with in Theorem (5.6.A), or to prove at once the second formulation by means of the more complete decomposition (6.5). Let us keep to the first procedure! It is clear how to construct the $A = A_f$ searched for. Any tensor F of rank f is split into $F^0 + \Phi$ according to Theorem (5.6.A), and its image $A(F)$ is defined to be $\bar{F}^0 + \bar{\Phi}$, where $\bar{F}^0 = A^0(F^0)$ and

$$\bar{\Phi} = \delta(i_1 i_2)\bar{F}^{12}(i_3 \cdots i_f) + \cdots$$

arises from (6.1) by applying the given A_{f-2} to the individual terms $F^{12}, \cdots$:

$$\bar{F}^{12} = A_{f-2}(F^{12}), \cdots.$$

The only trouble is to make certain that $\bar{\Phi}$ is unambiguously determined by Φ; for this we must show that $\Phi = 0$ implies $\bar{\Phi} = 0$.

From Theorem (5.6.A) we know that a tensor $\bar{\Phi}$ in $P_f^\dagger$ vanishes whenever all its (simple) traces $\bar{\Phi}_{\alpha\beta}$ ($\alpha < \beta$) do. Hence we try to show that the trace $\bar{\Phi}_{12}$ arises from the corresponding trace Φ_{12} of Φ by A_{f-2}. On forming the 12-trace we have to distinguish three types of terms in the sum (6.1):

a) $\delta(i_1 i_2)\bar{F}^{12}(i_3 \cdots i_f)$,

b) $\delta(i_1 i_\alpha)\bar{F}^{1\alpha}(i_2 \cdots \underset{\alpha}{|} \cdots i_f)$ $\{\alpha \neq 1, 2\}$,

c) $\delta(i_\alpha i_\beta)\bar{F}^{\alpha\beta}(i_1 \cdots \underset{\alpha}{|} \cdots \underset{\beta}{|} \cdots i_f)$ $\{\alpha \text{ and } \beta \neq 1, 2\}$.

(A stroke like $\underset{\alpha}{|}$ indicates that the argument i_α is missing.) The cases a) and b) are trivial. The 12-trace of a) is $n \cdot \bar{F}^{12}(i_3 \cdots i_f)$; the trace of b) is $\bar{F}^{1\alpha}(i_\alpha i_3 \cdots i_{\alpha-1} i_{\alpha+1} \cdots i_f)$ and since A_{f-2} is bisymmetric it makes no difference in the substitution if we rearrange the arguments in the order

$$i_3 \cdots i_{\alpha-1} i_\alpha i_{\alpha+1} \cdots i_f .$$

The 12-trace of c) is

$$\delta(i_\alpha i_\beta) \cdot \bar{F}_{12}^{\alpha\beta}(i_3 \cdots \underset{\alpha}{|} \cdots \underset{\beta}{|} \cdots i_f). \tag{6.8}$$

According to the equation $f - 2$ of the first set (2.2) the 12-trace $\bar{F}_{12}^{\alpha\beta}$ of $\bar{F}^{\alpha\beta}$ arises from $\bar{F}_{12}^{\alpha\beta}$ by the substitution A_{f-4} ; hence according to the equation $f - 2$ of the second set (2.2) the tensor (6.8) arises from the corresponding tedsor without the bar by A_{f-2} , as we claimed.

Our construction was such that the trace of Φ is changed into the trace of $\bar{\Phi}$ by A_{f-2} , and as the traces of F^0 and $\bar{F}^0 = A^0(F^0)$ are zero the trace of any tensor F passes into that of its image $\bar{F} = A(F)$ by the substitution A_{f-2} . Hence the substitution A_f we constructed has the following properties:

1) it is bisymmetric;
2) it coincides with A^0 within P_f^0 ;
3) it transforms the traces of tensors of rank f according to the given substitution A_{f-2} ;
4) it changes a tensor $\delta(i_1 i_2) F^{12}(i_3 \cdots i_f)$ into $\delta(i_1 i_2)\bar{F}^{12}(i_3 \cdots i_f)$ where $\bar{F}^{12}$ proceeds from F^{12} by A_{f-2} .

This is what we had to prove. While in the case of the full linear group the simultaneous consideration of the tensor spaces of lower rank along with P_f was perhaps an unnecessary complication, here we just cannot help combining the ranks $f - 2, f - 4, \cdots$ with f.

7. The irreducible representations of the full orthogonal group

In applying our general theory of Chapter III, B, to the case we encounter here, where γ is the symmetric group π_f , the vector space is the space P_f^0 of all tensors of rank f with vanishing traces, and the given representation of γ in P_f^0 is $F \to \mathbf{s}F$, we see how P_f^0 can be split into irreducibly invariant subspaces with respect to the algebra $\mathfrak{A}_f^0$ of the bisymmetric substitutions in P_f^0 . The invariant subspace

$$P_0(T) = P_0(f_1 f_2 \cdots)$$

corresponding to the diagram T whose rows are of lengths f_1 , f_2 , $\cdots$ respectively ($f_1 \geqq f_2 \geqq \cdots$) consists of the tensors $\mathbf{c}F$ where $\mathbf{c}$ is the Young symmetrizer of T, and F ranges over all tensors of rank $f = f_1 + f_2 + \cdots$ and of vanishing traces. The prevailing conditions, even as to the number of times each irreducible constituent appears, will be perfectly clear as soon as we know which portion of the ring $\mathfrak{r}$ of the symmetric group here plays the rôle of ρ_0. ρ_0 is defined as the linear closure of all quantities $F(\bullet; i_1 \cdots i_f)$ with the coefficients

$$F(s; i_1 \cdots i_f) = \mathbf{s}F(i_1 \cdots i_f)$$

obtained by letting $i_1 \cdots i_f$ range independently from 1 to n and F over all *tensors with vanishing traces.* This ρ_0 is more limited than the ρ_0 of the previous chapter, where F varied over *all* the tensors, and hence may be distinguished by the notation ρ_{00}. An irreducible invariant subspace of P_f^0 consists of all tensors of form $\mathbf{e}F$ where $\mathbf{e}$ is a primitive idempotent in the group ring of π_f and F ranges over P_f^0. In particular we may choose $\mathbf{e}$ as the Young symmetrizer corresponding to a diagram T or a partition of f. From this connection it is clear that when we first operate in the rational ground-field κ and determine the irreducible parts in κ these parts will be irreducible in any field k over κ. We shall know ρ_{00} when we know which Young symmetrizers $\mathbf{c}$ change every tensor of vanishing traces into zero: $\mathbf{c}F = 0$ for F in P_f^0.

THEOREM (5.7.A). $\mathrm{P}_0(T)$ *is empty unless the sum of the lengths of the first two columns of the symmetry scheme T is $\leqq n$.*

This proposition is an immediate consequence of

LEMMA (5.7.B). *A tensor*

$$F\begin{pmatrix} i_1 \cdots\cdots i_a \\ k_1 \cdots k_b \end{pmatrix}$$

skew-symmetric with respect to the arguments in each row lies in $\mathrm{P}_{a+b}^{\dagger}$ if $a + b$ surpasses n, and therefore vanishes if its traces vanish.

A multilinear form like

$$\sum_{i,k} F\begin{pmatrix} i_1 \cdots\cdots i_a \\ k_1 \cdots k_b \end{pmatrix} \xi_{i_1}^{(1)} \cdots \xi_{i_a}^{(a)} \eta_{k_1}^{(1)} \cdots \eta_{k_b}^{(b)},$$

skew-symmetric in the a vectors $\xi^{(\alpha)}$ as well as in the b vectors $\eta^{(\beta)}$, may be represented symbolically as

$$[\xi^{(1)} \cdots \xi^{(n)}]\cdot[\eta^{(1)} \cdots \eta^{(n)}] \tag{7.1}$$

by making use of $n - a$ symbolic vectors $\xi^{(a+1)}, \cdots, \xi^{(n)}$ and $n - b$ symbolic vectors $\eta^{(b+1)}, \cdots, \eta^{(n)}$. (7.1) is the determinant of the scalar products

$$(\xi^{(\alpha)}\eta^{(\beta)}) \qquad (\alpha, \beta = 1, \cdots, n).$$

Each term of the determinant is the product of n factors $(\xi\eta)$, and since the number of symbolic vectors

$$\xi^{(a+1)}, \cdots, \xi^{(n)}; \eta^{(b+1)}, \cdots, \eta^{(n)}$$

is $< n$, at least one factor in each term is a non-symbolical

$$(\xi^{(\alpha)}\eta^{(\beta)}) \qquad (\alpha = 1, \cdots, a; \beta = 1, \cdots, b).$$

Hence $F\begin{pmatrix} i_1 \cdots\cdots i_a \\ k_1 \cdots k_b \end{pmatrix}$ is a sum of ab terms each of which contains a factor $\delta(i_\alpha, k_\beta)$, q.e.d.

The proof evidently lends itself to the sharper statement that F lies in the space we have denoted by P_{a+b}^{a+b-n}. *Hence the vanishing of merely its* $(a + b - n)$-*fold traces is a sufficient condition for the vanishing of* F. The result could also be reached by a direct combinatorial approach.

Diagrams whose first two columns have a total length $\leqq n$ shall be called permissible diagrams. One sees at once that one can arrange them in pairs of "associate" diagrams T, T' such that the length of the first column in T is a number $m \leqq \frac{1}{2}n$ and in T' is $n - m$, while the lengths of the other columns coincide for T and T'. T is self-associate, $T = T'$, if m is exactly $= \frac{1}{2}n$, which can happen only in case of even dimensionality. It is therefore convenient to distinguish between odd and even dimensionality: $n = 2\nu + 1$ or $n = 2\nu$ respectively. To T containing $m \leqq \frac{1}{2}n$ rows one might ascribe exactly ν rows of lengths $f_1, \cdots, f_\nu$,

$$(7.2) \qquad f_1 \geqq \cdots \geqq f_\nu \geqq 0, \qquad f_1 + \cdots + f_\nu = f,$$

on allowing some of the f_i's to be zero. We use the notation

$$P_0(T) = P_0(f_1 \cdots f_\nu), \qquad P_0(T') = P_0'(f_1 \cdots f_\nu).$$

T consists of f fields and T' of a larger number, unless T is self-associate. With $f_1, \cdots, f_\nu$ varying over all integers satisfying (7.2) and with f taking on the values 0, 1, 2, $\cdots$, the associates

$$T = T(f_1, \cdots, f_\nu) \quad \text{and} \quad T'$$

exhaust all permissible diagrams. In case of odd dimensionality, each one is obtained exactly once, while in the even case T and T' coincide, and hence

$$P_0'(f_1 \cdots f_\nu) = P_0(f_1 \cdots f_\nu),$$

whenever T actually contains ν rows: $f_\nu > 0$. The converse of Theorem (5.7.A) is also true:

THEOREM (5.7.C). $P_0(T)$ *is not empty if* T *is a permissible diagram.*

Let T contain $m \leqq \nu$ rows, and denote the figures 1, 2, $\cdots$, n by $1, 1^*, \cdots, m, m^*, m + 1, \cdots, n - m$. We define a tensor $G_0(i_T)$ of the table of arguments

$$i_T = \begin{pmatrix} i_{11} & \cdots\cdots & i_{1f_1} \\ \cdots\cdots & \cdots & \\ i_{m1} & \cdots\cdots & \end{pmatrix}$$

as follows. A component $G_0(i_T)$ is zero unless a) all the arguments in the first row of T are $= 1$ or 1^*, in the second row $= 2$ or 2^*, $\cdots$, in the mth row $= m$ or m^*, and unless b) the number μ of arguments taking on starred values is even; the value of G_0 for arguments satisfying these two conditions shall be

$$(7.3) \qquad (-1)^{\mu/2}.$$

This tensor $G_0(i_T)$ is symmetric in the arguments of each row and its traces are evidently zero. One obtains a similar tensor G^0 by changing (7.3) into $(-1)^{(\mu-1)/2}$, requiring at the same time the number μ to be *odd*. Alternation with respect to the columns carries G_0 (and G^0) into a non-vanishing tensor F_0 (and F^0) of the desired kind. Verbal description of the simple picture which this tensor F_0 presents becomes a little bit clumsy. A component $F_0(i_T)$ is $= 0$ unless the values in the first column of T arise from $1, \cdots, m$ by permuting these figures and starring some of them, and unless similar conditions prevail for the other columns; besides, the number μ of arguments taking on starred values has to be even. The "leading" component

$$F_0\begin{pmatrix} 1 & 1 & \cdots\cdots & 1 \\ 2 & 2 & \cdots\cdot & 2 \\ \cdots & \cdots & \cdots & \\ m & \cdots & m & \end{pmatrix} \text{ equals } 1,$$

whereas each transposition of two arguments in the same column, or each simultaneous starring of two arguments, changes the value F_0 into $-F_0$.

In order to accomplish the same for the associated scheme T' we use

$$G_0(i_{T'}) = G_0(i_T \mid i_{m+1}, \cdots, i_{n-m}) = G_0(i_T)\cdot\varphi(i_{m+1}, \cdots, i_{n-m})$$

instead of $G_0(i_T)$, where

$$\varphi(i_{m+1}, \cdots, i_{n-m}) = \pm 1$$

according as $i_{m+1}, \cdots, i_{n-m}$ is an even or odd permutation of $m + 1, \cdots, n$, and 0 otherwise.

Later on we shall have occasion to make use of the following process: $F(i_1 \cdots i_\rho)$ being a given skew-symmetric tensor of rank $\rho \leqq n$ we define a "complementary" skew-symmetric tensor σF of rank $n - \rho$ by the equation

$$F(i_1 \cdots i_\rho) = \sigma F(i_{\rho+1} \cdots i_n) \tag{7.4}$$

holding whenever $i_1 \cdots i_\rho i_{\rho+1} \cdots i_n$ is an even permutation of $1 \cdots n$. (We do not care for the moment whether this process is orthogonally invariant.) When we apply it to the arguments in the first column of a tensor $F(i_{T'})$ skew-symmetric in the arguments of each column of T', we obtain a tensor $\sigma F(i_T)$ skew-symmetric in the arguments of each column of T:

$$F\begin{bmatrix} i_1 & i_{12} & \cdots & i_{1f_1} \\ \cdots & \cdots & \cdots & \\ i_m & i_{m2} & \cdots & \\ i_{m+1} & & & \\ \vdots & & & \\ i_{n-m} & & & \end{bmatrix} = \sigma F\begin{pmatrix} i_1^* & i_{12} & \cdots & i_{1f_1} \\ \cdots & \cdots & \cdots & \\ i_m^* & i_{m2} & \cdots & \end{pmatrix} \tag{7.5}$$

$$\{i_1 \cdots i_{n-m} i_1^* \cdots i_m^* \text{ any even permutation of } 1 \cdots n\}.$$

In particular our $F_0(i_{T'})$ changes by this process either into $\pm F_0(i_T)$ or into $\pm F^0(i_T)$ according as m is even or odd, a remark that should be kept in mind for a future purpose.

In the same way as Theorem (4.4.C) follows from Lemma (4.4.B) we now infer from Theorems (5.7.A and C):

THEOREM (5.7.D). *The sum*

$$i = \epsilon + \epsilon' + \cdots$$

extending over all permissible symmetry diagrams is the unit I *of* ρ_{00}.

Application of our general theory now culminates in the following

THEOREM (5.7.E). P_f^0 *splits under* $\mathfrak{A}_f^0$ *into irreducibly invariant subspaces* $\mathrm{P}_0(T)$ *each defined by a permissible symmetry diagram* T (and its corresponding symmetrizer c). *Different diagrams give rise to inequivalent subspaces. In the decomposition the diagram T shows up as often as the number g, defined in the previous chapter, formula* (4.4.5), *indicates. The partial spaces lie in* κ *but are irreducible in any field k over* κ.

Turning to the *total tensor space*, we first split P_f into the partial spaces $\mathrm{P}_f^0, \mathrm{P}_f^1, \cdots$ of valences $f, f-2, \cdots$, equation (6.5). The generic tensor of valence v is a sum of terms (6.4) in which φ ranges over the whole space P_v^0. This space decomposes into irreducibly invariant parts with respect to the algebra of all bisymmetric substitutions in P_v^0. The individual part consists of all tensors of form $e\psi$ where e is a primitive idempotent symmetry operator in v figures and ψ ranges over P_v^0. For

$$\alpha_1, \alpha_{1'} \mid \cdots \mid \alpha_r, \alpha_{r'} \mid \beta_1 \cdots \beta_v \tag{7.6}$$

we substitute all essentially different arrangements of $1\ 2 \cdots f$. In this manner P_f^r appears as the sum of a number of irreducibly invariant subspaces with respect to our algebra $\mathfrak{A}_f$ which in some order may be denoted by Σ_λ ($\lambda = 1, 2, \cdots$). We apply the often-used Lemma (3.2.B) for dropping those among them that are redundant and thus obtain a genuine decomposition of P_f into independent irreducibly invariant subspaces with respect to our algebra $\mathfrak{A}_f$.

THEOREM (5.7.F). P_f *is decomposable into irreducibly invariant subspaces under the algebra* $\mathfrak{A}_f$. *An individual part contains all tensors of form* (6.4), *with* φ *ranging over one of the irreducible subspaces* $\mathrm{P}_0(T) = \mathrm{P}_0(f_1f_2 \cdots)$ *of* P_v^0. *T is any permissible diagram in* $v = f_1 + f_2 + \cdots$ *figures, and v takes on the values* $f, f-2, \cdots$. *The irreducible parts in κ will stay irreducible in any field k over* κ.

THEOREM (5.7.G). *Any invariant subspace of* P_f *breaks up into irreducibly invariant parts each of which is similar to one of the spaces* $\mathrm{P}_0(f_1f_2 \cdots)$ *mentioned in the previous theorem.*

More generally: any representation of $\mathfrak{A}^{(f)}$ *is fully reducible, and if irreducible it is equivalent to a* $\langle \mathrm{P}_0(T) \rangle = \langle \mathrm{P}_0(f_1f_2 \cdots) \rangle$ *that comes from a permissible diagram T of* $f_1 + f_2 + \cdots = v \leqq f$ *figures.*

The fact that the parts (6.4) of F^r corresponding to all possible arrangements (7.6) are not linearly independent prevents us from explicitly predicting how many equivalent irreducible constituents of each sort $\langle \mathrm{P}_0(f_1 f_2 \cdots)\rangle$ will appear in the decomposition of P_f. The statement as given in Theorem (5.7.E) still holds good for the valence $v = f$, but not so for the lower valences $f - 2, \cdots$. In that respect our result is less complete than for the full linear group. One could think of decomposing P_f first with respect to the full algebra $\mathfrak{K}_f$ of bisymmetric substitutions before one splits it into the finer pieces according to our more limited algebra $\mathfrak{A}_f$. From (6.5) one derives

$$\mathbf{c}F = \mathbf{c}F^0 + \mathbf{c}F^1 + \cdots$$

where the right-hand summands again lie in $\mathrm{P}_f^0, \mathrm{P}_f^1, \cdots$. If $\mathbf{c}$ belongs to a permissible diagram this equation shows that the "rough" part consisting of all tensors $\mathbf{c}F$ contains the "finer" part of all tensors $\mathbf{c}F^0$ (F^0 in P_f^0) exactly once. That explains why the diagrams of valence f occur as often here as in the rough decomposition under the full algebra $\mathfrak{K}_f$, *as far as they occur at all.*

The theory of decomposition as developed in Part B of this chapter does not depend on the result of Part A stating that $\mathfrak{A}_f$ is the enveloping algebra of $\Pi_f(O)$. By now making use of this fact we conclude:[5]

THEOREM (5.7.H). *For each permissible diagram T the representation $\langle \mathrm{P}_0(T)\rangle$ of the orthogonal group is irreducible. Different diagrams, whether of the same or different numbers of fields, lead to inequivalent representations.*

Theorems (5.7.F *and* G) *still hold good when the algebra $\mathfrak{A}_f$ is replaced by the group* $\Pi_f(O)$.

Concerning two diagrams of different numbers of fields, f and $v < f$, one ought to observe again that the enveloping algebra $\mathfrak{A}^{(f)}$ contains an element which is the identical transformation in P_f^0 and the zero transformation in P_v^0. The last statement of Theorem (5.7.G) now refers to a representation $R(A)$ of $O(n)$ where the components of the representing matrix $R(A)$ are polynomials of the components $a(ik)$ of A of formal degree f.

Thus we finally return to the statement from which our whole investigation in this chapter took its start, namely that $\Pi_f(O)$ is fully reducible. However we are now in possession of an explicit description of its decomposition and have the additional information that the irreducible parts in κ are *absolutely* irreducible.

If one $\times$-multiplies two quantities of the types $\langle \mathrm{P}_0(f_1 f_2 \cdots)\rangle$ we have constructed here, the product splits into a number of independent primitive quantities each of which is again described as a $\langle \mathrm{P}_0(f_1 f_2 \cdots)\rangle$. Since there is no difference between covariant and contravariant vectors in the case of the orthogonal group, the representation $\langle \mathrm{P}_0(f_1 f_2 \cdots)\rangle$ is contragredient to itself. So we have here the same properties of closure we encountered in Chapter IV, §5 for the quantities of the full linear group.

The circuitous manner in which we proved the full reducibility of $\mathfrak{A}_f$ is indicated by the diagram

$$\begin{array}{ccc} \pi_v & \longrightarrow & \mathfrak{A}_v^0 \\ & & \downarrow \\ \omega_f^n & \longleftarrow & \mathfrak{A}_f \end{array} \qquad (v = f, f-2, \cdots)$$

where the arrow leading to ω_f^n may first be disregarded. But by applying on $\mathfrak{A}_f$ the same general Theorem (3.5.B) by which one could have established the upper horizontal arrow one would find that its commutator algebra, i.e. ω_f^n *in its concrete form*, is fully reducible. The concrete form being a faithful representation in the case $n \geqq 2f$, we infer from Theorem (3.5.C) that the abstract algebra ω_f^n, or rather its regular representation, is fully reducible in the rational groundfield κ if $n \geqq 2f$, and that the irreducible parts in κ are absolutely irreducible. I find it hard to believe that the magnitude of n will affect the structure of ω_f^n to such a degree as to cause the breakdown of our result for $n < 2f$, but the question must be left open. As far as full reduction is concerned, Theorem (3.5.B) allowed us to jump freely from a matric algebra to its commutator algebra as indicated by the horizontal arrows; could one make the vertical transition directly on the left side without jumping to and fro over the ditch one would probably be able to settle our question. As things stand now I could not allow the reader to abandon the subject without a prick of discontent in his heart.

C. The Proper Orthogonal Group

8. Clifford's theorem

From the full orthogonal group we now turn to the group $O^+(n)$ of all proper orthogonal transformations in n-space. The passage to this subgroup of index 2 is more easily carried out with the groups themselves than with the enveloping algebras, owing to a beautiful and general theorem which was discovered by A. H. Clifford in this context.[6]

Theorem (5.8.A). *Let a group* $\gamma = \{s\}$ *and an irreducible representation* $\mathfrak{A}(\gamma)$ *of* γ: $s \rightarrow A(s)$ *be given in a number field* k. *Let, furthermore,* $\gamma' = \{t\}$ *be a given invariant subgroup of* γ. *The representation*

$$\mathfrak{A}(\gamma') \text{ of } \gamma': t \rightarrow A(t)$$

breaks up into "conjugate" irreducible representations of γ' *of equal degrees. If the index* (γ/γ') *be finite, the number of irreducible components cannot surpass the index.*

The meaning of the adjective "conjugate" will be explained in the course of the demonstration. $\mathfrak{x}$ being a vector of the n-dimensional representation space P of $\mathfrak{A}(\gamma)$, $\mathfrak{x}' = s\mathfrak{x}$ signifies the image $\mathfrak{x}' = A(s)\mathfrak{x}$. The letter s, with or without index, always alludes to an element of γ and t to an element of γ'.

We choose an l-dimensional subspace Λ of P irreducibly invariant under γ', and as such the carrier of an irreducible representation $t \to B(t)$ of γ' of degree l. The letter $\mathfrak{x}$ henceforward shall be used for vectors in Λ only. With s being a fixed element in γ and $\mathfrak{x}$ varying in Λ, $s\mathfrak{x}$ varies over a subspace $s\Lambda$, likewise invariant under γ' as is shown by the equation

$$t(s\mathfrak{x}) = s(t'\mathfrak{x}) = s\mathfrak{x}'.$$

In writing $s^{-1}ts = t'$ we made use of the fact that γ' is an invariant subgroup of γ. $s\Lambda$ is the carrier of the *conjugate* representation:

$$t \to B(s^{-1}ts).$$

Λ^* being any subspace invariant under γ', those vectors $\mathfrak{x}$ in Λ for which $s\mathfrak{x}$ is in Λ^* form a subspace Λ_0 of Λ invariant in the same sense:

$$s\mathfrak{x} \text{ in } \Lambda^*, \quad \text{hence} \quad s(t\mathfrak{x}) = t''(s\mathfrak{x}) \text{ in } \Lambda^*.$$

Since Λ is irreducible, there are only two possibilities: $\Lambda_0 = 0$ or Λ; $s\Lambda$ is either linearly independent of Λ^* or contained in Λ^*.

In case of a finite index j, let

$$s_1\gamma' = \gamma', s_2\gamma', \cdots, s_j\gamma'$$

be the cosets of γ' in γ. To the row of subspaces

$$s_1\Lambda, s_2\Lambda, \cdots, s_j\Lambda \tag{8.1}$$

invariant under γ' we apply the argument of Lemma (3.2.B) and thus pick out a certain number among them,

$$s_1\Lambda, \cdots, s_e\Lambda,$$

which are linearly independent and in whose sum

$$s_1\Lambda + \cdots + s_e\Lambda$$

all the spaces (8.1) are contained. As this sum is then invariant under the entire group γ it must be the total space P, and the representation $\mathfrak{A}(\gamma')$ has been decomposed into e conjugate irreducible representations

$$t \to B(s_i^{-1}ts_i) \qquad (i = 1, \cdots, e).$$

e is certainly a divisor of n, $n = el$, and $\leq j$.

Without the assumption of a finite index, one proceeds as follows. If $s_1\Lambda = \Lambda$ is not yet the whole space, then there exists an element $s = s_2$ such that $s\Lambda$ is not contained in Λ; otherwise Λ would be invariant under all elements s of γ. We have proved that $s_2\Lambda$ is of necessity linearly independent of Λ. If the sum $s_1\Lambda + s_2\Lambda$ does not exhaust the entire P, there is an $s = s_3$ such that $s\Lambda$ is not contained in $s_1\Lambda + s_2\Lambda$; $s_3\Lambda$ is then linearly independent of that sum. And so on. The process must come to a stop after n/l steps.

The theorem applies in particular to a subgroup γ' in γ *of index* 2:

$$\gamma = \gamma' + u\gamma'.$$

A given irreducible representation

$$\mathfrak{A} = \mathfrak{A}(\gamma): s \rightarrow A(s)$$

either stays irreducible under restriction to γ' (first type), or it breaks up into two irreducible parts $\mathfrak{A}_1(\gamma')$, $\mathfrak{A}_2(\gamma')$ conjugate and of equal degrees (second type):

$$A(t) = \begin{Vmatrix} A_1(t) & 0 \\ 0 & A_2(t) \end{Vmatrix}. \tag{8.2}$$

In the second case, where P is decomposed into $\Lambda + u\Lambda$, the substitution u carries the subspace Λ into $u\Lambda$ and vice versa ($u^2 = c$ is an element of γ'). This means that u is represented in $\mathfrak{A}$ by a matrix of the form

$$A(u) = \begin{Vmatrix} 0 & C_1 \\ C_2 & 0 \end{Vmatrix}. \tag{8.3}$$

One may normalize the coördinate system in $u\Lambda$ in terms of the coördinate system in Λ such that C_2 becomes the unit matrix E; then

$$A_2(t) = A_1(t'), \qquad t' = u^{-1}tu.$$

To the element $u^2 = c$ there correspond in $\mathfrak{A}_1(\gamma')$ and $\mathfrak{A}_2(\gamma')$ the matrices C_1C_2 and C_2C_1 respectively.

The relations

$$s \rightarrow A(s) \quad \text{or} \quad s \rightarrow -A(s)$$

according as s is in γ' or in the coset $u\gamma'$, define another irreducible representation $\mathfrak{A}'(\gamma)$ of γ which we call *associated* with the given $\mathfrak{A}(\gamma)$. The decomposition

$$\mathfrak{A}(\gamma') = \mathfrak{A}_1(\gamma') + \mathfrak{A}_2(\gamma')$$

can occur only if the associated $\mathfrak{A}'(\gamma)$ is equivalent to $\mathfrak{A}(\gamma)$. Indeed, in the associated representation we have parallel to (8.2), (8.3)

$$t \rightarrow \begin{Vmatrix} A_1(t) & 0 \\ 0 & A_2(t) \end{Vmatrix}, \qquad u \rightarrow \begin{Vmatrix} 0 & -C_1 \\ -C_2 & 0 \end{Vmatrix},$$

and these change into (8.2), (8.3) by changing the signs of the coördinates in the second subspace. We summarize:

THEOREM (5.8.B). *Under limitation to a subgroup of index 2, a given irreducible representation either stays irreducible or breaks up into two irreducible conjugate parts of equal degree; the latter is possible only if the given representation is equivalent to its associate.*

The question of *equivalence* seems to be answerable in simple and general terms only in case of *absolute* irreducibility. Then the following two statements hold:

THEOREM (5.8.C). *Two absolutely irreducible representations* $\mathfrak{A}(\gamma)$, $\mathfrak{B}(\gamma)$ *of the* *first type lead to inequivalent representations $\mathfrak{A}(\gamma')$, $\mathfrak{B}(\gamma')$ of γ' provided $\mathfrak{B}(\gamma)$ is equivalent neither to $\mathfrak{A}(\gamma)$ nor to the associate $\mathfrak{A}'(\gamma)$.*

The parts $\mathfrak{A}_1(\gamma')$, $\mathfrak{A}_2(\gamma')$ of an absolutely irreducible $\mathfrak{A}(\gamma)$ of the second type are inequivalent.

Were $\mathfrak{A}(\gamma')$ and $\mathfrak{B}(\gamma')$ equivalent in the first part of the theorem, then we could assume that the matrices $A(t)$ and $B(t)$ in

$$\mathfrak{A}: s \to A(s) \quad \text{and} \quad \mathfrak{B}: s \to B(s)$$

coincide for elements t in γ'. The matrices U and V which correspond to u in $\mathfrak{A}$ and $\mathfrak{B}$ respectively, must satisfy the relations

$$U^{-1}A(t)U = A(t'), \qquad U^2 = A(c)$$

with

$$t' = u^{-1}tu, \qquad u^2 = c. \tag{8.4}$$

According to Schur's lemma, applicable to absolutely irreducible sets of matrices like $A(t)$, the first equation determines U except for a numerical factor; hence $V = \beta U$. The second equation then yields $\beta^2 = 1$: $V = U$ or $-U$. Therefore $\mathfrak{B}$ is either $\mathfrak{A}$ or its associate.

For the indirect proof of the second part, we assume that contrary to its statement $\mathfrak{A}_2(\gamma')$ coincides with $\mathfrak{A}_1(\gamma')$:

$$A(t) = \begin{Vmatrix} A_1(t) & 0 \\ 0 & A_1(t) \end{Vmatrix}. \tag{8.5}$$

With (8.4) one then finds

$$\begin{Vmatrix} 0 & C_1 \\ C_2 & 0 \end{Vmatrix}^{-1} \begin{Vmatrix} A_1(t) & 0 \\ 0 & A_1(t) \end{Vmatrix} \begin{Vmatrix} 0 & C_1 \\ C_2 & 0 \end{Vmatrix} = \begin{Vmatrix} A_1(t') & 0 \\ 0 & A_1(t') \end{Vmatrix}$$

or

$$A_1(t') = C_1^{-1}A_1(t)C_1\,, \qquad A_1(t') = C_2^{-1}A_1(t)C_2\,.$$

This leads to $C_2 = \mu C_1$ and, after multiplying the coördinates in the second subspace by the number μ, to $C_2 = C_1 = C$, without destroying the normal form (8.5). Denoting the coördinates in the two subspaces by

$$x_i\,, y_i \qquad (i = 1, \cdots, \nu;\ 2\nu = n),$$

$\mathfrak{A}$ represents

$$t \text{ by} \begin{cases} x' = A_1x \\ y' = A_1y \end{cases} \quad \text{and } u \text{ by} \begin{cases} x' = Cy \\ y' = Cx\,. \end{cases}$$

Hence the whole vector space breaks up into two ν-dimensional subspaces with the coördinates $x_i + y_i$ and $x_i - y_i$ respectively, which are invariant under the full group γ:

$$t \to A_1(t), \qquad u \to C$$

in the first, and

$$t \to A_1(t), \qquad u \to -C$$

in the second subspace. We thus arrive at a contradiction to the supposed irreducibility of $\mathfrak{A}(\gamma)$.

To the statements contained in Theorem (5.8.C) we may add the more trivial

THEOREM (5.8.D). *Under the hypothesis of absolute irreducibility, an* $\mathfrak{A}(\gamma')$ *proceeding from an* $\mathfrak{A}(\gamma)$ *of the first type is never equivalent to one of the two parts* $\mathfrak{B}_1(\gamma')$, $\mathfrak{B}_2(\gamma')$ *proceeding from a* $\mathfrak{B}(\gamma)$ *of the second type.*

If $\mathfrak{A}(\gamma)$, $\mathfrak{B}(\gamma)$ *are two inequivalent irreducible representations of the second type, then* $\mathfrak{A}_\alpha(\gamma')$ *is not equivalent to* $\mathfrak{B}_\beta(\gamma')$ $(\alpha, \beta = 1, 2)$.

PROOF. The fact that $\mathfrak{B}_2(\gamma')$ is not equivalent to $\mathfrak{B}_1(\gamma')$ implies the impossibility of extending the representation $t \to B_1(t)$ of γ' by a suitable correspondence $u \to C$ to become a representation of the whole γ, whereas the representation $\mathfrak{A}(\gamma')$ of the first type is extensible according to its provenance from $\mathfrak{A}(\gamma)$.

For the second type we can suppose

$$A_2(t) = A_1(t') \quad \text{and } u \to \begin{Vmatrix} 0 & C \\ E & 0 \end{Vmatrix}, \qquad C = A_1(u^2).$$

This shows at once that in an appropriate coördinate system the matrix corresponding to u is uniquely determined by $\mathfrak{A}(\gamma')$.

9. Representations of the proper orthogonal group

The equation (4.4) may be stated as follows as a proposition concerning the process σ introduced by (7.4). The proper orthogonal substitution $A = \| a(ik) \|$ induces the same transformation in the space of all skew-symmetric tensors F of rank ρ as in the space of the corresponding tensors σF of rank $n - \rho$, while for improper orthogonal A's the two transformations are opposite (i.e. the one equals minus the other). Let T, T' be two associate permissible diagrams, (7.2) being the lengths of the rows of T. If F runs over $\mathrm{P}_0(T')$, then σF as introduced by (7.5) will range over an invariant subspace $\sigma\mathrm{P}_0(T')$, and the corresponding representation $\langle\sigma\mathrm{P}_0(T')\rangle$ of the full orthogonal group is the associate of $\langle\mathrm{P}_0(T')\rangle$. We have observed that for at least one of the tensors F in $\mathrm{P}_0(T')$, namely $F_0(i_{T'})$, the corresponding σF lies in the subspace $\mathrm{P}_0(T)$ defined by the associated symmetry diagram T. Those F in $\mathrm{P}_0(T')$ for which σF lies in $\mathrm{P}_0(T)$ obviously form an invariant subspace of $\mathrm{P}_0(T')$; because of the irreducibility of $\mathrm{P}_0(T')$ this must be the whole space as it cannot be zero. In other words: the process σ maps $\mathrm{P}_0(T')$ upon $\mathrm{P}_0(T)$ in a one-to-one fashion, and the representations $\langle\mathrm{P}_0(T)\rangle$ and $\langle\mathrm{P}_0(T')\rangle$ corresponding to associated diagrams are themselves associated.

For even dimensionality $n = 2\nu$, T may be a self-associate diagram. For such diagrams it is convenient to modify the definition of the process σ, which now maps $\mathrm{P}_0(T)$ upon itself, by inserting the factor i^ν on the left side of the defining relation (7.4), $i = \sqrt{-1}$. Then σ becomes involutorial since the character "even or odd" of

$$\overset{*}{i_1} \cdots \overset{*}{i_\nu} i_1 \cdots i_\nu$$

is $(-1)^\nu$ times the character of

$$i_1 \cdots i_\nu \overset{*}{i_1} \cdots \overset{*}{i_\nu}.$$

We are now prepared to descend to the *proper* orthogonal group. According to the results of the last section, an irreducible representation $\mathfrak{A}^T$ of the full group remains irreducible under limitation to the proper rotations provided T is not self-associate. But $\mathfrak{A}^T$ and $\mathfrak{A}^{T'}$ corresponding to two associate diagrams T and T' now become equivalent. The $\mathfrak{A}^T$ of self-associate symmetry T break up into two parts. Indeed, every tensor F of the space $P_0(T)$ can be decomposed into an "even" and "odd" such tensor F_1, F_2 by the equation

$$F = \tfrac{1}{2}(F + \sigma F) + \tfrac{1}{2}(F - \sigma F); \qquad \sigma F_1 = F_1, \qquad \sigma F_2 = -F_2.$$

Both subspaces, that of even and that of odd tensors, are invariant under proper orthogonal transformations whereas any improper transformation, for instance J_n, interchanges them. Hence both subspaces are of equal dimensionality and neither is empty. From our general considerations follows the fact that they are irreducible and inequivalent under the proper group. No other equivalences besides those mentioned explicitly are created by descending to the proper group.

By the way, the case of odd dimensionality, where no reduction occurs, is capable of a much simpler treatment. For then $-E$ is an improper orthogonal transformation commuting with all elements of our group. Since the rational representations of the full group which we constructed are absolutely irreducible, this element must, on account of Schur's lemma, be represented in any of our representations by a multiple μ of the unit matrix, and the factor μ will be either $+1$ or -1 because $(-E)^2 = E$ leads to $\mu^2 = 1$. We summarize as follows.

THEOREM (5.9.A). *Under limitation to the proper orthogonal group each irreducible representation $\mathfrak{A}^T$ of the full orthogonal group stays irreducible unless T is self-associate; in this case $\mathfrak{A}^T$ breaks up into two irreducible parts of equal degree, with the proviso, however, that $\sqrt{-1}$ must be adjoined to the reference field if $n \equiv 2$ (mod 4).—Associated representations become equivalent but no other equivalences are created.*

At the same time this theorem settles the question of the inequivalent irreducible representations of the algebra $\mathfrak{A}_+^{(f)}$ defined by the equations (2.2) together with (4.6) which we found hard to attack directly.[7]

CHAPTER VI

THE SYMPLECTIC GROUP

1. Vector Invariants of the Symplectic Group*

The study of the symplectic group, which has a close analogy to the orthogonal group, will afford an opportunity for recapitulating in accelerated tempo our whole development spread over all the previous chapters.

Whereas the orthogonal group consists of all transformations leaving invariant a non-degenerate symmetric bilinear form (scalar product), the symplectic group $Sp(n)$ is defined as the set of all linear transformations under which a given non-degenerate skew-symmetric bilinear from $[xy]$ remains unaltered. We may assume this "skew product" of the two vectors

$$x = (x_1, x_1', x_2, x_2', \cdots, x_\nu, x_\nu') \text{ and } y$$

to be given in the normalized form

$$[xy] = (x_1y_1' - x_1'y_1) + \cdots + (x_\nu y_\nu' - x_\nu' y_\nu) \tag{1.1}$$

such that the $n = 2\nu$ fundamental vectors e_α, $e_\alpha'(\alpha = 1, \cdots, \nu)$ satisfy the relations

$$[e_\alpha e_\beta] = [e_\alpha' e_\beta'] = 0, \qquad [e_\alpha e_\beta'] = -[e_\alpha' e_\beta] = \delta_{\alpha\beta} \tag{1.2}$$

(symplectic coördinate system). The existence of a non-degenerate skew-symmetric form requires an *even* number $n = 2\nu$ of dimensions. Indeed, any given such form $[xy]$ is changed into (1.1) by a suitable choice of coördinates; and contrary to the similar normalization of the scalar product,

$$(xy) = x_1y_1 + \cdots + x_ny_n,$$

the procedure here is completely rational, not even requiring the adjunction of square roots. We start with an arbitrary vector $e_1 \neq 0$. $[xy]$ being non-degenerate, we may choose a second one e_1' such that $[e_1e_1'] \neq 0$ and then multiply it by a suitable numerical factor so as to make $[e_1e_1'] = 1$. e_1 and e_1' are linearly independent because of $[e_1e_1] = 0$, and the vectors x satisfying the simultaneous equations

$$[e_1x] = 0, \qquad [e_1'x] = 0$$

* The name "complex group" formerly advocated by me in allusion to line complexes, as these are defined by the vanishing of antisymmetric bilinear forms, has become more and more embarrassing through collision with the word "complex" in the connotation of complex number. I therefore propose to replace it by the corresponding Greek adjective "symplectic." Dickson calls the group the "Abelian linear group" in homage to Abel who first studied it.

form a subspace P_1 of two dimensions less. Every vector x whatsoever can be written in the form

$$x = \xi_1 e_1 + \xi'_1 e'_1 + x_*$$

where x_* is in P_1; one has simply to take

$$\xi_1 = [xe'_1], \qquad \xi'_1 = -[xe_1].$$

Our statement is now readily proved by induction with respect to the dimensionality $n = 2\nu$. At the same time we have shown the existence of a symplectic coördinate system whose first fundamental vector e_1 is an arbitrarily preassigned vector $\neq 0$—and this, to be sure, in any number field whatsoever (of characteristic 0).

Whereas in the case of the orthogonal group only the *square* of a bracket factor is expressible by its scalar products, we are here in the more fortunate position that the bracket factor itself, the determinant of n vectors $[x^1 \cdots x^n]$, is an aggregate of skew-products. Therefore no such distinction as between proper and improper transformations makes its appearance here, and every symplectic transformation is of determinant 1.

To prove these two statements we form the alternating sum

$$\frac{1}{\nu!2^\nu}\sum \pm [x^1x^2][x^3x^4] \cdots [x^{n-1}x^n] \tag{1.3}$$

extending to all $n!$ permutations of the n independent vectors $x^1, x^2, \cdots, x^n$. The factor $1/\nu!2^\nu$ has been added because each term occurs in $\nu!2^\nu$ equal copies arising from it by those permutations which do not sever any two arguments bracketed together by [ ] in that term. $[xy]$ may here at first designate an arbitrary skew-symmetric form

$$[xy] = \sum_{i,k} \gamma(ik)x_i y_k. \tag{1.4}$$

The canonical form (1.1) shall be denoted by $\sum \epsilon(ik)x_i y_k$ so that $I = \| \epsilon(ik) \|$ is the matrix

$$I = \begin{Vmatrix} 0 & 1 \\ -1 & 0 \end{Vmatrix} \dotplus \begin{Vmatrix} 0 & 1 \\ -1 & 0 \end{Vmatrix} \dotplus \cdots \dotplus \begin{Vmatrix} 0 & 1 \\ -1 & 0 \end{Vmatrix} \quad (\nu \text{ summands}).$$

(1.3) becomes

$$\sum_{i_1,\cdots,i_n} \gamma(i_1 i_2) \cdots \gamma(i_{n-1} i_n) \sum \pm x^1_{i_1} x^2_{i_2} \cdots x^n_{i_n}$$

with the inner sum running again over all permutations of $x^1, x^2, \cdots, x^n$. This inner sum vanishes unless $i_1, \cdots, i_n$ is a permutation s of $1, \cdots, n$, and it equals $\pm[x^1x^2 \cdots x^n]$ according as s is an even or odd permutation. We are thus led to introduce the "Pfaffian"

$$\text{Pf}\{\gamma(ik)\} = \frac{1}{\nu!2^\nu}\sum \pm \gamma(i_1 i_2) \cdots \gamma(i_{n-1} i_n) \tag{1.5}$$

in which the sum runs alternatingly over all permutations $i_1 i_2 \cdots i_{n-1} i_n$ of $1, \cdots, n$. The Pfaffian plays the same rôle for the anti-symmetric form (1.4) as the determinant plays for symmetric forms. Our result is the formula

$$\frac{1}{\nu! 2^\nu} \sum \pm [x^1 x^2] \cdots [x^{n-1} x^n] = \text{Pf } \{\gamma(ik)\} \cdot [x^1 x^2 \cdots x^n]. \tag{1.6}$$

When (1.4) goes over into

$$\sum_{i,k} \gamma'(ik) x'_i x'_k$$

by a linear transformation

$$x_i = \sum_j a_{ji} x'_j$$

(to be cogrediently performed on x and y), (1.6) leads at once to the relation

$$\text{Pf}\{\gamma'(ik)\} = \text{Pf}\{\gamma(ik)\} \cdot \det(a_{ij}). \tag{1.7}$$

When the canonical form (1.1) is adopted for $[xy]$, the Pfaffian $\text{Pf}\{\epsilon(ik)\}$ becomes $=1$. Hence (1.7) shows that a substitution A leaving this form $[xy]$ unaltered, must be of determinant 1. That was the second point. And the first point: on clinging to the same canonical form for the skew product, the equation (1.6) simplifies to

$$[x^1 x^2 \cdots x^n] = \frac{1}{\nu! 2^\nu} \sum \pm [x^1 x^2] \cdots [x^{n-1} x^n], \tag{1.8}$$

and thus gives the desired expression of the bracket factor in terms of the skew products.

THEOREM (6.1.A). (*First main theorem for the symplectic group.*) *All vector invariants of the symplectic group depending on an arbitrary number of covariant and contravariant vectors, $x \cdots$ and $\xi \cdots$, are expressible in terms of the basic invariants of type*

$$[xy], \qquad (\xi x), \qquad [\xi\eta]. \tag{1.9}$$

PROOF.[1] Let us first deal with covariant vectors $x, \cdots$ only. The proof may be carried through along exactly the same lines as for the orthogonal group —with the simplification arising from the redundancy of the bracket factor which eliminates at the same time the distinction of proper and improper transformations. There is, however, one further observation to be made lest the induction with respect to the dimensionality n be stopped. When considering an invariant f depending on $n - 1$ vectors $x, y, \cdots$ one introduces a new symplectic coördinate system such that the first components $x_1, y_1, \cdots$ of $x, y, \cdots$ vanish relatively to the new coördinate system. Here $x, y, \cdots$ are supposed to be numerically given and linearly independent. After thus being led to introduce the function

$$f_0\begin{pmatrix} x'_1, x_2, x'_2, \cdots \\ y'_1, y_2, y'_2, \cdots \\ \cdots\cdots\cdots \end{pmatrix} = f\begin{pmatrix} 0, x'_1, x_2, x'_2, \cdots \\ 0, y'_1, y_2, y'_2, \cdots \\ \cdots\cdots\cdots \end{pmatrix}$$

we should get into trouble if the arguments $x_1', y_1', \cdots$ did not disappear from f_0 along with $x_1, y_1, \cdots$; the induction from $2(\nu - 1)$ dimensions to 2ν would not work. Fortunately this obstacle can be overcome by the following simple reasoning. We shall not touch the components $x_2, x_2', \cdots$, while on the only two we put in evidence, x_1, x_1', we perform an arbitrary unimodular transformation

$$\begin{aligned} x_1 &\to \alpha x_1 + \beta x_1', \\ x_1' &\to \gamma x_1 + \delta x_1' \end{aligned} \qquad (\alpha\delta - \beta\gamma = 1).$$

As this is a symplectic transformation and f an invariant, we have

$$f(0, x_1'; 0, y_1'; \cdots) = f(\beta x_1', \delta x_1'; \beta y_1', \delta y_1'; \cdots). \tag{1.10}$$

(1.10) holds for arbitrary numbers β and δ if only $\delta \neq 0$. For then we may choose $\alpha = 1/\delta, \gamma = 0$. The argument of the algebraic irrelevance of inequalities establishes (1.10) as an identity in the variables β and δ. Consequently (1.10) remains true for the values $\beta = 0, \delta = 0$. The equation thus arising,

$$f(0, x_1'; 0, y_1'; \cdots) = f(0, 0; 0, 0; \cdots),$$

proves that $f_0(x_1'; y_1'; \cdots)$ is in fact independent of $x_1', y_1', \cdots$.

Introduction of contravariant beside the covariant argument vectors is hardly less trivial than in the case of the orthogonal group. Indeed, the relations

$$x_1 = \xi_1', \; x_1' = -\xi_1, \; \cdots, \qquad x_\nu = \xi_\nu', \; x_\nu' = -\xi_\nu \tag{1.11}$$

tie up a covariant vector $x = \xi'$ with a given contravariant ξ, because these equations can be summed up in the one relation

$$[xy] = (\xi y)$$

holding identically in the covariant argument y, and this relation is invariant under symplectic transformations. We observe that

$$[\xi' y] = (\xi y), \qquad [\xi'\eta'] = -[\xi\eta],$$

and we thus obtain the other two types in the table (1.9) besides $[xy]$.

THEOREM (6.1.B). (*Second Main Theorem for the symplectic group.*) *Every relation between skew products is an algebraic consequence of relations of the following ν types:*

$$\begin{aligned} J_1 &\equiv \sum \pm[x_0y_0][x_1x_2] \cdots [x_{n-1}x_n] = 0, \\ J_2 &\equiv \sum \pm[x_0y_0][x_1y_1][x_2y_2][x_3x_4] \cdots [x_{n-1}x_n] = 0, \\ &\cdot \quad \cdot \quad \cdot \quad \cdot \quad \cdot \quad \cdot \quad \cdot \quad \cdot \\ J_\nu &\equiv \sum \pm[x_0y_0][x_1y_1] \cdots [x_ny_n] = 0. \end{aligned}$$

The sum $\sum$ extends in each case alternatingly to all permutations of the $n + 1$ vectors $x_0, x_1, \cdots, x_n$. The left sides, being skew-symmetric multilinear forms with respect to these $n + 1$ vectors, are bound to vanish.

The proof is essentially like that given in Chapter II, §17, for the full orthogonal group. However, attention has to be paid to the possibility, now not excluded, that two of the "new symbols" x' may join in a single factor like $[x'_0x'_1]$; for this product is now skew-symmetric instead of symmetric, and hence not annihilated by alternation! This is the reason why, in addition to the relation J_ν that corresponds to the relation J in Theorem (2.17.A), the types $J_1, J_2, \cdots, J_{\nu-1}$ appear. Again, Capelli's congruence reduces the theorem under examination to the fact that no relation holds among the skew-products of n vectors x_i. This follows from the possibility of ascertaining n vectors x_i such that the matrix of their skew-products $[x_ix_k]$ coincides with an arbitrarily preassigned skew-symmetric matrix; the construction of the x_i can be accomplished in a purely rational way. But one may proceed also in a fashion analogous to the case of the orthogonal group.

2. Parametrization and unitary restriction

The first main theorem for symplectic invariants once established, we repeat the procedure followed in the case of the orthogonal group step by step, pointing out only such instances as demand modifications of a not altogether trivial sort.

According to Lemma (2.10.A) the Cayley parametrization

$$A = (E - S)(E + S)^{-1}, \qquad S = (E - A)(E + A)^{-1}, \tag{2.1}$$

applicable to non-exceptional matrices A (and S), changes the quadratic equation

$$A^*IA = I \tag{2.2}$$

into the linear

$$S^*I + IS = 0. \tag{2.3}$$

For the rest of this section S may always designate an *infinitesimal symplectic transformation*, i.e. a matrix satisfying (2.3). On arranging the indices in the order $1, \cdots, \nu, 1', \cdots, \nu'$ one readily verifies that

$$S = \begin{Vmatrix} s_{\alpha\beta} & t_{\alpha\beta} \\ t'_{\alpha\beta} & s'_{\alpha\beta} \end{Vmatrix} \qquad (\alpha, \beta = 1, \cdots, \nu), \tag{2.4}$$

where

$$\| t_{\alpha\beta} \|, \ \| t'_{\alpha\beta} \| \text{ are symmetric and } s'_{\alpha\beta} + s_{\beta\alpha} = 0; \tag{2.5}$$

the number of parameters on which S linearly depends amounts to

$$N = \nu(2\nu + 1) = \tfrac{1}{2}n(n + 1). \tag{2.6}$$

Until now the symplectic case seemed of decidedly simpler nature than the orthogonal; things that were of quadratic character or needed extractions of square roots there, became linear or rational here. But at this juncture, where we are about to turn to the study of exceptional symplectic A, we hit a snag.

The analogues of the real orthogonal transformations are in many respects the *unitary* rather than the real symplectic transformations. We operate in the field $K^\dagger$ of all complex numbers or more generally in a field $k^\dagger = (k, \sqrt{-1})$ arising from a real field k by adjunction of $i = \sqrt{-1}$. The numbers in k are the "real" numbers of $k^\dagger$. From now on $\bar{a}$ shall always designate the conjugate complex of the number a:

$$a = \alpha + i\beta, \qquad \bar{a} = \alpha - i\beta \qquad (\alpha, \beta \text{ in } k). \tag{2.7}$$

A form of the type

$$G(x, y) = \sum g_{ik}\bar{x}_i y_k \tag{2.8}$$

is linear in y and "antilinear" in x:

$$\left.\begin{array}{ll} G(x, y + y') = G(x, y) + G(x, y'), & G(x, \lambda y) = \lambda G(x, y) \\ G(x + x', y) = G(x, y) + G(x', y), & G(\lambda x, y) = \bar{\lambda} G(x, y) \end{array}\right\} \quad \{\lambda \text{ in } k^\dagger\}.$$

Its matrix $G = \| g_{ik} \|$ is changed into

$$G' = \bar{A}^*GA$$

by applying the same transformation $x \to Ax$ on x and y. Lemma (2.10.A) and its proof carry over to this case with the effect that (2.1) establishes a one-to-one correspondence between the non-exceptional A and S satisfying the conditions

$$\bar{A}^*GA = G, \qquad \bar{S}^*G + GS = 0, \tag{2.9}$$

respectively. (2.8) is called Hermitean if

$$G(y, x) = \overline{G(x, y)} \text{ or } g_{ki} = \bar{g}_{ik},$$

and

$$G(x, x) = \sum_{i,k} g_{ik}\bar{x}_i x_k \tag{2.10}$$

is then the corresponding *Hermitean form* whose values are real. On the unit form

$$(xx)_H = \bar{x}_1 x_1 + \cdots + \bar{x}_n x_n, \qquad (xy)_H = \bar{x}_1 y_1 + \cdots + \bar{x}_n y_n = \bar{x}^* y, \tag{2.11}$$

one can base in the complex field $k^\dagger$ a sort of modified Euclidean vector geometry by having $(xy)_H$ play the part of the scalar product. Unitary perpendicularity as defined by $(xy)_H = 0$ is a reciprocal relationship, because of

$$(yx)_H = \overline{(xy)_H}.$$

The *unitary transformations* A leaving (2.11) unchanged,

$$\bar{A}^*A = E \text{ and consequently } A\bar{A}^* = E,$$

form a group $U(n)$ which in this geometry is the analogue of the orthogonal group. A unitary coördinate system $e_1, \cdots, e_n$ satisfies the equations

$$(e_i e_k)_H = \delta_{ik}.$$

Any non-exceptional unitary A is expressed by (2.1) in terms of an "infinitesimal unitary" matrix, i.e. by a matrix $S = \| s_{ik} \|$ for which

(2.12) $$\bar{S}^* + S = 0 \quad \text{or} \quad \bar{s}_{ki} + s_{ik} = 0.$$

On account of the positive definite character of $(xx)_H$, the space P″ unitary-perpendicular to a given subspace P′ is complementary to P′ such that the whole vector space P splits into P′ + P″. The classical inductive construction of a unitary coördinate system works in the same fashion as in ordinary real Euclidean geometry. Lemma (5.2.A) is paralleled by

LEMMA (6.2.A). *Any set of unitary transformations in a field* $k^\dagger = (k, \sqrt{-1})$ *(k real) is fully reducible.*

We now consider the intersection $USp(n)$ of $U(n)$ and $Sp(n)$; the elements A of this group are at the same time symplectic and unitary. If A be non-exceptional, the substitution (2.1) carries it into an infinitesimal element S of the same group characterized by both relations (2.3) and (2.12). (2.12) imposes upon the parameters in (2.4) the restrictions

$$\bar{s}_{\alpha\beta} + s_{\beta\alpha} = 0 \quad (\bar{s}'_{\alpha\beta} + s'_{\beta\alpha} = 0) \quad \text{and} \quad \bar{t}_{\alpha\beta} + t'_{\alpha\beta} = 0,$$

or, on setting

(2.13) $$\begin{cases} t_{\alpha\beta} = v_{\alpha\beta} + \sqrt{-1}\, v'_{\alpha\beta}, \qquad t'_{\alpha\beta} = -v_{\alpha\beta} + \sqrt{-1}\, v'_{\alpha\beta} \\ \qquad\qquad\qquad\qquad\qquad\qquad (v_{\beta\alpha} = v_{\alpha\beta},\ v'_{\beta\alpha} = v'_{\alpha\beta}), \\ s_{\alpha\alpha} = \sqrt{-1} \cdot u_{\alpha\alpha}; \qquad s_{\alpha\beta} = u_{\alpha\beta} + \sqrt{-1}\, u'_{\alpha\beta}, \\ \qquad\qquad s_{\beta\alpha} = -u_{\alpha\beta} + \sqrt{-1}\, u'_{\alpha\beta} \quad (\text{for } \alpha < \beta), \end{cases}$$

the condition of *reality* upon the N parameters u, u', v, v'.

Under the unitary group a covariant vector $x = (x_i)$ gives rise to the contragredient $\xi = (\bar{x}_i)$. On combining this with (1.11) we see that the relations

(2.14) $$u_\alpha = -\bar{x}'_\alpha \qquad u'_\alpha = \bar{x}_\alpha \qquad (\alpha = 1, \cdots, \nu)$$

associate a (covariant) vector u with any such vector x in a manner invariant under $USp(n)$. Indeed one may unite (2.14) into the one identity

$$(uy)_H = [xy]$$

in y. We use the notation

$$u = \tilde{x}, \qquad x = -\tilde{u}$$

and then have

$$(\tilde{x}y)_H = [xy], \qquad (xy)_H = -[\tilde{x}y];$$

(2.15) $$(\tilde{x}y)_H = -(\tilde{y}x)_H \quad \text{or} \quad (\tilde{x}\tilde{y})_H = (yx)_H = \overline{(xy)_H}.$$

Given a vector $e \neq 0$, will it be possible to determine a vector basis e_α, e'_α which is at the same time symplectic and unitary,

$$(2.16)\qquad \begin{cases} [e_\alpha e_\beta] = [e'_\alpha e'_\beta] = 0, & [e_\alpha e'_\beta] = -[e'_\alpha e_\beta] = \delta_{\alpha\beta}; \\ (e_\alpha e_\beta)_H = (e'_\alpha e'_\beta)_H = \delta_{\alpha\beta}, & (e_\alpha e'_\beta)_H = (e'_\alpha e_\beta)_H = 0, \end{cases}$$

and such that e_1 coincides with e? Of course one will first have to normalize e such that $(ee)_H$ becomes $= 1$. $(ee)_H$ being a square sum in the real field k, since

$$\bar{a}a = \alpha^2 + \beta^2 \text{ for (2.7)},$$

this may be accomplished by a Pythagorean extension of k not destroying the reality of k. Let us therefore assume $(ee)_H = 1$. The two vectors

$$e_1 = e, \qquad e'_1 = \tilde{e}$$

then fulfill all the requirements (2.16) with $\alpha = \beta = 1$. Moreover the subspace P_1 of the vectors x satisfying

$$[e_1 x] = 0, \qquad [e'_1 x] = 0$$

or, what is the same,

$$(ex)_H = 0, \qquad (\tilde{e}x)_H = 0,$$

is closed with respect to the operation $\tilde{}$, as follows readily from (2.15). This enables us to carry on the inductive construction of the coördinate system which we wanted.

Let $A = \| a_{ik} \|$ now be an *exceptional* element of $USp(n)$. The subspace P^0 of the vectors x satisfying

$$x + Ax = 0$$

allows the operation $\tilde{}$ since $y = Ax$ entails $\tilde{y} = A\tilde{x}$. One easily sees that, granted the freedom of Pythagorean adjunctions to k, this fact enables one to choose a unitary sympletic coördinate system $e_1, \tilde{e}_1, \cdots, e_p, \tilde{e}_p$ in P^0 and to extend it to a similar system for the whole vector space P. As in the case of the orthogonal group, one then obtains the

LEMMA (6.2.B). *Any unitary symplectic A in $(k, \sqrt{-1})$ is the product of two commuting non-exceptional unitary symplectic matrices A_1, A_2.*

The notion of a *formal symplectic invariant* is introduced in the obvious manner. In order to prove that any such invariant is expressible in terms of skew-products, we operate in the Gaussian field $(\kappa, \sqrt{-1})$. Two remarks are essential. If $x^1, \cdots, x^{n-1}$ are $n - 1$ vectors with components in that field, one can determine a unitary symplectic coördinate system $e_1, e'_1, \cdots, e_\nu, e'_\nu$ whose first vector $e_1 = e$ satisfies all the $n - 1$ equations

$$[ex^1] = 0, \cdots, [ex^{n-1}] = 0.$$

The corresponding unitary symplectic transformation, simultaneously annulling the x'_1-component of the $n - 1$ arguments $x^1, \cdots, x^{n-1}$, will, generally speaking, demand Pythagorean extensions of κ to a larger real field k. The second remark refers to a form $f(x, x'; y, y'; \cdots)$ depending on *binary* vectors (x, x'), (y, y'), $\cdots$ and formally invariant under the two-dimensional symplectic group; and it states that f will be constant if it does not involve the second components x', y', $\cdots$. Indeed, provided $\alpha \neq 0$ and $\neq -1$,

$$\begin{Vmatrix} \alpha & 0 \\ \gamma & 1/\alpha \end{Vmatrix}$$

is a non-exceptional symplectic transformation carrying

$$f(x, 0; y, 0; \cdots) \quad \text{into} \quad f(\alpha x, \gamma x; \alpha y, \gamma y; \cdots).$$

This equation will be an identity in α and γ and will thus hold even for $\alpha = \gamma = 0$.

Queerly enough, a *real* field k would have served us in all the preceding considerations as well as $k^\dagger = (k, \sqrt{-1})$; the unitary symplectic matrices would then have been orthogonal and symplectic. (This group, the intersection of $O(n)$ and $Sp(n)$, perhaps deserves a little better than just this cursory mention.) For the determination of the enveloping algebra, however, it appears essential to use the *unitary trick*; its success depends on the simple observation that the unitary restriction for S, (2.4), amounts to a *reality* restriction for the parameters involved and hence is algebraically irrelevant. Indeed, according to Lemma (1.1.A) a given polynomial $\Phi(S)$ depending on the N parameters

$$t_{\alpha\beta},\ t'_{\alpha\beta}\ (\alpha \leqq \beta) \text{ and } s_{\alpha\beta}\ (\text{all } \alpha, \beta)$$

of S, or on the parameters u, u', v, v' as introduced by (2.13), will vanish identically if it vanishes for all real values of the latter (even all rational values would do) not annulling

$$\Delta(S) = \det (E + S).$$

3. Embedding algebra and representations of the symplectic group

S being the generic infinitesimal symplectic matrix (2.4), (2.5), with indeterminate elements, we now consider the transformation $\Pi_f(A)$ which is induced in tensor space P_f by the transformation

$$A = (E - S)(E + S)^{-1}. \tag{3.1}$$

We want to show that P_f is fully reducible with respect to $\Pi_f(A)$. We first operate in the ground field κ. Let Σ be an invariant subspace of P_f spanned by the linearly independent rational tensors $F_1, \cdots, F_p$. The subspace Σ' orthogonal to Σ in the sense of the metric (5.2.1) will be spanned by a number of rational tensors $F'_1, \cdots, F'_q$; $p + q = n^f$. We now pass to the Gaussian field $\kappa^\dagger = (\kappa, \sqrt{-1})$ so that Σ, Σ' consist of all linear combinations of $F_1, \cdots, F_p$ and $F'_1, \cdots, F'_q$ respectively with coefficients in $\kappa^\dagger$. Then each tensor F' in Σ' is also *unitary*-orthogonal to each tensor F in Σ:

$$\sum_{(i)} \bar{F}(i_1 \cdots i_f) F'(i_1 \cdots i_f) = 0.$$

The elements of the matrix describing the substitution $\Pi_f(A)$ relatively to the coördinate system $F_1, \cdots, F_p; F'_1, \cdots, F'_q$ are rational functions of S with the denominator $|E+S|^f$. Of its 2×2 parts into which it splits according to the decomposition $\mathrm{P}_f = \Sigma + \Sigma'$, the upper right rectangle is empty (invariance of Σ). We maintain that so is the lower left rectangle; in other words: Σ' is invariant too. To show this, we substitute for S an arbitrary numerical non-exceptional infinitesimal unitary symplectic matrix in $(\kappa, \sqrt{-1})$. The corresponding A is unitary, and *so is* $\Pi_f(A)$; hence by Lemma (6.2.A) the elements of said rectangle vanish after this substitution. The concluding remark of the last section allows us to infer therefrom their identical disappearance.

The road is now open for establishing all the theorems analogous to those proved for the orthogonal group.[2] The algebra $\mathfrak{A}^{(f)}$ within $\mathfrak{R}^{(f)}$ is to be described by the following equations for $v = 2, \cdots, f$:

$$\sum_{k_1, k_2} \epsilon(k_1k_2)a(i_1 \cdots i_v; k_1 \cdots k_v) = \epsilon(i_1i_2) \cdot a(i_3 \cdots i_v; k_3 \cdots k_v),$$

$$\sum_{i_1, i_2} \epsilon(i_1i_2)a(i_1 \cdots i_v, k_1 \cdots k_v) = \epsilon(k_1k_2) \cdot a(i_3 \cdots i_v; k_3 \cdots k_v)$$

imposed upon the row $A^{(f)} = (A_f, A_{f-1}, \cdots, A_0)$ of bisymmetric matrices

$$A_v = \| a(i_1 \cdots i_v, k_1 \cdots k_v) \|.$$

THEOREM (6.3.A). *$\mathfrak{A}^{(f)}$ is the enveloping algebra of the group of all $\Pi^{(f)}(A)$ induced by the symplectic transformations A. This holds in any number field k of characteristic zero; one may even confine oneself within $Sp(n)$ to rational non-exceptional A.*

Or, in another form:

THEOREM (6.3.B). *Let $A = \| a(ik) \|$ be the generic matrix consisting of n^2 independent variables $a(ik)$ and let S be the generic infinitesimal symplectic matrix. Every polynomial $\Phi(A)$ of formal degree f of the n^2 variables $a(ik)$ which vanishes formally by the substitution*

$$A = (E - S)(E + S)^{-1}$$

is a linear combination

$$\sum L_{i_1i_2}D_{i_1i_2} + \sum L^*_{k_1k_2}D^*_{k_1k_2}$$

of the particular forms

$$D_{i_1i_2} = \sum_{k_1, k_2} \epsilon(k_1k_2)a(i_1k_1)a(i_2k_2) - \epsilon(i_1i_2),$$

$$D^*_{k_1k_2} = \sum_{i_1, i_2} \epsilon(i_1i_2)a(i_1k_1)a(i_2k_2) - \epsilon(k_1k_2)$$

by means of polynomial coefficients L of formal degree $f - 2$.

This theorem determines a basis for the *symplectic prime ideal.* If one sets no store by the restriction concerning the formal degree, one of the two basic sets $D_{i_1i_2}, D^*_{k_1k_2}$ is redundant.

The (12)-trace of a tensor $F(i_1 i_2 i_3 \cdots i_f)$ is now defined by

$$\sum_{i_1, i_2} \epsilon(i_1 i_2) F(i_1 i_2 i_3 \cdots i_f) .$$

The tensors of rank f and of vanishing traces form the space P_f^0. By imposing upon them the symmetry corresponding to a given diagram T, one obtains a subspace $\mathrm{P}_0(T)$, irreducibly invariant under the algebra $\mathfrak{A}_f$ or under the group $\Pi_f(Sp(n))$. However $\mathrm{P}_0(T)$ will be empty *unless T consists of not more than $\nu = \frac{1}{2}n$ rows* (permissible diagrams). (This simplification as compared to the orthogonal case is due to the fact that a single bracket factor, rather than the product of two such, as in (5.7.1), is expressible in terms of skew products.) T will again be characterized by the lengths $f_1, \cdots, f_\nu$ of its rows:

$$f_1 \geqq f_2 \geqq \cdots \geqq f_\nu \geqq 0.$$

$\mathrm{P}_0(T) = \mathrm{P}_0(f_1 \cdots f_\nu)$ is the substratum of an irreducible representation $\langle \mathrm{P}_0(f_1 \cdots f_\nu) \rangle$ of $Sp(n)$. Any invariant subspace of P_f is fully reducible, and if irreducible similar to one of the subspaces $\mathrm{P}_0(f_1 \cdots f_\nu)$ of P_v^0 with

$$v = f_1 + \cdots + f_\nu = f \text{ or } f - 2 \text{ or } f - 4 \cdots .$$

The representation $\langle \mathrm{P}_0(f_1 \cdots f_\nu) \rangle$ is self-contragredient, as (1.11) allows immediate transition from covariant to contravariant quantities.

We leave it to the reader to formulate and prove in all details the same string of propositions as were established for the orthogonal group. Use of the same notations for the symplectic and the orthogonal group will not cause any confusion since it will be clear in each case with which of the two groups we are concerned.

CHAPTER VII

CHARACTERS

1. Preliminaries about unitary transformations

When full decomposition into absolutely irreducible constituents prevails, the characters may serve to uniquely characterize the representations (in the sense of equivalence). Indeed, under the assumption just mentioned, the characters $\chi(s)$, $\chi'(s)$, $\cdots$ of inequivalent irreducible representations are linearly independent, and if any representation $\mathfrak{A}$ of character $X(s)$ splits into m times the first irreducible constituent, m' times the second, $\cdots$, then we obtain the equation

$$X(s) = m\chi(s) + m'\chi'(s) + \cdots \tag{1.1}$$

Hence the character $X(s)$ unambiguously determines the coefficients $m, m', \cdots$ of the expansion (1.1) in terms of the primitive characters $\chi, \chi', \cdots$; and the multiplicities $m, m', \cdots$ fully describe the representation $\mathfrak{A}$ under consideration. Upon this remark is based a calculatory treatment of representations by means of their characters. The simplification effected by such a shift is obvious; in particular the formal processes $\dotplus$, $\times$ as applied to representations are reflected in ordinary addition and multiplication of characters. Let us therefore set out in the present chapter to compute the characters of those representations of the full linear, the orthogonal, and the symplectic group, which we obtained in the previous chapters by reducing tensor space. This procedure, despite its explicit algebraic nature, is far from yielding simple explicit formulas for the characters. In order to find such formulas we shall have to resort to transcendental methods, to processes of integration extending over the group manifold.

Of course this will be feasible only provided the underlying number field is a *continuum*. We shall use the continuum $K^\dagger$ of ordinary complex numbers. Nevertheless our results will prove to be valid in any number field k of characteristic zero, the reason being the same as before: the results are such as to be enunciable within the rational ground field κ which in its turn is embedded in the continuum $K^\dagger$. Our investigation will thus provide a new instance of the application of analysis to purely rational algebraic questions.

Integration over a manifold Γ is possible without any complicated restrictions concerning convergence if Γ is in the topological sense a closed or *compact* set. We therefore take refuge in what might be called the unitarian trick: each group is replaced by the *subgroup of those elements that are unitary transformations*. In the previous chapter we successfully applied this idea to the symplectic group. It was tacitly inherent in our treatment of the orthogonal group; indeed, the *unitary* are the *real* elements of $O(n)$. In the case of the full linear group we have so far gotten along without the unitarian device, but we shall now exemplify our method just by the group $GL(n)$, the simplest of them all. Its success is due to the fact that *nothing of algebraic import is lost by unitary restriction.*

Algebraic irrelevance of the unitary restriction. We consider any of the groups

$$\Gamma = GL(n),\ O(n),\ Sp(n)$$

and an arbitrary polynomial $\phi(A)$ depending on the n^2 variable components a_{ik} of the generic n-rowed matrix $A = ||a_{ik}||$.

LEMMA (7.1.A). *$\phi(A)$ vanishes for all elements A of Γ if it is annulled by its unitary elements.*

A is unitary if

$$\bar{A}^*A = E, \tag{1.2}$$

while for *infinitesimal* substitutions $S = ||s_{ik}||$ the unitary restriction is expressed by the relations

$$\bar{S}^* + S = 0 \qquad \text{or} \qquad \bar{s}_{ki} + s_{ik} = 0. \tag{1.3}$$

On setting

$$\begin{cases} s_{ii} = \sqrt{-1}\,\sigma_{ii}; \\ s_{ik} = \sigma_{ik} + \sqrt{-1}\,\sigma'_{ik}, \quad s_{ki} = -\sigma_{ik} + \sqrt{-1}\,\sigma'_{ik} \quad \text{for each pair } i < k, \end{cases} \tag{1.4}$$

(1.3) requires the parameters σ to be real. The generic infinitesimal element S of each of the groups Γ depends linearly on a certain number of parameters σ and thus varies within a linear set $\mathfrak{g}$, the infinitesimal group. By appropriate choice of its basis we can take care that the unitary S, (1.3), be those with real parameter values σ. The familiar equation

$$A = (E - S)(E + S)^{-1} \tag{1.5}$$

establishes a one-to-one correspondence between the non-exceptional elements A of the group Γ and the non-exceptional elements S of the infinitesimal group; on either side one is allowed to impose the unitary restriction (1.2) and (1.3) respectively. We invoke the following

LEMMA (7.1.B). *$\phi(A)$ vanishes for all elements A of Γ if it is annulled by the substitution* (1.5) *identically in the parameters σ of S (and by the one improper substitution J_n,* (2.9.3), *in the case $\Gamma = O(n)$).*

We proved this lemma explicitly for the orthogonal group by studying the orthogonal ideal. The same way is to be followed for $Sp(n)$. No proof is needed for $GL(n)$ as the only limiting condition imposed by the hypothesis of the lemma upon the n^2 arguments a_{ik} in $\phi(A) = 0$, namely det $(E + A) \neq 0$, is at once removed in the familiar fashion.

When $\phi(A)$ vanishes for all unitary elements A of Γ we have in particular the numerical equation

$$\phi\left(\frac{E - S}{E + S}\right) = 0, \tag{1.6}$$

together with $\phi(J_n) = 0$ when $\Gamma = O(n)$; holding for all *real* values σ for which $\Delta(\sigma) = \det(E + S) \neq 0$. But this implies that (1.6) holds identically in the variables σ, and hence Lemma (7.1.A) is obtained as a consequence of Lemma (7.1.B).

The algebraically ambitious will observe that the lemma holds good in any field over the Gaussian field $\kappa^\dagger = (\kappa, \sqrt{-1})$, even if the equation $\phi(A) = 0$ is assumed only for unitary matrices A of Γ *lying in* $\kappa^\dagger$. Equivalent to the lemma is the statement that the enveloping algebras $\mathfrak{A}_f$ and $\mathfrak{A}^{(f)}$ of the substitutions $\Pi_f(A)$ and $\Pi^{(f)}(A)$ induced by the elements A of the group Γ will not shrink if we impose the unitary restriction on A.

The parts $P(f_1, \cdots, f_n)$ into which we decomposed the tensor space P_f with respect to $GL(n)$ will therefore stay irreducible even under restriction of the full linear to the unitary group $U(n)$.

Compactness. A unitary matrix $A = \|a_{ik}\|$ in $K^\dagger$ satisfies the relations

$$\bar{A}^*A = E, \qquad A\bar{A}^* = E, \tag{1.7}$$

or

$$\sum_k \bar{a}_{ki}a_{kj} = \delta_{ij}, \qquad \sum_k a_{ik}\bar{a}_{jk} = \delta_{ij}. \tag{1.8}$$

Hence

$$\det \bar{A} \cdot \det A = 1,$$

or the determinant $\det A$ is of modulus 1. The equations (1.8) for $i = j$,

$$|a_{1i}|^2 + |a_{2i}|^2 + \cdots + |a_{ni}|^2 = 1,$$

show that each a_{ki} is of modulus $\leqq 1$, and by the well-known Weierstrass procedure this implies the *compactness* of the unitary group: if a sequence

$$A^{(p)} \qquad (p = 1, 2 \cdots)$$

of unitary matrices is given, then there exists a unitary (*limit*) matrix A such that in every neighborhood of A there lies at least one of the matrices

$$A^{(p)}, A^{(p+1)}, \cdots,$$

however large p is chosen.

Transformation on principal axes. Let there be given a unitary mapping $x \to x' = Ax$,

$$x'_i = \sum_k a_{ik}x_k,$$

of P upon itself. We maintain that in an appropriate unitary coördinate system y arising from the original one by a unitary transformation U: $x = Uy$, this mapping takes on the "diagonal" form

$$y'_i = \epsilon_i y_i$$

where the coefficients ϵ_i are of modulus 1. In other words

$$U^{-1}AU = \{\epsilon\} \tag{1.9}$$

becomes a diagonal unitary matrix

$$\{\epsilon\} = \left\| \begin{matrix} \epsilon_1 & 0 & \cdots & 0 \\ 0 & \epsilon_2 & \cdots & 0 \\ \cdots & \cdots & \cdots & \cdots \\ 0 & 0 & \cdots & \epsilon_n \end{matrix} \right\| = \{\epsilon_1, \epsilon_2, \cdots, \epsilon_n\}.$$

THEOREM (7.1.C). *A given unitary mapping $A: x \to x'$, when expressed in terms of appropriate unitary coördinates x_i, takes on the diagonal form*

$$x_i' = \epsilon_i x_i \qquad (|\epsilon_i| = 1).$$

Or: within the unitary group each element A is conjugate to a diagonal element $\{\epsilon\}$.

PROOF. Choose $\lambda = \epsilon$ as a root of the characteristic equation

$$|\lambda E - A| = 0.$$

Then their exists a non-vanishing vector $x = e$ such that

$$\epsilon e = Ae.$$

By the classical inductive construction we may ascertain a unitary coördinate system $e_1, e_2, \cdots, e_n$ whose first fundamental vector e_1, but for a positive numerical factor, coincides with e. We then have

$$\begin{aligned} Ae_1 &= \epsilon e_1, \\ Ae_2 &= \alpha_{12}e_1 + \cdots + \alpha_{n2}e_n, \\ &\cdots\cdots\cdots\cdots \\ Ae_n &= \alpha_{1n}e_1 + \cdots + \alpha_{nn}e_n; \end{aligned}$$

i.e., in the new coördinate system $e_1, \cdots, e_n$ the matrix $\|\alpha_{ik}\|$ of the mapping A has as its first column

$$(\alpha_{11}, \alpha_{21}, \cdots, \alpha_{n1}) = (\epsilon, 0, \cdots, 0).$$

The square sum of the moduli of its terms must equal 1; hence $|\epsilon| = 1$. At the same time the square sum of the moduli of the terms in the first row is to be $= 1$:

$$|\epsilon|^2 + |\alpha_{12}|^2 + \cdots + |\alpha_{1n}|^2 = 1, \quad \text{or}$$

$$|\alpha_{12}|^2 + \cdots + |\alpha_{1n}|^2 = 0.$$

This leads to

$$\alpha_{12} = \cdots = \alpha_{1n} = 0,$$

and consequently our matrix decomposes according to the scheme

$$\left\| \begin{matrix} \epsilon & 0 & \cdots & 0 \\ 0 & \alpha_{22} & \cdots & \alpha_{2n} \\ \cdots & \cdots & \cdots & \cdots \\ 0 & \alpha_{n2} & \cdots & \alpha_{nn} \end{matrix} \right\|.$$

The matrix $\|\alpha_{ik}\|$ where i, k run from 2 to n is an $(n - 1)$-dimensional unitary matrix. Our proposition for n-dimensional unitary matrices has thus been reduced to the corresponding theorem for $n - 1$ dimensions.

The elements ϵ_i of the diagonal matrix $\{\epsilon\}$ are obviously the roots of the characteristic equation

$$\det(\lambda E - A) = \det(\lambda E - \{\epsilon\}) = 0,$$

and are thus, but for their order, uniquely determined by A. The order remains arbitrary indeed, i.e., two diagonal unitary matrices

$$\{\epsilon\} = \{\epsilon_1, \cdots, \epsilon_n\} \text{ and } \{\epsilon'\} = \{\epsilon_1', \cdots, \epsilon_n'\}$$

are conjugate if (and only if) the ϵ_i' arise from the ϵ_i by permutation. On setting

$$\epsilon_k = e^{2\pi i\varphi_k} = e(\varphi_k),$$

we introduce the n real "angles" $\varphi_1, \cdots, \varphi_n$ of A. An angle has to be taken mod. 1, i.e. $\varphi \pm 1, \varphi \pm 2, \cdots$ all mean the same angle as φ.

Theorem (7.1.C) can be extended from a single unitary matrix to an arbitrary set $\mathfrak{A} = \{A\}$ of unitary matrices, all elements A of which commute with one another (commutative set):

THEOREM (7.1.D). *By means of an appropriate unitary coördinate system all unitary transformations of a commutative set may simultaneously be put in diagonal form.*

When we lump the "eigenvalues" or roots ϵ_i of the unitary mapping A together into groups of equal ones, then the proposition (7.1.C) may be stated thus: P is decomposable into a number of perpendicular subspaces $P_1 + P_2 + \cdots$ such that 1) each P_κ is invariant under A, and 2) the operation A in P_κ is a mere multiplication of all vectors by a certain number α_κ, the multipliers α_κ being distinct for the several P_κ. In P_κ lies *every* vector $\mathfrak{x}$ for which

$$A\mathfrak{x} = \alpha_\kappa \mathfrak{x}. \tag{1.10}$$

Let B be an arbitrary (unitary) operator commuting with A; I maintain that the spaces P_κ are also invariant with respect to B. Indeed, if $\mathfrak{x}$ lies in P_κ, i.e. satisfies (1.10), then the same holds for $B\mathfrak{x}$:

$$A(B\mathfrak{x}) = B(A\mathfrak{x}) = \alpha_\kappa \cdot B\mathfrak{x}.$$

This observation enables us to carry over our proposition to any commutative set $\mathfrak{A}$ of unitary operators A; in other words, P is decomposable into perpendicular subspaces P_κ 1) invariant under $\mathfrak{A}$, and 2) such that each operator A of $\mathfrak{A}$ is in P_κ a multiplication. Indeed, let P be decomposed into invariant subspaces P_κ, but assume that at least one operator A^0 in $\mathfrak{A}$ in one of the subspaces, e.g. in P_1, does not reduce to a mere multiplication. A^0 in P_1 is a unitary operator and hence P_1 may be split into perpendicular subspaces $P_1' + P_1'' + \cdots$ such that these are invariant under A^0 and that A^0 in each of them amounts to a mere multiplication (by distinct numbers $\alpha_1', \alpha_1'', \cdots$). Then according to our observation, which is to be applied to P_1 rather than to P, the parts P_1', P_1'', $\cdots$ are invariant with respect to all operators A of $\mathfrak{A}$, as these commute with A^0. The operator A^0 not being a mere multiplication in P_1, the number of summands in the decomposition

$$\mathrm{P}_1 = \mathrm{P}_1' + \mathrm{P}_1'' + \cdots$$

is at least 2. Our decomposition of P into invariant subspaces has thus been refined, a procedure that necessarily must come to a stop after at most n steps.

2. Character for symmetrization or alternation alone

The manifold $\Sigma(f)$ of the symmetric tensors of rank f is the substratum of a certain representation

$$\langle\Sigma(f)\rangle: \qquad A \rightarrow (A)_f$$

of the full linear group which may be defined as follows. While the coordinates x_i in the underlying n-dimensional vector space P undergo the linear transformation

$$A = \| a_{ik} \|: \qquad x_i' = \sum_k a_{ik} x_k, \tag{2.1}$$

all their monomials of degree f,

$$x_1^{r_1} \cdots x_n^{r_n} \qquad (r_1 + \cdots + r_n = f), \tag{2.2}$$

undergo the corresponding linear transformation $(A)_f$. It will be a good preparation for more difficult problems of similar nature ahead of us if we compute the character of this representation $\langle\Sigma(f)\rangle$.

One may say that the substratum of $\langle\Sigma(f)\rangle$ is the linear manifold of all forms of degree f of our variables x_i. This formulation at once evinces the fact that the choice of the coördinate system in P is immaterial for the determination of our character. Indeed, changing the coördinates x_i in P simply effects a change of the ("monomial") basis (2.2) for the forms of degree f, and one knows that the trace of a linear substitution is not affected by a change of basis.

In accordance with the general plan of our investigation, we first restrict ourselves to unitary matrices A. We then are allowed to assume A in its normal form $\{\epsilon\}$. Under the influence of A, each monomial (2.2) is multiplied by the factor

$$\epsilon_1^{r_1} \cdots \epsilon_n^{r_n},$$

i.e., the corresponding transformation $(A)_f$ is also in diagonal form, and the trace of this substitution or the character $\psi_f(A)$ is given by the sum

$$\sum \epsilon_1^{r_1} \cdots \epsilon_n^{r_n} \tag{2.3}$$

extending to all non-negative integers $r_1, \cdots, r_n$ whose sum $= f$. As is readily seen from the formula

$$1/(1 - \epsilon z) = \sum_{r=0}^{\infty} \epsilon^r z^r,$$

(2.3) is the coefficient of z^f in the Taylor expansion of

$$1/(1 - \epsilon_1 z) \cdots (1 - \epsilon_n z). \tag{2.4}$$

The expansion might be interpreted either in a formal way or as a numerical equation holding within the region of convergence $|z| < 1$ for the complex variable z. The denominator in (2.4) is the characteristic polynomial

$$\varphi(z) = \det(E - zA) = q_0 - q_1 z + q_2 z^2 - \cdots \pm q_n z^n \qquad (q_0 = 1), \tag{2.5}$$

and this observation at once allows us to go back from the normal form $\{\epsilon\}$, which our unitary A took on in a coördinate system *adapted to A*, to its matrix A in the original coördinate system common to all elements A of $U(n)$.

For any linear substitution $A = \| a_{ik} \|$ we therefore introduce the functions $p_f(A)$ by the formal expansion

$$1/\varphi(z) = 1/|E - zA| = \sum_{f=0}^{\infty} p_f \cdot z^f. \tag{2.6}$$

This means that for every $l \geqq 0$ we shall have the congruence

$$(1 - q_1 z + \cdots \pm q_n z^n)(p_0 + p_1 z + \cdots + p_l z^l) \equiv 1 \pmod{z^{l+1}},$$

which allows determination of the p_l's one after the other in a recurrent fashion:

$$p_0 = 1, \qquad p_l - q_1 p_{l-1} + \cdots \pm q_n p_{l-n} = 0 \qquad (l = 1, 2, \cdots).$$

The p with negative index, $p_{-1}, p_{-2}, \cdots$, are to be put $= 0$. From this determination it follows that $p_l(A)$ is a homogeneous polynomial of degree l of the quantities a_{ik}. The ensuing formula

$$\psi_f(A) = p_f(A) \tag{2.7}$$

was first proved only for unitary transformations A. However, according to their definition, both sides of the equation (2.7) are polynomials in the variable elements a_{ik} of A; consequently Lemma (7.1.A) carries the equality over to all elements A of $GL(n)$.

The method followed here will serve as a model for the future more complicated cases. But in view of the simple result one might ask oneself whether the formula (2.7) is not attainable without the detour via the unitary substitutions and their irrational normal form $\{\epsilon\}$ depending on the solution of the characteristic equation. Indeed it is easy enough to avoid the first step. By solving the characteristic equation for an arbitrary mapping A one brings its matrix into the recurrent or triangular form

$$\left\| \begin{matrix} \alpha_1 & * & \cdots & * \\ 0 & \alpha_2 & \cdots & * \\ \cdots & \cdots & \cdots & \cdots \\ 0 & 0 & \cdots & \alpha_n \end{matrix} \right\|$$

$(A)_f$ then takes on the same form and its trace is found to be

$$\sum \alpha_1^{r_1} \cdots \alpha_n^{r_n} \qquad (r_1 + \cdots + r_n = f).$$

To avoid the transformation of A into a handy normal form seems somewhat less easy. I give an analytic procedure complying with our present demand of using the substitution A in its original form (2.1). The coefficient $(r; \rho)$ of the linear transformation $(A)_f$:

$$x_1'^{r_1} \cdots x_n'^{r_n} = \sum_{\rho} (r; \rho) x_1^{\rho_1} \cdots x_n^{\rho_n}$$

can be computed as the integral

$$(r; \rho) = \frac{1}{(2\pi i)^n} \int \cdots \int \frac{x_1'^{r_1} \cdots x_n'^{r_n}}{x_1^{1+\rho_1} \cdots x_n^{1+\rho_n}} dx_1 \cdots dx_n,$$

each integration extending over the unit circle $|x_k| = 1$ in the complex x_k-plane. Hence the expansion

$$1/\prod_k (x_k - zx_k') = \frac{1}{x_1 \cdots x_n} \sum \left(\frac{x_1'}{x_1}\right)^{r_1} \cdots \left(\frac{x_n'}{x_n}\right)^{r_n} z^f \quad (r_1 + \cdots + r_n = f)$$

involving the auxiliary variable z leads by the same process of integration to the formula

$$(2.8) \qquad \frac{1}{(2\pi i)^n} \int \cdots \int \frac{dx_1 \cdots dx_n}{(x_1 - zx_1') \cdots (x_n - zx_n')} = \sum_{f=0}^{\infty} \psi_f(A) z^f.$$

If the x_k vary on the unit circle $|x_k| = 1$ the absolute values of the quantities

$$x_i' = \sum_k a_{ik} x_k$$

can not surpass a certain bound M, the biggest among the n numbers

$$|a_{i1}| + \cdots + |a_{in}| \qquad (i = 1, \cdots, n).$$

Convergence in (2.8) is secured if $|z| < 1/M$.

The desired equation

$$(2.9) \qquad \sum_{f=0}^{\infty} \psi_f(A) z^f = 1/|E - zA|$$

will be established within that circle by proving the following

LEMMA (7.2.A). *If*

$$A = \| \alpha_{ik} \| : \qquad y_i = \sum \alpha_{ik} x_k$$

is a linear substitution satisfying the inequalities

$$(2.10) \qquad \begin{cases} + |\alpha_{11}| - |\alpha_{12}| - \cdots - |\alpha_{1n}| > 0, \\ - |\alpha_{21}| + |\alpha_{22}| - \cdots - |\alpha_{2n}| > 0, \\ \cdots\cdots\cdots\cdots\cdots\cdots\cdots\cdots \\ - |\alpha_{n1}| - |\alpha_{n2}| - \cdots + |\alpha_{nn}| > 0. \end{cases}$$

then its determinant det A *is* $\neq 0$ *and*

$$\frac{1}{(2\pi i)^n}\int\cdots\int\frac{dx_1\cdots dx_n}{y_1\cdots y_n} = 1/\det A, \tag{2.11}$$

the integration extending over the unit circles $|x_k| = 1$.

We proceed by induction with respect to the dimensionality n and first make the somewhat vague hypothesis that A is sufficiently near to the unit matrix. Consider $1/y_1 \cdots y_n$ as a function of the variable x_1 for fixed values of $x_2, \cdots, x_n$ on their circles $|x_k| = 1$. Under our assumption the pole resulting from the equation $y_1 = 0$:

$$x_1 = -\frac{1}{\alpha_{11}}(\alpha_{12}x_2 + \cdots + \alpha_{1n}x_n) \tag{2.12}$$

will be small, whereas the other $n - 1$ poles annulling $y_2, \cdots, y_n$ are large. Hence none but the first one lies within the unit circle, and Cauchy's integral formula gives

$$\frac{1}{2\pi i}\int\frac{dx_1}{y_1\cdots y_n} = \frac{1}{\alpha_{11}}\cdot\frac{1}{y_2'\cdots y_n'}, \tag{2.13}$$

where

$$y_i' = \sum \alpha_{ik}'x_k \qquad (i, k = 2, \cdots, n)$$

are the linear forms arising from $y_2, \cdots, y_n$ by substituting for x_1 the value (2.12), or by subtracting from $y_2, \cdots, y_n$ those multiples, $\alpha_{21}/\alpha_{11}, \cdots, \alpha_{n1}/\alpha_{11}$, of y_1 which annihilate the coefficients of x_1. This subtraction does not change the determinant det A of the forms $y_1, \cdots, y_n$; hence

$$\det A = \alpha_{11}\cdot\det A'. \tag{2.14}$$

Observe that

$$A' = \|\alpha_{ik}'\| \qquad (i, k = 2, \cdots, n) \tag{2.15}$$

again is near to the identity. On comparing (2.14) with the integrated relation (2.13),

$$\frac{1}{(2\pi i)^n}\int\cdots\int\frac{dx_1\cdots dx_n}{y_1\cdots y_n} = \frac{1}{\alpha_{11}}\cdot\frac{1}{(2\pi i)^{n-1}}\int\cdots\int\frac{dx_2\cdots dx_n}{y_2'\cdots y_n'},$$

one proves inductively the equation (2.11).

Simple calculation shows that the inequalities (2.10) imply (1) the desired position of the x_1-zeros of $y_1 \mid y_2, \cdots, y_n$, namely inside and outside of the unit circle respectively, provided $x_2, \cdots, x_n$ lie on the unit circle, and (2) the same inequalities for A′. Hence we can claim the formula (2.11) under the more precise hypothesis of the lemma.[1]

After having thus determined in several ways the character of the representation $\langle\Sigma(f)\rangle$, we come nearer to our ultimate goal by dividing up the f arguments of a tensor of rank f into several rows of lengths $f_1, f_2, \cdots, f_r$ according to a symmetry diagram ($f_1 + f_2 + \cdots + f_r = f$) and by considering all tensors which are symmetric in the arguments of each row. They form an invariant manifold $\Sigma(f_1, \cdots, f_r)$ with its corresponding representation $\langle\Sigma(f_1, \cdots, f_r)\rangle$ of $GL(n)$. The representative of A being the Kronecker product

$$(A)_{f_1} \times (A)_{f_2} \times \cdots \times (A)_{f_r},$$

its character $\psi(f_1, \cdots, f_r)$ is given by

$$\psi(f_1, \cdots, f_r) = p_{f_1}(A)p_{f_2}(A) \cdots p_{f_r}(A). \tag{2.16}$$

Thus we have solved the problem of characters when a diagram $T(f_1, f_2, \cdots)$ is used as a basis for *symmetrization*

$$a = \sum p \qquad \text{(cf. Chapter IV, §2).}$$

Even simpler is the problem for *alternation*

$$b = \sum \delta_q \cdot q.$$

If we first limit ourselves to unitary transformations again, and use A in its normal form $\{\epsilon\}$, then the antisymmetric tensors of rank f form the substratum of the representation the character of which is equal to the fth elementary symmetric function of $\epsilon_1, \cdots, \epsilon_n$, or to the coefficient q_f in the characteristic polynomial (2.5). The same result may be obtained in an absolutely elementary way. For one sees at once that the trace of the substitution the skew-symmetric tensors undergo under the influence of A equals

$$\sum_{(i_1<i_2<\cdots<i_f)} \begin{vmatrix} a_{i_1i_1}, & a_{i_1i_2}, & \cdots, & a_{i_1i_f} \\ a_{i_2i_1}, & a_{i_2i_2}, & \cdots, & a_{i_2i_f} \\ \cdots & \cdots & \cdots & \cdots \\ a_{i_fi_1}, & a_{i_fi_2}, & \cdots, & a_{i_fi_f} \end{vmatrix} = q_f(A).$$

When the columns of the diagram are of lengths $f_1^*, f_2^*, \cdots$, the tensors which are antisymmetric in the arguments of each column form the substratum of a representation the character of which equals

$$q_{f_1^*} \cdot q_{f_2^*} \cdot \cdots. \tag{2.17}$$

The real problem however is to determine the character in case both processes, symmetrization with respect to the rows, and alternation with respect to the columns, are applied one after the other. This task we shall now approach by an essentially more transcendental method: quite independently of all of the foregoing investigations we shall try to compute the characters corresponding to all irreducible continuous representations of the unitary group.

3. Averaging over a group

For the study of a finite group γ the process of *averaging over all group elements* s is a powerful instrument yielding unexpectedly far-reaching results. We encountered this method before in Chapter III where we derived from it the existence of generating idempotents and the full reducibility of the regular and hence of every representation of a finite group. In order to familiarize ourselves with the process we give some further examples.

Let there be given in the n-dimensional affine point space a finite group γ of affine transformations s. *I maintain that the group has a fixed point that is left invariant by each transformation of the group.* If P is the center of gravity of some positive masses $\mu_1, \cdots, \mu_h$ at the points $P_1, \cdots, P_h$ and s is any affine transformation, then sP is the center of gravity of the same masses at the points $sP_1, \cdots, sP_h$. We now take an arbitrary point P and form the center of gravity P_0 of the h equivalent points sP which originate from P by the h transformations s of the given group, attaching the same mass or weight 1 to each of these points:

$$P_0 = \frac{1}{h} \cdot \sum_s sP.$$

P_0 is a fixed point. Indeed, if a be any transformation of our group, we have

$$aP_0 = \frac{1}{h} \sum_s asP = \frac{1}{h} \sum s'P$$

with

$$s' = as.$$

But s' ranges with s over all group elements; hence the last sum again equals P_0.—

Another important example much nearer to our present interests is concerned with *invariants* $F(x, y, \cdots)$ of a given abstract finite group γ where the arguments $x, y, \cdots$ are generic vectors in the representation spaces of a number of given representations of γ (cf. Chapter I, §5). Let F be any polynomial of such vectors, homogeneous in the components of each of them. Again we form the average

$$[F] = \frac{1}{h} \cdot \sum_s sF \tag{3.1}$$

on the group γ. *Then $[F]$ is an invariant.* This averaging process [], which carries every polynomial into an invariant, is firstly a *linear* process and secondly leaves F unaltered if F itself is an invariant. Moreover, J being an invariant, one has

$$[JF] = J[F]. \tag{3.2}$$

We shall see (Chapter VIII, §14) that this process provides a simple proof for the first invariant-theoretic main theorem in the case of finite groups.—

As a third example I prove

THEOREM (7.3.A). *Any finite group of homogeneous linear transformations leaves invariant a certain positive definite Hermitean form.*

PROOF. A Hermitean form

$$G(x) = \sum g_{ik} \bar{x}_i x_k \qquad (g_{ki} = \bar{g}_{ik})$$

is positive definite if $G(x) > 0$ except for $x_1 = \cdots = x_n = 0$. One starts with an arbitrary such form G, e.g. with the unit form

$$\bar{x}_1 x_1 + \cdots + \bar{x}_n x_n . \tag{3.3}$$

The average

$$G_0 = \frac{1}{h} \sum_s sG$$

satisfies our requirements.

By a suitable choice of coördinates, any positive definite Hermitean form $G(x)$ may be changed into the unit form. This is accomplished by the classical inductive construction of a unitary coördinate system, taking

$$G(x, y) = \sum g_{ik} \bar{x}_i y_k$$

as the scalar product in our unitary geometry. On applying this remark to G_0 one obtains this important

COROLLARY (7.3.B). *Any finite group of homogeneous linear transformations is equivalent to a group of unitary transformations.*

Taking into account Lemma (6.2.A), one derives anew the full reducibility of the representations of finite groups. Although this is a very quick way of establishing that fundamental truth, it has its disadvantages as compared with our former procedure in Chapter III. For while the latter went through in any number field, the proof here given operates in the ordinary field of all complex numbers. Even with the utmost generosity its scope may not be extended beyond the fields of type $k^\dagger = (k, \sqrt{-1})$, where k is real.

These examples may suffice. But what we really have in mind is to apply the averaging process to a *compact continuous group* rather than to a finite group. Let us suppose we are given a continuous group γ, i.e. one whose elements s form a continuous manifold in the sense of topology. We shall naturally assume that the composite st of two elements s and t depends *continuously* on both arguments s, t, and that s^{-1} is a continuous function of s. We shall even suppose that we can apply the differential geometric notion of *line elements* to our group manifold; in that case we speak of a *Lie group*. Nobody so far has succeeded in stating in a natural and satisfactory way the intrinsic requirements a manifold must meet so as to allow application of the idea of line elements and thus of Calculus ("differentiable manifold"); however, for all practical purposes this is what it amounts to. Corresponding to each point of the manifold we have a class of admissible coördinate systems $(s_1, \cdots, s_r)$; the transformation functions linking two admissible coördinate systems are not only continuous, but have continuous first derivatives and a non-vanishing functional determinant (in a certain neighborhood of the point). In this way a number of dimensions r is at once attached to the manifold. The "parameters" s_i are assumed to vary in the domain of real numbers. For a more careful formulation we refer the reader to Veblen and Whitehead, *The Foundations of Differential Geometry*, Cambridge Tracts No. 29, 1932. In the case of a group, one needs to make these requirements only for the unit point I. By means of the "left translation" a:

$$s \to s' = as \tag{3.4}$$

we can transport the neighborhood of I into the neighborhood of the arbitrary point a. We must of course now assume that, for any two elements s and t in a sufficiently small neighborhood of I, the parameters of st are functions of the parameters of s and t with continuous first derivatives.

The line elements $\delta s = (\delta s_1, \cdots, \delta s_r)$ at the unit point I are the infinitesimal elements of the group; they form the r-dimensional tangent plane of the group manifold at I. r such line elements $\delta' s, \cdots, \delta^{(r)} s$ at I span an infinitesimal parallelepipedic volume element; as its volume we are to consider the absolute value of the determinant

$$\begin{vmatrix} \delta' s_1, & \cdots, & \delta' s_r \\ \cdots & \cdots & \cdots \\ \delta^{(r)} s_1, & \cdots, & \delta^{(r)} s_r \end{vmatrix}.$$

If we change the system of parameters $s_1, \cdots, s_r$ covering the neighborhood of I, all these volumes are multiplied by one and the same positive constant. Hence they are independent of the parameter system except for the choice of a unit.

The process of averaging over a compact Lie group presupposes our ability to compare volume elements *at different points* of the group manifold. We must find the analogue of the equal weights attached to the several group elements in the case of a finite group. Our examples reveal at once the necessary condition which such a "good" volume measure is to satisfy: it must be invariant with respect to all left translations (3.4). But after defining the volume for elements at I this requirement is just sufficient to carry over the measure from the central bureau of standards at the point I to any other point: the transport takes place by *left translation*. A volume element $d\omega_a$ at a which arises from the volume element $d\omega$ at I by the left translation a shall by definition have the same measure as $d\omega$. A line element ds at s leads from s to an infinitely near point $s + ds$; by the left translation s^{-1} it goes over into a line element δs at I defined by

$$\mathsf{I} + \delta s = s^{-1}(s + ds). \tag{3.5}$$

According to our definition these δs are used for computing the volume of an infinitesimal parallelepipedon $d\omega_s$ at s (by means of the absolute determinant of the components of r such δs which correspond to r line elements ds at s spanning $d\omega_s$).

Using the volume measure thus defined we can average any continuous function on a compact Lie group γ over the whole group.[2] If we like we may fix the arbitrary unit for the measurement of volumes so as to make the volume of the total group $= 1$. We are now in a position to carry over from finite to compact Lie groups all the examples considered in this section. In particular any continuous representation of such a group leaves invariant a certain positive definite Hermitean form, and hence is equivalent to a unitary representation. In its unitary normalization the representing matrix $U(s) = \| u_{ik}(s) \|$ satisfies the equations

$$u_{ki}(s^{-1}) = \hat{u}_{ik}(s) = \bar{u}_{ik}(s),$$

and for the character $\chi(s)$ we therefore have, under all circumstances,

$$\chi(s^{-1}) = \bar{\chi}(s).$$

Any representation is decomposable into irreducible ones.

I. Schur's proof of the orthogonality relations, Chapter IV, §1, goes through as well.[3] In particular, we obtain for the primitive characters the orthogonality relations:

$$\mathfrak{M}_s\{\bar{\chi}(s)\chi'(s)\} = 1 \text{ or } 0,$$

according as the two irreducible representations of the characters χ and χ' are equivalent or not. The multiplicities $m, m', \cdots$ in the expansion of an arbitrary character $X(s)$, equation (1.1), may therefore be determined by the mean values

$$m = \mathfrak{M}_s\{X(s)\bar{\chi}(s)\}, \cdots. \tag{3.6}$$

Moreover one finds

$$\mathfrak{M}_s\{X(s)\bar{X}(s)\} = m^2 + m'^2 + \cdots. \tag{3.7}$$

This shows that *a representation of character* $X(s)$ *is irreducible if and only if the mean value of* $|X(s)|^2$ *equals* 1, a criterion often used by Frobenius.

The *completeness relation* for a full set of inequivalent irreducible representations of a compact Lie group was established by F. Peter and the author; the construction, as it was known for a finite group, had to be twisted around somewhat in adapting it to the compact continuous groups.[4]

Sometimes we should be severely handicapped if our volume measure were not invariant with respect to right-hand as well as left-hand translations, i.e. under the operations

$$s \to s' = sa$$

We carry a volume element $d\omega$ at I to a by the left-hand translation a, and then back to I by the right-hand translation a^{-1}. The question is whether the resulting operation, the "conjugation" a:

$$s' = asa^{-1}, \tag{3.8}$$

leaves the volume of $d\omega$ unaltered. Associating the operation (3.8) with the group element a establishes a realization of the group, the so-called *adjoint realization* whose point field is the group manifold itself. Since (3.8) leaves I fixed, it carries an infinitesimal group element $\mathsf{I} + \delta s$ into an infinitesimal one $\mathsf{I} + \delta' s$, and the transition from the line element δs to $\delta' s$ is naturally a linear transformation $K(a)$; the correspondence $a \to K(a)$ is the *adjoint representation.* We raised the question as to whether this representation is quasi-unimodular, that is to say whether

$$|\det K(a)| = 1$$

for all elements a. It is answered in the affirmative by the general

LEMMA (7.3.C). *A continuous representation of a compact group is necessarily quasi-unimodular.*

PROOF. Consider any element a of the compact group γ and its representative $T(a)$. Suppose that

$$|\det T(a)| > 1.$$

The sequence $a^n (n = 1, 2, \cdots)$ has a limit point b on the group manifold. Because of

$$T(a^n) = (T(a))^n,$$

$|\det T(a^n)|$ tends towards ∞ as $n \to \infty$. This is in contradiction to the fact that for certain arbitrarily high n this number must come as close as one wishes to the finite value $|\det T(b)|$. In the same manner one discards the possibility $|\det T(a)| < 1$ by considering the sequence a^{-n} $(n = 1, 2, \cdots)$.

The substitution

$$s' = s^{-1}$$

changes the line element δs at I into $-\delta s$ and consequently leaves unaltered the volume of any infinitesimal parallelepiped at I. The same remains true for a volume element at any point a. For while s is submitted to the left-hand translation a: $s \to as$, the inverse element s^{-1} undergoes the right-hand translation a^{-1}:

$$s^{-1} \to s^{-1}a^{-1}.$$

Thus our statement is an immediate consequence of the fact that volume is invariant under both right-hand and left-hand translations.

In the case of a group of linear transformations or matrices, the infinitesimal elements of the group form an r-dimensional linear set of matrices

$$\delta A = \delta s_1 \cdot K_1 + \cdots + \delta s_r \cdot K_r,$$

of which $K_1, \cdots, K_r$ is an arbitrarily chosen basis. If A and $A + dA$ are two infinitely near matrices of the group, $\delta A = A^{-1} \cdot dA$ is the infinitesimal element defined by (3.5), and hence the components δs_i as introduced by

$$A^{-1} \cdot dA = \delta s_1 \cdot K_1 + \cdots + \delta s_r \cdot K_r \tag{3.9}$$

serve for computing the volume of an infinitely small parallelepipedon at A spanned by r line elements dA.

The group $U(n)$ of all unitary transformations is an n^2-parameter compact Lie group. An infinitesimal unitary matrix $\delta A = \| \delta a_{ik} \|$ satisfies the conditions

$$\delta a_{ki} = -\overline{\delta a_{ik}},$$

and its n^2 real parameters δs may be introduced by the substitution (1.4):

$$\delta a_{ii} = \sqrt{-1} \cdot \delta s_{ii} \qquad (\text{for each } i = 1, \cdots, n);$$

$$\delta a_{ik} = \delta s_{ik} + \sqrt{-1}\, \delta s'_{ik}, \qquad \delta a_{ki} = -\delta s_{ik} + \sqrt{-1}\, \delta s'_{ik}$$

$$(\text{for each pair } i, k \text{ with } i < k).$$

This shows at once that in computing the volume of a volume element, the complex parameters δa_{ik} may serve as well as the real ones δs_{ik}.

The statement that the adjoint representation is quasi-unimodular, deduced before by a simple topological consideration for any abstract compact Lie group, is capable of an algebraic demonstration for Lie groups of unitary linear transformations.

LEMMA (7.3.D). *Let U be a unitary matrix and $\mathfrak{K}$ a linear set of order r of matrices K:*

$$K = \lambda_1 K_1 + \cdots + \lambda_r K_r \tag{3.10}$$

invariant under the transformation

$$K \to K' = UKU^{-1}: \tag{3.11}$$

$$UK_\alpha U^{-1} = \sum_\beta c_{\beta\alpha} K_\beta \qquad (\alpha, \beta = 1, \cdots, r).$$

Then this linear transformation $\| c_{\alpha\beta} \|$ leaves invariant a certain positive definite Hermitean form. (Hence by proper choice of the basis K_α of $\mathfrak{K}$ the matrix $\| c_{\alpha\beta} \|$ will become unitary and certainly have a determinant of modulus 1.)

Every matrix $A = \| a_{ik} \|$ has a *"norm"*

$$n(A) = \operatorname{tr}(\bar{A}^* A) = \sum_{i,k} | a_{ik} |^2,$$

positive except for $A = 0$. The norm is not altered by changing A into $B = UAU^{-1}$ provided U be unitary:

$$\bar{U}^* = U^{-1}, \qquad \bar{B}^* B = \bar{U}^{*-1} \bar{A}^* \bar{U}^* U A U^{-1} = U(\bar{A}^* A) U^{-1}.$$

Consequently

$$n(K) = \sum_{\alpha,\beta} g_{\alpha\beta}\bar{\lambda}_\alpha\lambda_\beta$$

is for the generic K, (3.10), a positive definite Hermitean form of the parameters λ which is invariant under the substitution (3.11):

$$\lambda'_\alpha = \sum_\beta c_{\alpha\beta}\lambda_\beta .$$

The simplest compact group is the *group of plane rotations* (or what is the same, the unitary group in 1 dimension). It is an Abelian 1-parameter group. If φ be the angle of rotation, then for any integer m,

$$\varphi \rightarrow e^{2\pi i m\varphi} = e(m\varphi)$$

is obviously a unitary representation of degree 1. I maintain that these are the only irreducible continuous representations, and hence

$$\chi(\varphi) = e(m\varphi) \tag{3.12}$$

the only primitive characters. This is a very natural approach to the theory of Fourier series. In particular, the Parseval equation

$$\int_0^1 | f(\varphi) |^2 \, d\varphi = \sum_{m=-\infty}^{+\infty} \left| \int_0^1 f(\varphi)e(-m\varphi) \, d\varphi \right|^2,$$

holding for any continuous function $f(\varphi)$ of period 1, comes out as a special case of the group-theoretical completeness relation. To prove our statement, we first observe that any representation is equivalent to a unitary one. Because of the Abelian nature of our group and Theorem (7.1.D), the unitary representation breaks up into parts of degree 1. Therefore we have to look for continuous solutions $\chi(\varphi)$ of the following functional equations:

$$\chi(\varphi + \varphi') = \chi(\varphi)\cdot\chi(\varphi'), \qquad | \chi(\varphi) | = 1. \tag{3.13}$$

Given such a function χ we introduce the real function $g(\varphi)$ by

$$\chi(\varphi) = e(g(\varphi));$$

it will be uniquely determined not only mod 1, but even absolutely if we require $g(0) = 0$ and $g(\varphi)$ to vary continuously with φ. The congruence

$$g(\varphi + \varphi') \equiv g(\varphi) + g(\varphi') \pmod 1$$

may at once be replaced by the equation

$$g(\varphi + \varphi') = g(\varphi) + g(\varphi'). \tag{3.14}$$

For the difference of left- and right-side is continuous in φ, always an integer and vanishes for $\varphi = 0$. (3.14) implies

$$g(k\varphi) = k\cdot g(\varphi) \tag{3.15}$$

for any positive integer k and

$$(3.16) \qquad g(0) = 0, \qquad g(-\varphi) = -g(\varphi).$$

As $e(g(1)) = 1$, $g(1)$ must be a certain integer m. We deduce the equation

$$g(\varphi) = m\varphi$$

first for any aliquot part $\varphi = 1/k$ of the full rotation by setting $\varphi = 1/k$ in (3.15), and then from the same equation and (3.16) for any multiple k' (positive, zero or negative) of this φ, i.e. for any rational angle k'/k, and finally by continuity for all. We thus arrive at the desired equation (3.12).

Harald Bohr's theory of *almost-periodic functions* may be looked upon as the theory of the Abelian group of one-dimensional translations, i.e., of the group of real numbers with addition as their composition. Bohr discovered that the essential facts of the theory of periodic functions, orthogonality and completeness, go through provided one restricts the notion of function in an appropriate manner. The author showed that the group theoretical interpretation leads to a much simpler deduction of the central theorem of completeness.[5]

By an ingenious device A. Haar succeeded in defining a "good" volume measure on every locally Euclidean compact group.[6] In other words, he got rid of the awkward assumptions of differentiability involved in the notion of a Lie group.

With the theory of compact groups and Bohr's example of a non-compact group before his eyes, J. von Neumann established the theory of "almost-periodic" representations, their orthogonality and completeness, for any group whatsoever.[7] For this generality, to be sure, he had to pay heavily in limiting the notion of function to the often very narrow domain of almost-periodic functions. I shall try to give in a few words his fundamental idea. Even if our group should happen to be a topological group, we malevolently disregard its topology. Instead we introduce an artificial topology relative to a given function $f(s)$ on the given group by defining: s has a distance $\leqq \epsilon$ from s_0 provided the inequalities

$$|f(st) - f(s_0t)| \leqq \epsilon, \qquad |f(ts) - f(ts_0)| \leqq \epsilon$$

hold for all group elements t. f is called almost-periodic if the group endowed with this topology is compact or finite, which means that however small ϵ may be, the manifold may be covered by a finite number of circular disks of radius ϵ. Von Neumann shows how to form the *mean value* of such a function. This once accomplished, the theory proceeds along ready-made channels. His theory is undeniably the culminating point of this whole trend of ideas, though by no means "the end of every man's desire," as is shockingly revealed by this remark: on the group $GL(n)$ of all non-singular real transformations, the only almost-periodic function is the *constant*. Hence the simple representations $\langle P(f_1f_2 \cdots)\rangle$ of $GL(n)$ which resulted from the decomposition of tensor space, nay, the representation of the group by itself: $s \to s$, lie beyond the scope of von Neumann's theory! We should be satisfied if the almost-periodic functions would allow distinguishing between distinct points; i.e., if for any two distinct elements there exists an almost-periodic function which takes on different values at these two points. But the only groups on which the almost-periodic functions "satisfy" in this sense are the direct products of compact groups with a number of one-dimensional translations (H. Bohr's case). This result, due to H. Freudenthal,[8] clearly indicates the limits of any "almost-periodic" theory. I should venture to say that only in combination with the unitarian trick can almost-periodicity give anything like a satisfactory solution of the problem.

The group γ may be replaced by any point field on which the given group is realized by one-to-one transformations. The point field is homogeneous with respect to this group if any two points go over into each other by a suitable transformation of the group. What has been said about functions on a group remains true for this more general situation of *functions on a homogeneous manifold,* for which the spherical harmonics constitute the most representative and classical instance.[9]

The completeness of the primitive characters of Abelian groups has recently found important applications in *general topology.*[10] The ideas we discussed in this section seem to form a crossing point of a number of recent advances in different fields of mathematics.

4. The volume element of the unitary group

Let us study somewhat more closely the fundamental equation:

$$A = U\{\epsilon\}U^{-1}; \tag{4.1}$$

$$\{\epsilon\} = \text{diag. matr. } \{\epsilon_1, \cdots, \epsilon_n\}, \qquad \epsilon_k = e(\varphi_k),$$

(Theorem 7.1.C) holding for the elements A of the unitary group. We first infer from it:

THEOREM (7.4.A). *The unitary group is a connected compact manifold; it consists of one piece.*

Indeed, we keep U in (4.1) fixed while replacing the angles φ_k in $\{\epsilon\}$ by $\tau\varphi_k$ where τ is a real parameter. The element A_τ thus resulting varies from $A_{\tau=0} = E$ to the given $A_{\tau=1} = A$ when τ varies from 0 to 1.

The diagonal elements $\{\epsilon\}$ form an n-parameter Abelian subgroup Λ of the unitary group. On using the angles φ_k as parameters, the combination of two elements (φ_k) and (φ_k') yields the element $(\varphi_k + \varphi_k')$.

In formula (4.1) the generic $\{\epsilon\}$ depends on n real parameters, U on n^2 parameters. One might therefore expect that the resulting A, (4.1), would depend on $n + n^2$ parameters, whereas the correct number is merely n^2. How is this apparent contradiction dissolved? In (4.1) U may be replaced by $U\{\rho\}$, where $\{\rho\}$ is any element of Λ, without altering the resulting A. It is therefore convenient to identify two U's, U, U_1, which are right equivalent mod Λ or belong to the same right co-set mod Λ, i.e. for which $U^{-1}U_1$ is in Λ. This process of identification changes the n^2-dimensional manifold $U(n)$ into an $(n^2 - n)$-dimensional one $[U(n)]$. U, as an element of $[U(n)]$, is denoted by $[U]$. We shall use a suggestive geometric nomenclature by saying that two such elements U, U_1 which are right equivalent mod Λ lie on the same *vertical*.

When we passed from an element U to a nearby element $U + dU$ we attached to this transition the infinitesimal element $\delta U = U^{-1}dU$. In studying the process on $[U(n)]$ we are free to replace

$$U + dU \quad \text{by} \quad (U + dU)(1 + 2\pi i\{d\rho\}),$$

where the second factor with the real infinitely small parameters $d\rho$ is any infinitesimal element of Λ. Thus dU is changed into

$$d'U = dU + 2\pi i U\{d\rho\}$$

and δU into

$$\delta' U = \delta U + 2\pi i\{d\rho\}.$$

Since the diagonal components δu_{ii} of δU are pure imaginary, one may assign to them by proper choice of the $d\rho_i$ any such values as one wants, whereas the lateral components are not affected by the arbitrariness of the $d\rho$. In particular, one may choose $\delta u_{ii} = 0$; we shall then call δU a *horizontal* transition from the vertical $[U]$ to the infinitely near vertical $[U + dU]$ at the altitude U. A parallelepipedic fiber of verticals spanned by $n^2 - n$ line elements at $[U]$ on $[U(n)]$ has a cross section determined by $n^2 - n$ such horizontal δU at U. As the volume of this cross section, let us consider the absolute determinant of the lateral components δu_{ik} $(i \neq k)$ of these δU's. When we cut across the same fiber at another altitude, we have to replace U, $U + dU$ by

$$UR \quad \text{and} \quad (U + dU)R$$

respectively, where

$$R = \{\rho\} = \{\rho_1, \cdots, \rho_n\}, \qquad \rho_k = e(\psi_k)$$

is any element of Λ. Hence $\delta U = U^{-1}dU$ is replaced by

$$\delta' U = (UR)^{-1} \cdot dU \cdot R = R^{-1} \cdot \delta U \cdot R$$

with the components

$$\delta' u_{ik} = \frac{\rho_k}{\rho_i}\,\delta u_{ik}. \tag{4.2}$$

If δU is horizontal so is $\delta' U$. Since the linear substitution (4.2) for the $n^2 - n$ lateral components δu_{ik} is unimodular, the factors ρ_k/ρ_i, ρ_i/ρ_k of two associated pairs (i, k) and (k, i) canceling each other, the volume of the cross-section is independent of the altitude, and we have thus introduced a reasonable volume measure on the $(n^2 - n)$-dimensional manifold $[U(n)]$.

Is it true that no U leads to the same A in (4.1) unless it is right equivalent mod Λ to the original U? In replacing U by UR the matrix A will stay unaltered if and only if $R = \| \rho_{ik} \|$ commutes with $\{\epsilon\}$:

$$\epsilon_i \rho_{ik} = \rho_{ik} \epsilon_k \quad \text{or} \quad \rho_{ik}(\epsilon_i - \epsilon_k) = 0.$$

Hence if all eigenvalues ϵ_k are distinct, R has to be diagonal and the answer to our question is affirmative. An element A for which two eigenvalues ϵ_k coincide, for example $\epsilon_1 = \epsilon_2$, may be called *singular*. *The singular elements on the n^2-dimensional unitary group form a manifold, not of one, as one might expect, but of three dimensions less.* Indeed, our R is then allowed to be of the form

$$\left\| \begin{array}{cc|ccc} \rho_{11} & \rho_{12} & & & \\ \rho_{21} & \rho_{22} & & & \\ \hline & & \rho_3 & & \\ & & & \ddots & \\ & & & & \rho_n \end{array} \right\|$$

and since a two-dimensional unitary transformation involves 4 parameters, it depends on $4 + (n - 2) = n + 2$ real parameters. The number of essential parameters in U (modulo the subgroup of these R) is thus reduced to $n^2 - (n + 2)$ while the generic $\{\epsilon\}$ with $\epsilon_1 = \epsilon_2$ depends on $n - 1$ parameters. The sum is $n^2 - 3$.

We have given in §1 a direct algebraic proof for the equation (4.1). One could think of an analytic proof by the method of continuity: one would try to follow a given infinitesimal variation of A by corresponding increments of U and $\{\epsilon\}$. In more general cases, in the theory of semi-simple groups, this procedure is forced upon us; but even here it is worth while to carry through the calculation:

$$AU = U\{\epsilon\} \quad \text{implies} \quad dA \cdot U + A \cdot dU = dU \cdot \{\epsilon\} + U\{\epsilon\} \cdot 2\pi i \cdot \{d\varphi\}.$$

Multiply with $(AU)^{-1} = U^{-1}A^{-1}$ on the left side and with the equal $\{\epsilon\}^{-1}U^{-1}$ on the right side:

$$U^{-1} \cdot \delta A \cdot U + \delta U = \{\epsilon\}^{-1} \delta U \{\epsilon\} + 2\pi\sqrt{-1}\{d\varphi\}.$$

For

$$\delta B = U^{-1} \cdot \delta A \cdot U = \| \delta b_{ik} \|$$

we obtain the formulas

$$(4.3) \quad \left\{ \begin{array}{l} \delta b_{ii} = 2\pi\sqrt{-1}\, d\varphi_i, \\[2mm] \delta b_{ik} = \left(\dfrac{\epsilon_k}{\epsilon_i} - 1\right) \delta u_{ik} \qquad (i \neq k). \end{array} \right.$$

Supposing δA to be given, the first set of equations determines uniquely the increments $d\varphi_i$, the second set the increments of the lateral components

δu_{ik} $(i \neq k)$, while the diagonal components δu_{ii} remain free. These statements, bound to the assumption that A is non-singular, are in complete agreement with our previous considerations. In varying A it is possible to avoid the singular points, which form a manifold of three dimensions less.

When one considers that the transition from

$$\delta A = \| \delta a_{ik} \| \quad \text{to} \quad \delta B = U^{-1} \cdot \delta A \cdot U = \| \delta b_{ik} \|$$

is a quasi-unimodular transformation, one derives from (4.3) the following formula for the volume element $d\omega_A$ over which (4.1) varies when $[U]$ on the manifold $[U(n)]$ ranges over the volume element $[d\omega_U]$ and when the angles φ_k vary between φ_k and $\varphi_k + d\varphi_k$:

$$d\omega_A = (2\pi)^n \cdot \left| \prod_{i \neq k} \left(\frac{\epsilon_k}{\epsilon_i} - 1 \right) \right| \cdot [d\omega_U] \cdot d\varphi_1 \cdots d\varphi_n .$$

$d\omega_A$ and $[d\omega_U]$ stand here for the volumes of the respective elements. The factors

$$(\epsilon_k / \epsilon_i) - 1$$

occur in associated pairs:

$$(\epsilon_k/\epsilon_i) - 1 \quad \text{and} \quad (\epsilon_i/\epsilon_k) - 1 = (\bar{\epsilon}_k/\bar{\epsilon}_i) - 1,$$

and by joining them one obtains

THEOREM (7.4.B). *If A is defined in terms of $[U]$ and $\{\epsilon\}$ by (4.1), the volumes of corresponding infinitesimal parts are related by the formula*

$$d\omega_A = (2\pi)^n [d\omega_U] \cdot \Delta\bar{\Delta} d\varphi_1 \cdots d\varphi_n ,$$

where

$$\Delta = \prod_{i<k} (\epsilon_i - \epsilon_k)$$

is the difference product $D(\epsilon_1, \cdots, \epsilon_n)$ of the ϵ_i .

After integrating over the whole $[U(n)]$ and fixing the unit in a convenient manner, one arrives at the following fundamental equation for the density of classes in the $(\varphi_1, \cdots, \varphi_n)$-space:

THEOREM (7.4.C). *The volume of that part of the unitary group whose elements have their angles between the limits φ_k and $\varphi_k + d\varphi_k$ is given by*

$$\Delta\bar{\Delta} \cdot d\varphi_1 \cdots d\varphi_n . \tag{4.4}$$

The formula is not very surprising after all that went before. The singular elements for which $\epsilon_1 = \epsilon_2$ form a manifold of three dimensions less; consequently they are like the center of polar coördinates in three-dimensional space. The formula for the volume element of three-space in terms of polar coördinates contains the factor r^2 which vanishes in second order at the origin. For the same reason the density here must vanish in second order with $\epsilon_1 - \epsilon_2$, i.e.

it must contain the factor $|\epsilon_1 - \epsilon_2|^2$. The same holds for all other pairs ϵ_i, ϵ_k. In the simplest fashion, equation (4.4) takes care of this requirement.

Any *class function* $f(A)$ on the unitary group is a symmetric function $F(\varphi_1, \cdots, \varphi_n)$ of the angles φ_k, periodic with the period 1 in all n arguments. Our result permits this new formulation:

THEOREM (7.4.D). *The mean value of any class function F is given by*

$$\frac{1}{\Omega} \cdot \int_0^1 \cdots \int_0^1 F \cdot \Delta\bar{\Delta}\, d\varphi_1 \cdots d\varphi_n,$$

where

$$\Omega = \int_0^1 \cdots \int_0^1 \Delta\bar{\Delta}\, d\varphi_1 \cdots d\varphi_n.$$

5. Computation of the characters

After these preparations we can accomplish the computation of the characters of the unitary group $U(n)$ in a few strokes. It rests upon three observations.[11] Let us consider any continuous representation of degree N of $U(n)$ and its character χ.

1) As a class function, χ will be *a continuous symmetric periodic function* of the n angles φ_k. In this way the formula (4.1) bears upon our problem.

2) χ is the trace of the matrix representing the diagonal element $\{\epsilon\}$. Hence we may limit ourselves to the group Λ of the diagonal elements, which is a compact Abelian group of very simple structure. Its given representation is equivalent to a unitary one, and hence according to Theorem (7.1.D) breaks up into N unitary representations $E_K(\varphi_1, \cdots, \varphi_n)$ of degree 1 ($K = 1, \cdots, N$), each of which satisfies the functional equations

$$(5.1) \qquad |E(\varphi_1, \cdots, \varphi_n)| = 1,$$

$$(5.2) \qquad E(\varphi_1 + \varphi_1', \cdots, \varphi_n + \varphi_n') = E(\varphi_1, \cdots, \varphi_n)\cdot E(\varphi_1', \cdots, \varphi_n').$$

On putting

$$E(\varphi, 0, \cdots, 0) = f_1(\varphi), \qquad E(0, \varphi, \cdots, 0) = f_2(\varphi), \cdots,$$

one obtains from (5.2):

$$E(\varphi_1, \cdots, \varphi_n) = f_1(\varphi_1) \cdots f_n(\varphi_n).$$

Each of the n functions f of a single variable is a solution of the functional equation (3.13), and hence is of the form $e(m\varphi)$, where m is an integer. Consequently

$$(5.3) \qquad E(\varphi_1, \cdots, \varphi_n) = e(m_1\varphi_1 + \cdots + m_n\varphi_n) = \epsilon_1^{m_1} \cdots \epsilon_n^{m_n}.$$

The trace, the sum of the N quantities E_K, or our character χ, is therefore *a finite Fourier series with non-negative integral coefficients.*

We always arrange the monomials

$$\epsilon_1^{m_1} \cdots \epsilon_n^{m_n} = e(m_1\varphi_1 + \cdots + m_n\varphi_n)$$

in lexicographic order so that the term with exponents $m_1, \cdots, m_n$ precedes that with the exponents $m_1', \cdots, m_n'$ if the first of the non-vanishing differences

$$m_1 - m_1', \cdots, m_n - m_n'$$

is positive.

3) The form (4.4) of the volume element occurring in the orthogonality relations for the primitive characters induces us to associate the function

$$\chi \cdot \Delta = \xi \tag{5.4}$$

with every character χ. This ξ will be an *antisymmetric* periodic function and again a finite Fourier series with integral coefficients. The coefficient of the highest term is the same as in χ and therefore positive.

The simplest antisymmetric periodic functions are the "*elementary sums*"

$$\xi(l_1, \cdots, l_n) = \sum \pm e(l_1\varphi_1 + \cdots + l_n\varphi_n), \tag{5.5}$$

extending alternatingly to all permutations of $\varphi_1, \cdots, \varphi_n$ (or of $l_1, \cdots, l_n$). l_i are integers in the order

$$l_1 > l_2 > \cdots > l_n. \tag{5.6}$$

One may write (5.5) as a determinant

$$| \epsilon^{l_1}, \cdots, \epsilon^{l_n} |$$

of n rows arising from the one written down on replacing ϵ successively by $\epsilon_1, \cdots, \epsilon_n$.

If

$$c \cdot e(l_1\varphi_1 + \cdots + l_n\varphi_n) \qquad (c > 0) \tag{5.7}$$

is the highest term in the antisymmetric ξ, then ξ must contain all terms

$$\pm\, c \cdot e(l_1'\varphi_1 + \cdots + l_n'\varphi_n)$$

in which $l_1', \cdots, l_n'$ is any permutation of $l_1, \cdots, l_n$. Hence the terms with the $-$ sign, in particular those resulting from a transposition, must be actually lower than (5.7), or

$$l_1 > l_2 > \cdots > l_n.$$

After subtracting $c \cdot \xi(l_1, \cdots, l_n)$ from ξ one can apply the same argument to the remainder and thus find an expansion proceeding to lower and lower terms of the kind:

$$\xi = c \cdot \xi(l_1 \cdots l_n) + c' \cdot \xi(l_1' \cdots l_n') + \cdots \tag{5.8}$$

where the coefficients $c, c', \cdots$ are integers and

$$c > 0. \tag{5.9}$$

From

$$\int_0^1 \cdots \int_0^1 e(m_1\varphi_1 + \cdots + m_n\varphi_n) \cdot \bar{e}(m_1'\varphi_1 + \cdots + m_n'\varphi_n)\, d\varphi_1 \cdots d\varphi_n = 1 \text{ or } 0$$

according as

$$m_1 = m_1', \cdots, m_n = m_n'$$

or not, one readily deduces the relation

$$\int_0^1 \cdots \int_0^1 \xi(l_1 \cdots l_n)\bar{\xi}(l_1' \cdots l_n')\, d\varphi_1 \cdots d\varphi_n = n! \text{ or } 0 \tag{5.10}$$

according as

$$l_1 = l_1', \cdots, l_n = l_n'$$

or not. We observe that Δ itself is an elementary sum, namely

$$\Delta = \xi(n - 1, \cdots, 1, 0);$$

hence

$$\Omega = \int_0^1 \cdots \int_0^1 \Delta\bar{\Delta}\, d\varphi_1 \cdots d\varphi_n = n!$$

On applying the "orthogonality relations" (5.10) to the expansion (5.8), we therefore get

$$\mathfrak{M}_s\{\chi(s)\bar{\chi}(s)\} = \frac{1}{\Omega}\int_0^1 \cdots \int_0^1 \chi\bar{\chi}\Delta\bar{\Delta}\, d\varphi_1 \cdots d\varphi_n = c^2 + c'^2 + \cdots.$$

If χ is primitive this mean is to equal 1. Consequently the expansion consists of the first term only and $c = \pm 1$, or, on account of (5.9), $c = 1$.

THEOREM (7.5.A). *Any primitive character of the unitary group is of the form*

$$\frac{|\epsilon^{l_1}, \epsilon^{l_2}, \cdots, \epsilon^{l_n}|}{|\epsilon^{n-1}, \cdots, \epsilon, 1|} \tag{5.11}$$

where $l_1, \cdots, l_n$ are descending integers.

The leading term of this finite Fourier series (5.11) is

$$\epsilon_1^{f_1} \cdots \epsilon_n^{f_n},$$

where

$$f_1 = l_1 - (n - 1), \cdots, f_{n-1} = l_{n-1} - 1, \qquad f_n = l_n - 0 \tag{5.12}$$

and therefore

$$f_1 \geqq f_2 \geqq \cdots \geqq f_n. \tag{5.13}$$

We denote the function (5.11) by $\chi(f_1 \cdots f_n)$.

To find the degree N we have to put $\varphi_1 = \cdots = \varphi_n = 0$. This cannot be done immediately because it would result in the quotient 0/0. Therefore we first set

$$\varphi_1 = (n-1)\varphi, \cdots, \varphi_{n-1} = 1 \cdot \varphi, \qquad \varphi_n = 0 \cdot \varphi,$$

whereby the numerator of (5.11) changes into the difference product of the numbers $e(l_i\varphi)$, and since for small φ in first approximation

$$e(l_i\varphi) - e(l_k\varphi) \sim 2\pi\sqrt{-1} \cdot (l_i - l_k)\varphi,$$

one obtains

$$N = N(f_1 \cdots f_n) = \frac{D(l_1, \cdots, l_{n-1}, l_n)}{D(n-1, \cdots, 1, 0)}. \tag{5.14}$$

So much for the transcendental part. In Chapter IV, §5, we bored the tunnel from the other side, showing that the algebraically constructed irreducible representation of signature $(f_1, \cdots, f_n)$—which is $\langle \mathrm{P}(f_1 \cdots f_n) \rangle$ if $f_n \geqq 0$—has as its character χ a polynomial

$$\sum k_{m_1 \cdots m_n} \epsilon_1^{m_1} \cdots \epsilon_n^{m_n}$$

with non-negative integral coefficients k, the highest term of which is

$$1 \cdot \epsilon_1^{f_1} \cdots \epsilon_n^{f_n}.$$

Hence χ can be nothing else than our $\chi(f_1 \cdots f_n)$.

THEOREM (7.5.B). *The irreducible representation of the unitary group of signature* $(f_1, \cdots, f_n)$ *has the character*

$$\chi(f_1 \cdots f_n) = \frac{|\epsilon^{l_1}, \cdots, \epsilon^{l_n}|}{|\epsilon^{n-1}, \cdots, \epsilon^0|} \tag{5.15}$$

and its degree is

$$N(f_1 \cdots f_n) = D(l_1 \cdots l_n)/D(n-1, \cdots, 0), \tag{5.14}$$

where

$$l_1 = f_1 + (n-1), \cdots, l_{n-1} = f_{n-1} + 1, \qquad l_n = f_n + 0. \tag{5.16}$$

Theorem (7.5.A) then allows us to conclude:

THEOREM (7.5.C). *There are no other continuous irreducible representations of the unitary group beside those of signature*

$$\boxed{(f_1, \cdots, f_n), \qquad f_1 \geqq \cdots \geqq f_n.}$$

What we called quantics are the only primitive quantities for the unitary group.

6. The characters of $GL(n)$. Enumeration of covariants

From now on we make the harmless restriction $f_n \geqq 0$ so that the representation of signature $(f_1, \cdots, f_n)$ is $\langle \mathrm{P}(f_1 \cdots f_n) \rangle$.

$\chi(f_1 \cdots f_n)$ is a symmetric polynomial of the variables $\epsilon_1, \cdots, \epsilon_n$ and must hence be expressible in terms of the elementary symmetric functions, i.e., of the coefficients of the characteristic polynomial

$$\varphi(z) = \prod_i (1 - z\epsilon_i) = \det(E - zA) = 1 - q_1 z + \cdots \pm q_n z^n.$$

The explicit computation rests on Cauchy's

LEMMA (7.6.A).

$$\det\left(\frac{1}{1 - x_i y_k}\right) = \frac{D(x_1 \cdots x_n)\, D(y_1 \cdots y_n)}{\prod (1 - x_i y_k)} \qquad (i, k = 1, \cdots, n).$$

Proof by induction with respect to n. Subtract the first row from the second, $\cdots$, n^{th} row of the determinant on the left side:

$$\frac{1}{1 - x_2 y_k} - \frac{1}{1 - x_1 y_k} = \frac{x_2 - x_1}{1 - x_1 y_k} \cdot \frac{y_k}{1 - x_2 y_k}.$$

One obtains

$$(6.1) \qquad \frac{(x_2 - x_1) \cdots (x_n - x_1)}{\prod_{k=1}^{n} (1 - x_1 y_k)} \begin{vmatrix} 1, & 1, & \cdots, & 1 \\ \dfrac{y_1}{1 - x_2 y_1}, & \dfrac{y_2}{1 - x_2 y_2}, & \cdots, & \dfrac{y_n}{1 - x_2 y_n} \\ \cdots & \cdots & \cdots & \cdots \end{vmatrix}.$$

Subtract the first column from the second, $\cdots$, n^{th}: this changes the first row into $|\, 1, 0, \cdots, 0 \,|$ while the fate of the others may be read from the equation

$$\frac{y_k}{1 - x_2 y_k} - \frac{y_1}{1 - x_2 y_1} = \frac{y_k - y_1}{1 - x_2 y_1} \cdot \frac{1}{1 - x_2 y_k}.$$

Therefore the determinant in (6.1) changes into

$$\frac{(y_2 - y_1) \cdots (y_n - y_1)}{(1 - x_2 y_1) \cdots (1 - x_n y_1)} \begin{vmatrix} 1 & 0 & \cdots & 0 \\ * & \dfrac{1}{1 - x_2 y_2} & \cdots & \dfrac{1}{1 - x_2 y_n} \\ \cdots & \cdots & \cdots & \cdots \\ * & \dfrac{1}{1 - x_n y_2} & \cdots & \dfrac{1}{1 - x_n y_n} \end{vmatrix},$$

thus giving the desired result:

$$\det\left(\frac{1}{1 - x_i y_k}\right)_{i,\,k=1,\cdots,n} =$$

$$\frac{(x_2 - x_1) \cdots (x_n - x_1) \cdot (y_2 - y_1) \cdots (y_n - y_1)}{\prod_{(i \text{ or } k=1)} (1 - x_i y_k)} \cdot \det\left(\frac{1}{1 - x_i y_k}\right)_{i,\,k=2,\cdots,n}$$

We take $x_i = \epsilon_i$ and find

$$(6.2) \qquad \frac{D(\epsilon_1 \cdots \epsilon_n) \cdot D(z_1 \cdots z_n)}{\varphi(z_1) \cdots \varphi(z_n)} = \left| \frac{1}{1 - \epsilon_i z_k} \right|.$$

By developing

$$1/(1 - \epsilon z) = 1 + \epsilon z + \epsilon^2 z^2 + \cdots$$

one sees that the expansion of the right side in powers of the z_k contains the monomial

$$(6.3) \qquad z_1^{l_1} \cdots z_n^{l_n}$$

multiplied by the coefficient

$$| \epsilon^{l_1}, \cdots, \epsilon^{l_n} |.$$

Thus we get for the right side the sum

$$\sum | \epsilon^{l_1}, \cdots, \epsilon^{l_n} | \cdot | z^{l_1}, \cdots, z^{l_n} |$$

extending over all non-negative integers l satisfying the inequalities (5.6). $\chi(f_1 \cdots f_n)$ is therefore the coefficient of (6.3) in the expansion of

$$\frac{| z^{n-1}, \cdots, z, 1 |}{\varphi(z_1) \cdots \varphi(z_n)}.$$

By introducing as before the polynomials $p_0, p_1, \cdots$ by

$$(6.4) \qquad 1/\varphi(z) = 1/| E - zA | = p_0 + p_1 z + p_2 z^2 + \cdots \qquad (p_{-1} = p_{-2} = \cdots = 0)$$

one arrives at the formula[12]

$$(6.5) \qquad \chi(f_1 \cdots f_n) = | p_{l-(n-1)}, \cdots, p_{l-1}, p_l |.$$

The determinant on the right side is to be interpreted as consisting of n rows arising from the one written down in replacing l successively by $l_1, l_2, \cdots, l_n$.

At this juncture we can *return from the unitary to the full linear group.* The character $\chi(f_1 \cdots f_n)$ of the representation of $GL(n)$ which springs from the diagram $T(f_1 \cdots f_n)$ is obviously a polynomial in the n^2 components a_{ik} of the generic element $A = \| a_{ik} \|$ of the group. So is the right side of the equation (6.5). According to Lemma (7.1.A) the equation therefore holds good for all A; and it lies in the nature of the formula to be valid in any field of characteristic 0.

THEOREM (7.6.B). *The character $\chi(f_1 \cdots f_n)$ of the representation $\langle P(f_1 \cdots f_n) \rangle$ of $GL(n)$ arising from the partition $f = f_1 + \cdots + f_n$ is given by the formula*[13]

$$(6.5) \qquad \chi(f_1 \cdots f_n) = | p_{l-(n-1)}, \cdots, p_l |,$$

where l in the several rows of the determinant is to be replaced by

$$l_1 = f_1 + (n - 1), \cdots, \qquad l_{n-1} = f_{n-1} + 1, \qquad l_n = f_n,$$

and the p_f are the coefficients of the Taylor expansion:

$$1/|E - zA| = p_0 + p_1 z + p_2 z^2 + \cdots.$$

This result is only slightly less simple than the formula (2.16) for the character of the representation $\langle\Sigma(f_1 f_2 \cdots f_r)\rangle$. We proceed next to the decomposition of the latter, in which the number r is arbitrary.

Let $f_1, \cdots, f_r$ be any integers satisfying

$$f_1 \geqq \cdots \geqq f_r \geqq 0, \qquad f_1 + \cdots + f_r = f, \tag{6.6}$$

and let the l's be defined by

$$l_1 = f_1 + (r - 1), \cdots, l_{r-1} = f_{r-1} + 1, \qquad l_r = f_r + 0.$$

The equation (6.5) prompts us to study the determinant

$$\chi(f_1, \cdots, f_r) = | p_{l-(r-1)}, \cdots, p_{l-1}, p_l | \tag{6.7}$$

of signature $(f_1, \cdots, f_r)$. In considering the case $r > n$ we want to reduce the length r of our symbols χ to n. First, when $f_{r+1} = 0$, we have the obvious and simple reduction:

$$\chi(f_1 \cdots f_r\, 0) = \chi(f_1 \cdots f_r).$$

If, however, $f_{r+1} > 0$ then I maintain that

$$\chi(f_1 \cdots f_{r+1}) = 0.$$

Indeed, let us write $\varphi(z)$ as a polynomial of formal degree r:

$$\varphi(z) = c_0 + c_1 z + \cdots + c_r z^r \qquad (c_0 = 1). \tag{6.8}$$

(Of course, with the notation (2.5), we have $c_i = (-1)^i q_i$ for $i = 0, 1, \cdots, n$ and $c_i = 0$ for $i > n$.) The recursive relations

$$c_0 p_l + c_1 p_{l-1} + \cdots + c_r p_{l-r} = \begin{cases} 0 & \text{for } l > 0 \\ 1 & \text{for } l = 0 \end{cases} \tag{6.9}$$

are just another form of the defining equation

$$\sum_{l=0}^{\infty} p_l z^l \cdot \varphi(z) = 1.$$

By taking $l = l_1, \cdots, l_{r+1}$ in (6.9) we obtain $r + 1$ homogeneous linear equations with a non-vanishing solution $c_0, c_1, \cdots, c_r$; hence their determinant $\chi(f_1 \cdots f_{r+1})$ must be zero.

All the symbols of length n are linearly independent, as is borne out by our expression (5.15) in terms of a diagonal matrix A:

$$a_{ik} = 0 \text{ for } i \neq k, \qquad a_{ii} = \epsilon_i.$$

The ϵ_i are here no longer subject to the condition $|\epsilon_i| = 1$, meaningless in an arbitrary field, but are rather to be considered as independent variables.

RULE OF REDUCTION (7.6.C). *For* $r \geqq n$ *we have*

$$\chi(f_1 \cdots f_r f_{r+1}) = 0 \qquad \textit{if } f_{r+1} > 0,$$

$$\chi(f_1 \cdots f_r\, 0) = \chi(f_1 \cdots f_r) \qquad \textit{if } f_{r+1} = 0.$$

The symbols $\chi(f_1 \cdots f_n)$ *of length* n *are linearly independent.*

We now engage in a purely combinatorial argument in which $p_0, p_1, \cdots$ are looked upon as independent quantities. The signatures $(f_1, \cdots, f_r)$ of given rank f,

$$f_1 + \cdots + f_r = f, \qquad f_1 \geqq f_2 \geqq \cdots \geqq f_r \geqq 0,$$

may be arranged in lexicographic order starting with the highest one $(f, 0, \cdots, 0)$. The determinant $\chi(f_1 \cdots f_r)$ is an aggregate of $r!$ terms

$$p_{f_1'} \cdots p_{f_r'} = \psi(f_1' \cdots f_r') \qquad (f_1' \geqq \cdots \geqq f_r' \geqq 0) \tag{6.10}$$

whose signatures $(f_1', \cdots, f_r')$ have the same rank f. The leading term is $\psi(f_1 \cdots f_r)$, and I maintain that all other terms stand higher. Hence the linear relations linking the $\chi(f_1 \cdots f_r)$ to the $\psi(f_1 \cdots f_r)$ of a given rank f are of the arithmetic recursive type. Equations between two sets of variables x_α, y_α with indices forming an ordered set,

$$y_\alpha = \sum k_{\beta\alpha} x_\beta,$$

are of this type if the integral coefficients $k_{\alpha\beta}$ satisfy the conditions

$$k_{\alpha\alpha} = 1, \qquad k_{\beta\alpha} = 0 \qquad \text{for } \beta < \alpha.$$

Such a substitution has an inverse of the same type. Therefore:

THEOREM (7.6.D). *On considering the* p_f *as independent variables, the products* $p_{f_1} \cdots p_{f_r}$ *of a preassigned total rank* $f_1 + \cdots + f_r = f$ *arise from the determinants* $\chi(f_1 \cdots f_r)$ *by a linear substitution of the arithmetic recursive type:*

$$p_{f_1} \cdots p_{f_r} = \sum \mu \cdot \chi(f_1' \cdots f_r'). \tag{6.11}$$

Our statement concerning the recursive character is not quite as trivial as it may seem. By expanding the determinant (6.7) and writing the factors in each term (6.10) in the order in which they come from the successive rows, it is clear indeed that $(f_1' \cdots f_r')$ is higher than $(f_1 \cdots f_r)$, i.e. that the first non-vanishing difference $f_i' - f_i$ $(i = 1, \cdots, r)$ is positive. However, the numbers $f_1', \cdots, f_r'$ might not occur in their proper order $f_1' \geqq \cdots \geqq f_r'$. But when we rearrange them in proper order: $g_1 \geqq g_2 \geqq \cdots \geqq g_r$, one readily sees that $(g_1, g_2, \cdots, g_r)$ cannot be lower than $(f_1', f_2', \cdots, f_r')$.

We now express the p_f again by (6.4) in terms of an arbitrary n-rowed matrix $A = \| a_{ik} \|$. If $n = r$ the formula (6.11) exhibits the decomposition of $\langle \Sigma(f_1 \cdots f_r) \rangle$ into its irreducible parts $\langle \mathrm{P}(f_1' \cdots f_r') \rangle$; hence the coefficients μ

in (6.11) must be $\geqq 0$. (6.11) has the same significance for $n < r$ after our rule of reduction has been applied:

THEOREM (7.6.E). *In the recursive formula* (6.11) *the coefficients* μ *are non-negative. The formula exhibits the decomposition of* $\langle\Sigma(f_1 \cdots f_r)\rangle$ *into its irreducible parts, immediately when* $n = r$, *and after due reduction according to Rule* (7.6.C) *when* $n < r$.

Hence the decomposition is *cum grano salis* independent of the dimensionality n, and we can compute the multiplicities μ with which each constituent appears by a simple combinatory device, namely by the formulas (6.11) resulting from the inversion of the equations (6.7) which define $\chi(f_1 \cdots f_r)$ in terms of products $p_{f_1'} \cdots p_{f_r'}$.

The fact that $\langle\Sigma(e_1 \cdots e_r)\rangle$ contains $\mu\begin{pmatrix} e_1 \cdots e_r \\ f_1 \cdots f_n \end{pmatrix}$ times the irreducible component $\langle\mathrm{P}(f_1 \cdots f_n)\rangle$ as expressed by our formula

$$p_{e_1} \cdots p_{e_r} = \sum_f \mu\begin{pmatrix} e_1 \cdots e_r \\ f_1 \cdots f_n \end{pmatrix} \cdot \chi(f_1 \cdots f_n) \tag{6.12}$$

means at the same time the existence of as many linearly independent covariant quantities of the type $\langle\mathrm{P}(f_1 \cdots f_n)\rangle$ which depend in the degrees $e_1, \cdots, e_r$ on r argument vectors. We may arrange the multiplicities μ in an (infinite) matrix where $(e_1 \cdots e_r)$ indicates the rows, $(f_1 \cdots f_n)$ the columns. When for a given type $\langle\mathrm{P}(f_1 \cdots f_n)\rangle$ we wish to ascertain the numbers of covariant quantities simultaneously for all possible degrees $(e_1, \cdots, e_r)$, we must compute the $(f_1 \cdots f_n)$-column of the μ-matrix while decomposition of $\langle\Sigma(e_1 \cdots e_r)\rangle$ requires the knowledge of its $(e_1 \cdots e_r)$-row. In a more explicit way than by our combinatorial device both problems may be solved through *generating functions.* Let us start with the first question and therefore ask for the "column-wise generating function"

$$\Phi_{f_1 \cdots f_n}(z_1 \cdots z_r) = \sum_{(e)} \mu\begin{pmatrix} e_1 \cdots e_r \\ f_1 \cdots f_n \end{pmatrix} z_1^{e_1} \cdots z_r^{e_r}, \tag{6.13}$$

a formal power series of r auxiliary variables $z_1, \cdots, z_r$.

We consider, for $r \geqq n$, the function

$$H(z_1 \cdots z_r) = D(z_1 \cdots z_r)/\varphi(z_1) \cdots \varphi(z_r),$$

and first develop a simple recursive formula for the transition $r \to r + 1$. In the determinant

$$| z^r, z^{r-1}, \cdots, 1 |$$

forming the numerator of $H(z_1 \cdots z_{r+1})$ we multiply the first r columns by the coefficients $c_r, \cdots, c_1$ introduced in (6.8) and add them to the last column. After the last column has thus changed into $\varphi(z)$ we expand in terms of that column and obtain (for $r \geqq n$):

$$H(z_1 \cdots z_{r+1}) = H(z_1 \cdots z_r) \cdot (z_1 \cdots z_r) - + \cdots . \tag{6.14}$$

The sum contains $r + 1$ alternating terms in which $z_{r+1}, z_r, \cdots, z_1$ are successively absent.

Next we observe that for $r = n$

$$1/\varphi(z_1) \cdots \varphi(z_n) = \sum_l p_{l_1} \cdots p_{l_n} z_1^{l_1} \cdots z_n^{l_n};$$

hence

$$\begin{aligned} H(z_1 \cdots z_n) &= \sum |p_{l-(n-1)}, \cdots, p_l| z_1^{l_1} \cdots z_n^{l_n} \\ &= \sum_{(l_1 > \cdots > l_n \geq 0)} |p_{l-(n-1)}, \cdots, p_l| \cdot |z^{l_1}, \cdots z^{l_n}| \\ &= \sum_{(f_1 \geq \cdots \geq f_n \geq 0)} |z^{f_1+(n-1)}, \cdots, z^{f_n+0}| \cdot \chi(f_1 \cdots f_n). \end{aligned}$$

We therefore find an expansion of $H(z_1 \cdots z_r)$ in terms of the characters $\chi(f_1 \cdots f_n)$:

$$H(z_1 \cdots z_r) = \sum_{(f)} L_{f_1 \cdots f_n}(z) \cdot \chi(f_1 \cdots f_n) \tag{6.15}$$

where the L's for each signature $(f_1 \cdots f_n)$ obey the same recursion rule (6.14) as H and for $r = n$ reduce to:

$$L_{f_1 \cdots f_n}(z_1 \cdots z_n) = |z^{f_1+(n-1)}, \cdots, z^{f_n+0}|.$$

One readily verifies that the expressions

$$L_{f_1 \cdots f_n}(z_1 \cdots z_r) = |z^{f_1+(r-1)}, \cdots, z^{f_n+(r-n)}\ z^{r-n-1}, \cdots, z, 1| \tag{6.16}$$

fulfil both conditions and consequently they are the sought-for coefficients in (6.15). The generating functions Φ are by definition the coefficients, not depending on A, in the expansion

$$1/\varphi(z_1) \cdots \varphi(z_r) = \sum_{(f)} \Phi_{f_1 \cdots f_n}(z) \cdot \chi(f_1 \cdots f_n),$$

and therefore come out as L/D. Since L is skew-symmetric in its r arguments $z_1 \cdots z_r$, the quotient L/D is a polynomial, clearly of degree $f_1 + \cdots + f_n$.

THEOREM (7.6.F). *The generating function Φ, (6.13), giving the numbers μ of linearly independent covariants of type $(f_1 \cdots f_n)$ depending in all possible degrees $(e_1 \cdots e_r)$ on a given number r of argument vectors, is the quotient of the antisymmetric polynomial* (6.16) *by the difference product $D(z_1 \cdots z_r)$.*

We have thus arrived at a very substantial, neat formula. In particular, on putting $f_1 = \cdots = f_n = g$ it yields the *numbers of linearly independent vector invariants of a given weight g* and of any degrees $e_1, \cdots, e_r$. The generating function for invariants of weight g *depending on n vectors* $x^1, \cdots, x^n$ turns out to be

$$(z_1 \cdots z_n)^g.$$

In other words: there is no such invariant unless the degrees are $e_1 = \cdots = e_r = g$, and then there is exactly *one* (namely the gth power of the bracket factor

$[x^1 \cdots x^n]$). In this way our formula yields a proof of the first invariant-theoretic main theorem for n vectors from which Capelli's general identity allows inference of the general case. According to that theorem one obtains a linear basis for the invariants of r vectors of given degrees $e_1, \cdots, e_r$ by forming all possible "monomials," i.e. all products of bracket factors in which the arguments $x^1, \cdots, x^r$ occur $e_1, \cdots, e_r$ times respectively. However, there will in general prevail linear relations among these monomials of which the second main theorem takes care. Nevertheless it does not enable one to predict just how many of the monomials, in view of those relations, will stay linearly independent. Thus the two main theorems on the one side, and our present formula on the other side, are, so to speak, the known ends of a chain, the intermediate links of which remain in the dark.[14]

The column-wise generating function was obtained from the expressions (6.5) of the characters. The row-wise generator is at once gathered from (6.12) by using the expressions (5.15). In considering $\epsilon_1, \cdots, \epsilon_n$ as n auxiliary variables, we write for clarity's sake $p_f(\epsilon)$ for the symmetric functions of the ϵ defined by

$$1/\prod_i (1 - \epsilon_i z) = p_0(\epsilon) + p_1(\epsilon)z + \cdots.$$

We then have

$$D(\epsilon_1, \cdots, \epsilon_n) \cdot p_{e_1}(\epsilon) \cdots p_{e_r}(\epsilon) = \sum_f \mu\begin{pmatrix} e_1 \cdots e_r \\ f_1 \cdots f_n \end{pmatrix} \cdot | \epsilon^{f_1+(n-1)}, \cdots, \epsilon^{f_n} |,$$

and in this sense the polynomial of the indeterminates ϵ on the left side is indeed the sought-for row-wise generating function: the μ put in evidence is the coefficient of

$$\epsilon_1^{l_1} \cdots \epsilon_n^{l_n} \qquad (l_1 = f_1 + (n-1), \cdots, l_n = f_n + 0)$$

in that skew-symmetric polynomial.

For the following cases, the symplectic and the orthogonal group, we shall treat only the column-wise generators, omitting the row-wise arrangements as well as the combinatorial procedure (with the p_f as indeterminates and a subsequent "rule of reduction").

Our results apply almost without modification to the group $SL(n)$. The limitation to unimodular transformations has the effect that the character (5.15) depends merely on the differences of the f_i, i.e., two signatures like

$$(f_1, \cdots, f_n) \quad \text{and} \quad (f_1 + e, \cdots, f_n + e)$$

are to be considered the same.

7. A purely algebraic approach

Our final result (6.5) is so simple that it should be possible to reach the goal by a shorter road. Here is one that follows closely an ingenious paper by G. Frobenius.[15]

Considering the complete reciprocity between $GL(n)$ and the symmetric group π_f, we try to determine the characters of the symmetric group, which we denote by the same letters as for $GL(n)$. If distinction be needed, we add the index L for $GL(n)$ and π for π_f.

As in Chapter IV, let

$$a = \hat{a} = \sum p \tag{7.1}$$

be the operator of symmetrization belonging to the partition $f = f_1 + \cdots + f_n$. The invariant subspace of all symmetry quantities of form xa may be designated by $\pi(f_1 \cdots f_n)$, the corresponding representation of π_f by $\langle\pi(f_1 \cdots f_n)\rangle$. From formula (3.7.14) we see that its character equals

$$\frac{1}{f_1!\,f_2!\cdots} \sum_t a(t^{-1}st) = \psi(f_1 \cdots f_n; s),$$

or the number of elements t for which

$$t^{-1}st \text{ is a } p \tag{7.2}$$

divided by $f_1!\,f_2!\cdots$. The permutation s, when decomposed into disjoint cycles, may consist of α_1 cycles of length 1, α_2 cycles of length 2, . . .:

$$1\alpha_1 + 2\alpha_2 + \cdots = f.$$

The numbers $\alpha_1, \alpha_2, \cdots$ characterize the class $\mathfrak{f}$ of the element; we therefore write for any class function $\psi(s)$:

$$\psi(s) = \psi(\mathfrak{f}) = \psi(\alpha_1, \alpha_2, \cdots).$$

The equation (7.2) will have no solution t unless we can compose lines of lengths $f_1, \cdots, f_n$ out of the cycles of s. Let α_{i1} cycles of length 1, α_{i2} cycles of length 2, $\cdots$, contribute to the line of length f_i. We then must have

$$1\alpha_{i1} + 2\alpha_{i2} + \cdots = f_i \quad \text{and} \tag{7.3}$$

$$\sum_i \alpha_{i1} = \alpha_1, \qquad \sum_i \alpha_{i2} = \alpha_2, \quad \cdots. \tag{7.4}$$

The α_1 cycles of length 1 can be split in

$$\alpha_1!/\alpha_{11}!\,\alpha_{21}!\cdots\alpha_{n1}!$$

different manners into groups of $\alpha_{11}, \alpha_{21}, \cdots$ cycles respectively; the same for the other lengths. The line of f_i figures shall be made up by first writing down the α_{i1} cycles of length 1 in their natural or some fixed order, then the cycles of length 2 in the same manner, etc. t will be a solution of (7.2) if by the substitution t the figures in the first row of length f_1 change into the figures $1, 2, \cdots, f_1$ *in any of their orders*, the figures in the second row change into any arrangement of $f_1 + 1, \cdots, f_1 + f_2$, etc. Hence the number of solutions t of (7.2) equals $f_1!\,f_2!\cdots$ times

$$\sum \frac{\alpha_1!}{\alpha_{11}!\,\alpha_{21}!\cdots} \cdot \frac{\alpha_2!}{\alpha_{12}!\,\alpha_{22}!\cdots} \cdots, \tag{7.5}$$

the sum extending to all non-negative solutions $\alpha_{i1}, \alpha_{i2}, \cdots$ of the simultaneous equations (7.3), (7.4). The sum (7.5) itself is the character $\psi(f_1 \cdots f_n; \alpha_1, \alpha_2, \cdots)$. By introducing n indeterminates $\epsilon_1, \cdots, \epsilon_n$ we condense our combinatorial result into the more manageable formula

$$\sigma(\mathfrak{f}) = \sigma_1^{\alpha_1}\sigma_2^{\alpha_2} \cdots = \sum_{(f)} \psi(f_1 \cdots f_n; \alpha_1\alpha_2 \cdots)\epsilon_1^{f_1} \cdots \epsilon_n^{f_n}, \tag{7.6}$$

where $\sigma_1, \sigma_2, \cdots$ are the power sums of the ϵ's:

$$\sigma_r = \epsilon_1^r + \cdots + \epsilon_n^r.$$

What is of chief importance to us is the fact that the coefficients of the expansions of the power products $\sigma(\mathfrak{f})$ in terms of the variables ϵ are *characters*.

We now form

$$\sigma(\mathfrak{f})\cdot|\epsilon^{n-1}, \cdots, \epsilon, 1| = \sum_{(f_1 \geqq \cdots \geqq f_n)} \omega_{f_1\cdots f_n}(\mathfrak{f})\cdot|\epsilon^{l_1}, \cdots, \epsilon^{l_n}| \tag{7.7}$$

where

$$\omega_{f_1\cdots f_n} = \sum \pm \psi(l_1 - r_1, \cdots, l_n - r_n), \tag{7.8}$$

the sum extending alternatingly to all permutations $r_1, \cdots, r_n$ of $n-1, \cdots 1, 0$.

$$\omega_{f_1\cdots f_n}(\mathfrak{f}) = \omega(s)$$

is, like the ψ's, a linear combination of primitive characters $\chi, \chi', \cdots$ with integral coefficients:

$$\omega(s) = m\cdot\chi(s) + m'\cdot\chi'(s) + \cdots; \tag{7.9}$$

but some of the coefficients might be negative, which would prevent ω itself from being a character. Our next step is to show by direct calculation that the ω's satisfy the same orthogonality relations as the primitive characters χ. This is done by Cauchy's lemma.

We use a second set of variables $z_1, \cdots, z_n$ with the power sums $\tau_1, \tau_2, \cdots$ and start with Cauchy's relation (certainly valid for $|\epsilon_i| < 1, |z_i| < 1$)

$$\sum_{(l_1 > \cdots > l_r \geqq 0)} \frac{|\epsilon^{l_1}, \cdots, \epsilon^{l_n}|}{|\epsilon^{n-1}, \cdots 1|}\cdot\frac{|z^{l_1}, \cdots, z^{l_n}|}{|z^{n-1}, \cdots, 1|} = \frac{1}{\prod_{i,k}(1 - \epsilon_i z_k)}. \tag{7.10}$$

The logarithm of

$$1/\prod_i (1 - \epsilon_i z) \text{ equals } \sum_i \sum_{r=1}^{\infty} \frac{\epsilon_i^r z^r}{r} = \frac{\sigma_1 z}{1} + \frac{\sigma_2 z^2}{2} + \cdots; \tag{7.11}$$

therefore the logarithm of the right side of (7.10) is

$$\frac{\sigma_1\tau_1}{1} + \frac{\sigma_2\tau_2}{2} + \cdots.$$

Hence the terms of degree $f_1 + \cdots + f_n = f$ in the product itself are the terms of that degree in the expansion of

$$\exp\left(\frac{\sigma_1\tau_1}{1} + \frac{\sigma_2\tau_2}{2} + \cdots\right),$$

or the sum

$$\sum \frac{\sigma_1^{\alpha_1}\sigma_2^{\alpha_2}\cdots\cdot\tau_1^{\alpha_1}\tau_2^{\alpha_2}\cdots}{1^{\alpha_1}2^{\alpha_2}\cdots\alpha_1!\,\alpha_2!\cdots}$$

extending over all $\alpha_1, \alpha_2, \cdots$ satisfying

$$1\alpha_1 + 2\alpha_2 + \cdots = f.$$

As one readily sees,

$$f!/1^{\alpha_1}2^{\alpha_2}\cdots\alpha_1!\,\alpha_2!\cdots$$

is the number $n(\mathfrak{k})$ of group elements in the class $\mathfrak{k}$. Thus we find this form of Cauchy's relation:

$$\sum \frac{|\epsilon^{l_1}, \cdots, \epsilon^{l_n}|}{|\epsilon^{n-1}, \cdots, 1|}\cdot\frac{|z^{l_1}, \cdots, z^{l_n}|}{|z^{n-1}, \cdots, 1|} = \frac{1}{f!}\sum_{\mathfrak{k}} n(\mathfrak{k})\sigma(\mathfrak{k})\tau(\mathfrak{k}), \tag{7.12}$$

where the sum on the left ranges over

$$f_1 \geqq f_2 \geqq \cdots \geqq f_n \geqq 0, \qquad f_1 + \cdots + f_n = f,$$

on the right over all classes $\mathfrak{k}$ of π_f.

On the other hand, if one multiplies the equation (7.7) for ϵ with the corresponding one for the z's, one finds for the right side of (7.12) an expansion in terms of the products

$$\frac{|\epsilon^{l_1}, \cdots, \epsilon^{l_n}|}{|\epsilon^{n-1}, \cdots, 1|}\cdot\frac{|z^{l'_1}\cdots, z^{l'_n}|}{|z^{n-1}, \cdots, 1|},$$

where this product bears the coefficient

$$\frac{1}{f!}\sum_{\mathfrak{k}} n(\mathfrak{k})\omega_{f_1\cdots f_n}(\mathfrak{k})\omega_{f'_1\cdots f'_n}(\mathfrak{k}) = \mathfrak{M}_s\{\omega(s)\omega'(s)\}.$$

Hence

$$\mathfrak{M}_s\{\omega(s)\omega'(s)\} = 1 \text{ or } 0 \tag{7.13}$$

according as ω and ω' are associated with the same or different signatures $(f_1, \cdots, f_n)$.

$$\mathfrak{M}_s\{\omega^2(s)\} = 1,$$

together with (7.9) and the orthogonality relations for the primitive characters $\chi(s), \chi'(s), \cdots$, leads to the result

$$m^2 + m'^2 + \cdots = 1$$

which shows that either $+\omega(s)$ or $-\omega(s)$ must be a primitive character. Moreover one infers from (7.13) that the $\pm\omega_{f_1\cdots f_n}$ for different signatures $(f_1 \cdots f_n)$ are characters of *inequivalent* irreducible representations.

There remain two things to prove:

1) The $+$ sign prevails for each signature;

2) $\omega_{f_1\cdots f_n}$ is the character of the representation $\langle\rho(f_1 \cdots f_n)\rangle$ corresponding to the partition $f = f_1 + \cdots + f_n$.

1) The inverse of the equations (7.8) reads:

$$\psi(f_1 \cdots f_n) = \sum \mu \cdot \omega_{f'_1\cdots f'_n} = \omega_{f_1\cdots f_n} + \cdots \tag{7.14}$$

with the same coefficients μ as in Theorem (7.6.D). The dots stand here for a linear combination of terms $\omega_{f'_1\cdots f'_n}$ of higher rank $(f'_1 \cdots f'_n)$ than the leading term $\omega_{f_1\cdots f_n}$. This equation exhibits the decomposition of $\langle\pi(f_1 \cdots f_n)\rangle$ into its irreducible components. The leading term $\omega_{f_1\cdots f_n}$ makes it impossible that $-\omega_{f_1\cdots f_n}$ should be the character because in terms of the *characters* the coefficients in (7.14) cannot be negative. Thus we conclude two things: the character is $+\omega_{f_1\cdots f_n}$ and the coefficients μ are $\geqq 0$.

2) In Chapter IV, §3 we incidentally observed, equation (4.3.4), that

$$c'xa = 0 \tag{7.15}$$

if the diagram T' stands lower than T. If we write this in the form

$$ax\hat{c}' = 0,$$

it means that $\langle\pi(f_1 \cdots f_n)\rangle$ contains no irreducible $\langle\rho(f'_1 \cdots f'_n)\rangle$ of lower rank than $(f_1 \cdots f_n)$; but it certainly contains $\langle\rho(f_1 \cdots f_n)\rangle$ since, in contrast to (7.15),

$$ca = f_1!\, f_2! \cdots c \neq 0.$$

Let us now assume that our statement about $\omega_{f_1\cdots f_n}$ being the character of $\langle\rho(f_1 \cdots f_n)\rangle$ had been proved for all partitions higher than the one $f_1 + \cdots + f_n$ under consideration. Then the formula (7.14) shows that $\omega_{f_1\cdots f_n}$ must correspond to the one $\langle\rho\rangle$ that, besides those of higher rank than $(f_1 \cdots f_n)$, is certainly part of $\langle\pi(f_1 \cdots f_n)\rangle$, namely to $\langle\rho(f_1 \cdots f_n)\rangle$.

The representation of $GL(n)$ which corresponds to the representation $\langle\pi(f_1, \cdots, f_n)\rangle$ of π_f is that induced in the space of all tensors of rank f of the form aF, so that the character of $GL(n)$ corresponding to $\psi_\pi(f_1, \cdots, f_n)$ is given by (2.16). Hence if we shift to the left side of (7.8) all terms occurring with a $-$ sign on the right, pass to the corresponding equation for the characters of $GL(n)$, and then shift these terms back, we find that the character of the representation $\langle \mathrm{P}(f_1, \cdots, f_n)\rangle$ is

$$\sum \pm\, p_{l_1-r_1} \cdots p_{l_n-r_n},$$

or the determinant (6.5). We thus arrive at our former result. Besides this we have won the following simple *formula for the computation of the primitive characters* $\chi(f_1 \cdots f_n)$ *of the symmetric group*:

THEOREM (7.7.A).

$$\sigma(\mathfrak{k})\cdot|\epsilon^{n-1}, \cdots, 1| = \sum \chi(f_1 \cdots f_n; \mathfrak{k})\,|\epsilon^{l_1}, \cdots, \epsilon^{l_n}|, \tag{7.16}$$

the sum on the right side ranging over

$$f_1 \geqq \cdots \geqq f_n \geqq 0, \qquad f_1 + \cdots + f_n = f.$$

No more powerful instrument could be devised for that purpose; it yields $\chi(f_1 \cdots f_n\,;\,\mathfrak{k})$ as the coefficient of

$$\epsilon_1^{l_1} \cdots \epsilon_n^{l_n} \qquad (l_i = f_i + (n - i); \qquad l_1 > \cdots > l_n \geqq 0) \tag{7.17}$$

in the expansion of

$$\sigma(\mathfrak{k})\cdot|\,\epsilon^{n-1}, \cdots, 1\,|.$$

We give two easily obtainable consequences:

1) The degree $g = g(f_1 \cdots f_n)$ of $\langle\rho(f_1 \cdots f_n)\rangle$ is the coefficient of $\epsilon_1^{l_1} \cdots \epsilon_n^{l_n}$ in the expansion of

$$(\epsilon_1 + \cdots + \epsilon_n)^f\cdot|\,\epsilon^{n-1}, \cdots, 1\,|.$$

If we take the term

$$\pm\epsilon_1^{r_1} \cdots \epsilon_n^{r_n}$$

of the second factor, $r_1, \cdots, r_n$ being a permutation of $n - 1, \cdots, 0$, we must select the term

$$\frac{f!}{(l_1 - r_1)!(l_2 - r_2)!\cdots}\,\epsilon_1^{l_1 - r_1}\epsilon_2^{l_2 - r_2}\cdots$$

of the first factor in order to obtain a contribution to the monomial (7.17) in the product. Consequently

$$g = f!\,|\,1/(l - n + 1)!, \cdots, 1/(l - 1)!, 1/l!\,|$$

$$= \frac{f!}{l_1!\cdots l_n!}\,|\cdots, l(l - 1), l, 1\,|.$$

The last determinant is

$$|\,l^{n-1}, \cdots, l, 1\,| = D(l_1, \cdots, l_n).$$

Therefore:

THEOREM (7.7.B). *The degree $g(f_1 \cdots f_n)$ of the irreducible representation $\langle\rho(f_1 \cdots f_n)\rangle$ of the symmetric group of $f = f_1 + \cdots + f_n$ figures is*

$$= f!\,\frac{D(l_1, \cdots, l_n)}{l_1!\cdots l_n!},$$

where, as always,

$$l_1 = f_1 + (n - 1), \cdots, \qquad l_n = f_n + 0.$$

2) Let us suppose the permutation s to contain a cycle of length v ($\alpha_v \geqq 1$). By dropping it we reduce s to a permutation of $f - v$ figures whose class $\mathfrak{k}'$

is characterized by the same numbers $\alpha_1, \alpha_2, \cdots$ as the class $\mathfrak{k}$ of s, except that α_v is reduced to $\alpha_v - 1$; whence the simple relation

$$\sigma(\mathfrak{k}) = \sigma(\mathfrak{k}')\cdot\sigma_v = \sigma(\mathfrak{k}')(\epsilon_1^v + \cdots + \epsilon_n^v).$$

On writing for the moment the right side of (7.16) in the form

$$\sum \{l_1, \cdots, l_n\}_{\mathfrak{k}}\, \epsilon_1^{l_1} \cdots \epsilon_n^{l_n},$$

the sum ranging over all integers $l_1, \cdots, l_n$, we obtain

$$\{l_1, l_2, \cdots, l_n\}_{\mathfrak{k}} =$$

$$\{l_1 - v, l_2, \cdots, l_n\}_{\mathfrak{k}'} + \{l_1, l_2 - v, \cdots, l_n\}_{\mathfrak{k}'} + \cdots + \{l_1, l_2, \cdots, l_n - v\}_{\mathfrak{k}'}.$$

Even if the $l_1, \cdots, l_n$ follow the proper order

$$l_1 > l_2 > \cdots > l_n \geqq 0$$

this may not be so for some of the brackets on the right side, e.g. for

$$\{l_1, l_2 - v, \cdots, l_n\}.$$

If $l_2 - v$ is $= l_3$ the term is to be dropped; if $l_2 - v < l_3$ we interchange these two arguments:

$$\{l_1, l_2 - v, l_3, \cdots\} = -\{l_1, l_3, l_2 - v, \cdots\}.$$

When necessary we repeat this process until $l_2 - v$ finds its proper place. Should $l_2 - v$ be <0 it will be pushed to the end of the row and the term is likewise to be canceled. On going back from the l's to the f's we obtain the following rule for the recurrent calculation of characters:

THEOREM (7.7.C). *If the class $\mathfrak{k}$ contains a cycle of length v and if $\mathfrak{k}'$ is the class arising by canceling this cycle, then*

$$\chi(f_1 \cdots f_n; \mathfrak{k}) = \chi(f_1 - v, f_2, \cdots, f_n; \mathfrak{k}') + \chi(f_1, f_2 - v, \cdots, f_n; \mathfrak{k}') + \cdots.$$

While $f_1 \geqq f_2 \geqq \cdots \geqq f_n \geqq 0$, the indices of a χ on the right side may deviate from this proper order at one place. For a $\chi(f_1 \cdots f_n)$ where this happens, the following reduction is to be performed:

(1) *If the discrepancy occurs at the last place, $f_n < 0$, then χ is to be canceled.*

(2) *If it occurs at an earlier place,*

$$f_1 \geqq \cdots \geqq f_{i-1} \geqq f_{i+1} \geqq \cdots \geqq f_n, \text{ but } f_i < f_{i+1},$$

the same holds, provided the gap $f_{i+1} - f_i = 1$;

(3) *If, however, the gap $f_{i+1} - f_i$ is $\geqq 2$, one replaces*

$$\chi(\cdots, f_i, f_{i+1}, \cdots) \text{ by } -\chi(\cdots, f_{i+1} - 1, f_i + 1, \cdots)$$

(f_i is increased by 1, f_{i+1} is lowered by 1, and the order exchanged). *The discrepancy is thus either removed or moved on to the next place (with a lower gap).*

This rule has often been used before for $v = 1$. In particular the recursive equation holds:

$$g(f_1, f_2, \cdots) = g(f_1 - 1, f_2, \cdots) + g(f_1, f_2 - 1, \cdots) + \cdots, \tag{7.18}$$

where terms on the right side are to be dropped, whose arguments do not keep proper order. The general rule was only recently pointed out by Professor F. D. Murnaghan[16] who found it very useful for the actual computation of characters.

If one multiplies (7.16) by

$$n(\mathfrak{k}) \cdot \chi(f_1 \cdots f_n; \mathfrak{k})$$

and sums over $\mathfrak{k}$, making use of the orthogonality relations of the characters χ, one arrives at the equation

$$\frac{|\epsilon^{l_1}, \cdots, \epsilon^{l_n}|}{|\epsilon^{n-1}, \cdots, 1|} = \frac{1}{f!} \sum_{\mathfrak{k}} n(\mathfrak{k})\sigma(\mathfrak{k})\chi_\pi(f_1 \cdots f_n; \mathfrak{k}).$$

On the left side we have the character $\chi_L(f_1 \cdots f_n)$ of $GL(n)$. This connection,

$$\chi_L = \frac{1}{f!} \sum n(\mathfrak{k})\sigma(\mathfrak{k})\chi_\pi(\mathfrak{k}), \tag{7.19}$$

evidently carries over from the irreducible representations to any linear combinations of them, and hence to *all* representations. It is indeed nothing else than the expression in terms of characters of the general reciprocity discussed in Chapter III, B. In this way, as has been done by the author,[17] the equation may be directly established through the same combinatorial considerations as led to Frobenius's equation (7.6). A particular case is the equation

$$p_f = \frac{1}{f!} \sum n(\mathfrak{k})\sigma(\mathfrak{k}) = \sum \frac{1}{\alpha_1! \, \alpha_2! \cdots} \left(\frac{\sigma_1}{1}\right)^{\alpha_1} \left(\frac{\sigma_2}{2}\right)^{\alpha_2} \cdots$$

$$(1\alpha_1 + 2\alpha_2 + \cdots = f)$$

which may also be derived from (7.11):

$$\sum_{f=0}^{\infty} p_f z^f = 1/\prod_i (1 - \epsilon_i z) = \exp\left(\frac{\sigma_1 z^1}{1} + \frac{\sigma_2 z^2}{2} + \cdots\right).$$

If one is concerned with the symmetric group only, one might to some advantage give this whole development the following turn.[18] By means of independent variables $\sigma_1, \sigma_2, \cdots$ one associates with each class $\mathfrak{k}$ whose permutations consist of α_1 cycles of length 1, α_2 cycles of length 2, $\cdots$, the monomial

$$\sigma(\mathfrak{k}) = \sigma_1^{\alpha_1}\sigma_2^{\alpha_2} \cdots .$$

With each character $\chi(\mathfrak{k})$ of the symmetric group one associates the following polynomial $\Psi(\sigma_1, \sigma_2, \cdots)$ of the variables $\sigma_1, \sigma_2, \cdots$ which Schur calls the characteristic:

$$\Psi(\sigma_1, \sigma_2, \cdots) = \frac{1}{f!} \sum_{\mathfrak{k}} n(\mathfrak{k})\chi(\mathfrak{k})\sigma(\mathfrak{k}).$$

The characteristic of the representation $s \to 1$ whose character is $\chi(\mathfrak{k}) = 1$ may be denoted by p_f :

$$p_f = \frac{1}{f!} \sum n(\mathfrak{k})\sigma(\mathfrak{k}).$$

The introduction of characteristics might formally be justified by observing that the product of two characteristics of π_f and $\pi_{f'}$ is a characteristic of $\pi_{f+f'}$. This statement takes the place of the combinatorial argument by which we derived Frobenius's formula (7.6), and is proved in similar fashion. [For us who are initiated and know that the characteristic is the corresponding character of $GL(n)$, all this is plain enough.] Hence the products

$$p_{f_1} p_{f_2} \cdots$$

associated with arbitrary partitions $f = f_1 + f_2 + \cdots + f_r$ are characteristics; and so are the determinants

$$| p_{l-(r-1)}, \cdots, p_{l-1}, p_l | = \frac{1}{f!} \sum_{\mathfrak{k}} n(\mathfrak{k})\chi(f_1 \cdots f_r; \mathfrak{k})\sigma(\mathfrak{k})$$

$$(l_1 = f_1 + (r - 1), \cdots, \qquad l_r = f_r + 0),$$

and their coefficients $\chi(\mathfrak{k})$ are characters, at least in Frobenius's extended sense, meaning linear combinations of primitive characters with integral, not necessarily positive, coefficients. By means of the orthogonality relations and a suitable form of Cauchy's lemma, one realizes that the $\chi(f_1 \cdots f_r; \mathfrak{k})$ themselves are primitive characters.

The absence of the number n alien to the symmetric group π_f may be counted as an advantage of this procedure. But whichever way one turns this theory, it always depends on three arguments, viz. the typical combinatorial consideration, Cauchy's lemma, and the orthogonality relations. And the determinant comes in, one does not know whence, as a *deus ex machina*. Not so in the analytic method where Δ springs from the volume element of the unitary group. Thus, after due respect has been paid to the algebraic verifications in their various shapes, I am still convinced against all Puritan doctrines that the analytic method is the least artificial, affording the deepest insight and best in keeping with our program: to solve concrete problems by means of general ideas which shed light upon a much wider range of mathematical facts than were needed for our immediate purpose. One of the most conspicuous advantages of our analytic procedure is its being capable of immediate generalization to the symplectic and the orthogonal group.

8. Characters of the symplectic group

For the symplectic group $Sp(n)$ the dimensionality n is even, $= 2\nu$. The vector components are designated as

$$x_1, x_1'; x_2, x_2', \cdots, x_\nu, x_\nu'.$$

Lemma (7.1.A) justifies limitation to the *unitary* symplectic transformations which form the group $USp(n)$. Similar to Theorem (7.1.C) one has:

Theorem (7.8.A). *Within $USp(n)$ each element A is conjugate to a diagonal element with the components*

$$\epsilon_1, \epsilon_1', \cdots, \epsilon_\nu, \epsilon_\nu' \tag{8.1}$$

where $|\epsilon_i| = 1$ *and* $\epsilon_i' = 1/\epsilon_i = \bar{\epsilon}_i$ $(i = 1, \cdots, \nu)$.

Proof. From the theory of unitary matrices we know that the roots ϵ of the characteristic equation

$$|\lambda E - A| = 0$$

are of modulus 1, and that the eigenspaces $\mathrm{P}(\epsilon)$ which consist of the vectors x satisfying the equation

$$\epsilon x = Ax \tag{8.2}$$

are unitary perpendicular to each other for numerically different roots ϵ, and yield as their sum the full space P. To each vector x there corresponds the vector denoted in Chapter VI, §2, by $\tilde{x}$, and the equation (8.2) entails

$$\bar{\epsilon}\tilde{x} = A\tilde{x}.$$

We now determine the basis in each $\mathrm{P}(\epsilon)$ in a particular way, and we distinguish two cases:

1) $\epsilon \neq \pm 1$. In $\mathrm{P}(\epsilon)$ we choose as before an arbitrary unitary-orthogonal basis $e_1, \cdots, e_\mu$. The vectors $\tilde{x}$ corresponding to the x in $\mathrm{P}(\epsilon)$ form the eigenspace $\mathrm{P}(\bar{\epsilon})$ and as its basis we choose

$$e_1' = \tilde{e}_1, \cdots, e_\mu' = \tilde{e}_\mu.$$

Thus the eigenvalues $\epsilon \neq \pm 1$ occur in pairs ϵ, $\bar{\epsilon} = 1/\epsilon$ of equal multiplicity; and the construction just indicated is carried through for each of these pairs independently.

2) $\epsilon = \pm 1$. The case $\epsilon = -1$ was treated in Chapter VI, §2 and we succeeded in constructing a basis of $\mathrm{P}(-1)$ which is at the same time unitary-orthogonal and symplectic. The same construction goes through for $\epsilon = +1$.

If we combine the bases thus constructed for the different eigenspaces, we obtain a basis of the whole n-dimensional space which is at the same time unitary-orthogonal and symplectic. In this coördinate system the transformation takes on the diagonal form with the components (8.1).

We put

$$\epsilon_i = e(\varphi_i), \qquad \bar{\epsilon}_i = e(-\varphi_i) \qquad (i = 1, \cdots, \nu)$$

and call $\varphi_1, \cdots, \varphi_\nu$ again the angles of A. They are uniquely determined mod. 1 except for their order and their signs. We use the notation

$$c(\varphi) = e(\varphi) + e(-\varphi), \qquad s(\varphi) = e(\varphi) - e(-\varphi).$$

THEOREM (7.8.B). *The volume of that part of $USp(n)$ the elements of which have angles within the infinitely near limits*

$$(\varphi_1, \varphi_1 + d\varphi_1), \cdots, (\varphi_\nu, \varphi_\nu + d\varphi_\nu)$$

is given by

$$\Delta\bar{\Delta}\, d\varphi_1 \cdots d\varphi_\nu,$$

where

$$\Delta = \prod_i s(\varphi_i) \cdot \prod_{i<k} \{c(\varphi_i) - c(\varphi_k)\} \qquad (i, k = 1, \cdots, \nu). \tag{8.3}$$

The proof is similar to that of Theorem (7.4.C).

Each character of $USp(n)$ is a class function and hence a function of the angles $\varphi_1, \cdots, \varphi_\nu$ periodic with period 1 in each argument and invariant under the "octahedral" group Q_ν of order $2^\nu \cdot \nu!$, which consists of permutations of the arguments $\varphi_1, \cdots, \varphi_\nu$ combined with arbitrarily distributed changes of signs. Δ is antisymmetric with respect to that group. The simplest antisymmetric functions are the elementary sums

$$\xi(l_1 \cdots l_\nu) = \sum \pm e(l_1\varphi_1 + \cdots + l_\nu\varphi_\nu),$$

where the l's are integers satisfying the inequalities

$$l_1 > l_2 > \cdots > l_\nu > 0$$

and the sum extends alternatingly over the group Q_ν.

$$\xi(l_1, \cdots, l_\nu) = | s(l_1\varphi), \cdots, s(l_\nu\varphi) |.$$

Δ is the lowest of these elementary sums:

$$\Delta = \xi(l_1^0, \cdots, l_\nu^0), \qquad l_1^0 = \nu, \qquad l_2^0 = \nu - 1, \cdots, l_\nu^0 = 1. \tag{8.4}$$

THEOREM (7.8.C). *Each primitive character $\chi(f_1 \cdots f_\nu)$ of $USp(n)$ is given by a quotient the numerator of which is*

$$\xi(l_1 \cdots l_\nu) = | \epsilon^{l_1} - \epsilon^{-l_1}, \cdots, \epsilon^{l_\nu} - \epsilon^{-l_\nu} | \tag{8.5}$$

and the denominator of which is $\xi(\nu, \nu - 1, \cdots, 1)$.

Its highest term is $\epsilon_1^{f_1} \cdots \epsilon_\nu^{f_\nu}$, where

$$l_1 - l_1^0 = f_1, \cdots, l_\nu - l_\nu^0 = f_\nu.$$

The degree is determined by first setting

$$\varphi_1 = \nu\varphi, \qquad \varphi_2 = (\nu - 1)\varphi, \cdots, \varphi_\nu = 1\varphi,$$

whereby the numerator (8.5) changes into

$$\prod_i s(l_i\varphi) \cdot \prod_{i<k} \{c(l_i\varphi) - c(l_k\varphi)\},$$

and then letting φ tend to zero:

$$N(f_1 \cdots f_\nu) = P(l_1 \cdots l_\nu)/P(l_1^0 \cdots l_\nu^0),$$

$$P(l_1 \cdots l_\nu) = \prod_i l_i \cdot \prod_{i<k} (l_i - l_k)(l_i + l_k).$$

The identification with the algebraic construction in Chapter V reveals that (8.5) corresponds to the linear space $P_0(f_1 \cdots f_\nu)$ cut out from P_f^0 $(f = f_1 + \cdots + f_\nu)$ by the symmetry diagram $T(f_1 \cdots f_\nu)$.

THEOREM (7.8.D). *With $f_1, \cdots, f_\nu$ ranging over all integers in the order $f_1 \geqq \cdots \geqq f_\nu \geqq 0$ the representations $\langle P_0(f_1 \cdots f_\nu)\rangle$ constitute a full set of inequivalent continuous irreducible representations of the unitary symplectic group.*

We write Cauchy's lemma in the form

$$\left|\frac{1}{x_i - y_k}\right| = \frac{|1, \cdots, x^{\nu-1}| \cdot |y^{\nu-1}, \cdots, 1|}{\prod (x_i - y_k)} \qquad (i, k = 1, \cdots, \nu)$$

and now put

$$x_i = z_i + z_i^{-1}, \qquad y_i = \epsilon_i + \epsilon_i^{-1}.$$

Then

$$x_i - y_k = \frac{1}{z}(1 - \epsilon z)\left(1 - \frac{1}{\epsilon} z\right) \quad \text{for} \quad z = z_i, \ \epsilon = \epsilon_k.$$

We obtain

$$\left|\frac{1}{(1 - \epsilon_k z_i)(1 - \epsilon_k^{-1} z_i)}\right| = \frac{|\epsilon^{\nu-1} + \epsilon^{-(\nu-1)}, \cdots, 1| \cdot |z^{\nu-1}, z^{\nu-2} + z^{\nu}, \cdots, 1 + z^{2\nu-2}|}{\prod (1 - \epsilon_k z_i)(1 - \epsilon_k^{-1} z_i)}.$$

$$\frac{1}{(1 - \epsilon z)(1 - \epsilon^{-1} z)} \text{ is } \sum_{l=1}^{\infty} \{\epsilon^{l-1} + \epsilon^{l-3} + \cdots + \epsilon^{-(l-1)}\} z^{l-1} = \sum_{l=1}^{\infty} \frac{\epsilon^{l} - \epsilon^{-l}}{\epsilon - \epsilon^{-1}} \cdot z^{l-1}.$$

Hence with the abbreviations (8.4) and

$$\varphi(z) = \prod_k (1 - \epsilon_k z)(1 - \epsilon_k^{-1} z) = |E - zA|:$$

$$\frac{\Delta \cdot |z^{\nu-1}, \cdots, 1 + z^{2\nu-2}|}{\varphi(z_1) \cdots \varphi(z_\nu)} = \sum |\epsilon^{l_1} - \epsilon^{-l_1}, \cdots, \epsilon^{l_\nu} - \epsilon^{-l_\nu}| z_1^{l_1-1} \cdots z_\nu^{l_\nu-1}.$$

When we again introduce the p_f by (6.4) we see that the character $\chi(f_1 \cdots f_\nu)$ equals the coefficient of $z_1^{l_1-1} \cdots z_\nu^{l_\nu-1}$ in

$$|z^{\nu-1}, z^{\nu-2} + z^{\nu}, \cdots, 1 + z^{2\nu-2}| / \varphi(z_1) \cdots \varphi(z_\nu),$$

or

$$|p_{l-\nu}, \quad p_{l-\nu+1} + p_{l-\nu-1}, \cdots, p_{l-1} + p_{l-2\nu+1}|.$$

Now we may cast off the unitarian fetters. Wishing to supply a uniform notation for $GL(n)$ and $Sp(n)$ we replace the symbol l here by $l + 1$.

THEOREM (7.8.E). *The character of the irreducible representation $\langle P_0(f_1 \cdots f_\nu)\rangle$ of the symplectic group is*

(8.6) $$\chi(f_1, \cdots, f_\nu) = |p_{l-\nu+1}, p_{l-\nu+2} + p_{l-\nu}, \cdots, p_l + p_{l-2\nu+2}|,$$

the p_f being defined in terms of the arbitrary symplectic transformation A by

$$1/|E - zA| = \sum p_f z^f$$

and

$$l_1 = f_1 + (\nu - 1), \cdots, l_\nu = f_\nu + 0. \tag{8.7}$$

Nowhere in this sweeping advance do we encounter the least resistance.[19] The characteristic polynomial

$$\varphi(z) = | E - zA |$$

of a symplectic transformation A:

$$A^*IA = I,$$

has the property

$$z^n \cdot \varphi(1/z) = \varphi(z). \tag{8.8}$$

Indeed

$$(E - zA^*)IA = I(A - zE)$$

implies

$$| E - zA^* | = | A - zE | \quad \text{or} \quad | E - zA | = | zE - A |$$

because $| A | = 1$. We put

$$\psi_r(z) = \varphi(z) \text{ for } r = \nu \text{ and}$$

$$\psi_r(z) = \{1 + z^{2(r-\nu)}\}\varphi(z) \text{ for } r > \nu.$$

The polynomial ψ_r of degree $2r$ satisfies the equation analogous to (8.8):

$$z^{2r} \cdot \psi_r(1/z) = \psi_r(z),$$

and is therefore of the form

$$c_0(1 + z^{2r}) + c_1(z + z^{2r-1}) + \cdots + c_r z^r \qquad (c_0 = 1). \tag{8.9}$$

Hence on forming

$$\nabla(z_1 \cdots z_r) = | z^{r-1}, z^{r-2} + z^r, \cdots, 1 + z^{2(r-1)} | \tag{8.10}$$

the function

$$H(z_1 \cdots z_r) = \nabla(z_1 \cdots z_r)/\varphi(z_1) \cdots \varphi(z_r)$$

will satisfy, for $r \geqq \nu$, the recursive equation

$$H(z_1 \cdots z_{r+1}) = H(z_1 \cdots z_r)z_1 \cdots z_r\{1 + z_{r+1}^{2(r-\nu)}\} - + \cdots,$$

the factor $\{1 + z_{r+1}^{2(r-\nu)}\}$ in the lowest case $r = \nu$ to be replaced by 1. Moreover

$$H(z_1 \cdots z_\nu) = \sum z_1^{l_1} \cdots z_\nu^{l_\nu} \, | p_{l-(\nu-1)}, \cdots, p_l + p_{l-2(\nu-1)} |$$

$$= \sum_{(f_1 \geqq \cdots \geqq f_\nu \geqq 0)} | z^{f_1+(\nu-1)}, \cdots, z^{f_\nu} | \cdot \chi(f_1 \cdots f_\nu).$$

Hence by induction with respect to r starting with $r = \nu$:

$$H(z_1 \cdots z_r) = \sum_{(f)} L_{f_1 \cdots f_\nu}(z_1 \cdots z_r) \cdot \chi(f_1 \cdots f_\nu)$$

with

$$L_{f_1 \cdots f_\nu}(z_1 \cdots z_r) = \tag{8.11}$$
$$| z^{f_1+(r-1)}, \cdots, z^{f_\nu+(r-\nu)} \mid z^{r-\nu-1}, z^{r-\nu-2} + z^{r-\nu}, \cdots, 1 + z^{2(r-\nu-1)} |.$$

The determinant

$$\nabla = (z_1 \cdots z_r)^{r-1} \mid 1, z + z^{-1}, \cdots, z^{r-1} + z^{-r+1} \mid$$

is the difference product of the $z_i + z_i^{-1}$ multiplied by $(z_1 \cdots z_r)^{r-1}$. As

$$(z_k + z_k^{-1}) - (z_i + z_i^{-1}) = (z_i - z_k)\left(\frac{1}{z_iz_k} - 1\right) = (z_i - z_k)(1 - z_iz_k)/z_iz_k,$$

one finds

$$\nabla(z_1 \cdots z_r) = D(z_1 \cdots z_r) \prod_{i<k}(1 - z_iz_k).$$

It is convenient to introduce the intermediate *polynomials*

$$L_{f_1 \cdots f_\nu}(z_1 \cdots z_r)/D(z_1 \cdots z_r) = \Lambda_{f_1 \cdots f_\nu}(z). \tag{8.12}$$

Then the result is:

Theorem (7.8.F). *The generating functions*

$$\Phi_{f_1 \cdots f_\nu}(z) = \Lambda_{f_1 \cdots f_\nu}(z)/\prod_{i<k}(1 - z_iz_k) \qquad (i, k = 1, \cdots, r)$$
$$= \sum \mu\begin{pmatrix} e_1 \cdots e_r \\ f_1 \cdots f_\nu \end{pmatrix} z_1^{e_1} \cdots z_r^{e_r}$$

as defined by (8.11), (8.12), *on the one hand describe the decomposition of* $\langle \Sigma(e_1 \cdots e_r)\rangle$ *into irreducible parts* $\langle P_0(f_1 \cdots f_\nu)\rangle$:

$$\langle \Sigma(e_1 \cdots e_r)\rangle \sim \sum_{(f)} \mu\begin{pmatrix} e_1 \cdots e_r \\ f_1 \cdots f_\nu \end{pmatrix} \cdot \langle P_0(f_1 \cdots f_\nu)\rangle,$$

on the other hand determine the numbers of independent covariants of the prescribed type $\langle P_0(f_1 \cdots f_\nu)\rangle$ *depending in arbitrary degrees* $e_1, \cdots, e_r$ *on* r *argument vectors.*

It is not possible to condense in a neater form the abundance of information this formula contains. For $f_1 = \cdots = f_\nu = 0$ it gives the numbers of linearly independent *vector invariants*. Up to $r = 2\nu$, the numerator $L_{00 \cdots 0}$ permits the simplified expression

$$| z^{r-1}, \cdots, z^{\nu} \mid z^{\nu-1}, \cdots, 1 |;$$

hence, for $r = n$,

$$\Phi_{0 \cdots 0}(z_1 \cdots z_n) = 1/\prod_{i<k}(1 - z_iz_k) \qquad (i, k = 1, \cdots, n). \tag{8.13}$$

Let $x^1, \cdots, x^n$ be the n argument vectors and $[x^i x^k]$ $(i < k)$ their $\frac{1}{2}n(n-1)$ skew-products. (8.13) indicates that there are as many linearly independent invariants as "monomials"

$$\prod_{i<k} [x_i x_k]^{e_{ik}} \qquad (e_{ki} = e_{ik}) \tag{8.14}$$

of the preassigned degrees e_i, $e_i = \sum_k e_{ik}$. This of course is in perfect agreement with the statements of the first and second main theorems, that all invariants are expressible in terms of skew-products, and that there is no algebraic relation among the skew-products of n vectors. On assuming either of these two propositions, our formula would establish the other. We see now what the denominator $\prod_{i<k} (1 - z_i z_k)$ in our formulas means; for the purpose of enumeration it takes account of the obvious fact that multiplication of any covariant by an arbitrary monomial (8.14) leads to a covariant of the same type again.

We formerly advocated the procedure of descending from $GL(n)$ to the subgroup $Sp(n)$ after having first decomposed $\langle \Sigma(e_1 \cdots e_r) \rangle$ into its irreducible constituents $\langle \mathrm{P}(e_1 \cdots e_n) \rangle$ under the regime of the full linear group:

$$\langle \mathrm{P}(e_1 \cdots e_n) \rangle \sim \sum_{(f)} \mu^* \begin{pmatrix} e_1 & \cdots & e_n \\ f_1 & \cdots & f_\nu \end{pmatrix} \cdot \langle \mathrm{P}_0(f_1 \cdots f_\nu) \rangle. \tag{8.15}$$

Considering the previous relations

$$D(z_1 \cdots z_n)/\varphi(z_1) \cdots \varphi(z_n) = \sum_{(e)} | z^{e_1+(n-1)}, \cdots, z^{e_n} | \cdot \chi(e_1 \cdots e_n)$$

$$= \sum_{(f)} \frac{L_{f_1 \cdots f_\nu}(z_1 \cdots z_n)}{\prod_{i<k} (1 - z_i z_k)} \cdot \chi^S(f_1 \cdots f_\nu),$$

where χ and χ^S indicate the characters of $GL(n)$ and $Sp(n)$ respectively, one arrives at this formula:

THEOREM (7.8.G). *The Taylor expansion of*

$$L_{f_1 \cdots f_\nu}(z_1 \cdots z_n)/\prod_{i<k} (1 - z_i z_k) \qquad (i, k = 1, \cdots, n) \tag{8.16}$$

is skew-symmetric in the n variables $z_1, \cdots, z_n$ and may hence be written in the form

$$\sum_{(e_1 \geqq \cdots \geqq e_n \geqq 0)} \mu^* \begin{pmatrix} e_1 & \cdots & e_n \\ f_1 & \cdots & f_\nu \end{pmatrix} \cdot | z^{e_1+(n-1)}, \cdots, z^{e_n} |.$$

As a generating function in this sense, (8.16) *exhibits the multiplicities μ^*,* (8.15), *with which the irreducible parts $\langle \mathrm{P}_0(f_1 \cdots f_\nu) \rangle$ occur in $\langle \mathrm{P}(e_1 \cdots e_n) \rangle$.*

9. Characters of the orthogonal group

It was not without purpose that this time we gave the symplectic group precedence over the orthogonal group. For the latter the situation is considerably complicated by the distinction between proper and improper substitutions.[20] For the analytic investigation it proves convenient to separate the cases of odd and even dimensionality, $n = 2\nu + 1$ and $n = 2\nu$, and to assume as the fundamental quadratic form left invariant by the orthogonal transformations:

$$\begin{cases} 2(x_1x_1' + \cdots + x_\nu x_\nu') + x_0^2 & (\text{for } n = 2\nu + 1), \\ 2(x_1x_1' + \cdots + x_\nu x_\nu') & (\text{for } n = 2\nu) \end{cases} \tag{9.1}$$

rather than

$$(x_1^2 + x_1'^2) + \cdots + (x_\nu^2 + x_\nu'^2) \{+ x_0^2\}. \tag{9.2}$$

As the one changes into the other by means of the substitution

$$x_k \to \frac{1}{\sqrt{2}}(x_k + ix_k'), \qquad x_k' \to \frac{1}{\sqrt{2}}(x_k - ix_k') \qquad (k = 1, \cdots, \nu)$$

which is itself unitary, the group $UO(n)$ of unitary transformations leaving (9.1) invariant is equivalent under this substitution to the group of unitary, i.e. real, transformations with the invariant form (9.2). With the fundamental form (9.1) the scalar product of two vectors x, y equals

$$(xy) = (x_1y_1' + x_1'y_1) + \cdots + (x_\nu y_\nu' + x_\nu'y_\nu) \{+ x_0y_0\},$$

and thus the whole treatment will resemble more that of the symplectic group with the invariant skew product

$$[xy] = (x_1y_1' - x_1'y_1) + \cdots + (x_\nu y_\nu' - x_\nu'y_\nu).$$

The one improper orthogonal transformation J_n which we used in Chapter V for breaking up $O(n)$ into $O^+(n)$ and its coset $O^-(n)$ will now be defined by:

$$\begin{cases} x_1 \to x_1, x_1' \to x_1'; \cdots; x_\nu \to x_\nu, x_\nu' \to x_\nu'; x_0 \to -x_0 & (n = 2\nu + 1); \\ x_1 \to x_1, x_1' \to x_1'; \cdots; x_{\nu-1} \to x_{\nu-1}, x_{\nu-1}' \to x_{\nu-1}'; x_\nu \to x_\nu', x_\nu' \to x_\nu & (n = 2\nu). \end{cases} \tag{9.3}$$

Let us first study the odd case $n = 2\nu + 1$ in a manner needing but little modification when we afterwards pass to the even case. Within $UO(n)$ each proper or improper element is conjugate to a diagonal element

$$x_1 \to \epsilon_1 x_1, \quad x_1' \to \epsilon_1' x_1', \cdots, x_\nu \to \epsilon_\nu x_\nu, \quad x_\nu' \to \epsilon_\nu' x_\nu', \quad x_0 \to \pm x_0 \tag{9.4}$$

respectively, where $\epsilon_i' = \bar{\epsilon}_i = 1/\epsilon_i$. We introduce the ν angles φ_i by $\epsilon_i = e(\varphi_i)$. Each class function, in particular each character χ, is therefore a periodic function $\chi_+(\varphi_1, \cdots, \varphi_\nu)$ on the proper, another $\chi_-(\varphi_1, \cdots, \varphi_\nu)$ on the improper part, which are invariant under the octahedral group $Q = Q_\nu$ of order $2^\nu \cdot \nu!$. The diagonal elements of form (9.4) constitute an Abelian group and consequently we are able to bring their representing matrices in a given representation simultaneously into diagonal form. We know that for the proper elements the matrix will consist of N terms

$$E_K^+ = e(m_1\varphi_1 + \cdots + m_\nu\varphi_\nu)$$

along the diagonal. Let the one improper element J_n, (9.3), be represented by the diagonal matrix $\{a_K\}$; because of $J_n^2 = E$ we obtain

$$a_K^2 = 1, \qquad a_K = \pm 1,$$

and then for the improper (9.4):

$$E_K^- = a_K \cdot e(m_1\varphi_1 + \cdots + m_\nu\varphi_\nu).$$

Hence $\chi_+ = \sum E_K^+$ and $\chi_- = \sum E_K^-$ are finite Fourier series with integral coefficients:

$$\text{(9.5)} \qquad \begin{cases} \chi_+ = \sum k^+_{m_1 \cdots m_\nu}\, e(m_1\varphi_1 + \cdots + m_\nu\varphi_\nu), \\ \chi_- = \sum k^-_{m_1 \cdots m_\nu}\, e(m_1\varphi_1 + \cdots + m_\nu\varphi_\nu). \end{cases}$$

The k^+ are non-negative and the k^- satisfy the conditions

$$\text{(9.6)} \qquad |\, k^-_{m_1 \cdots m_\nu} \,| \leqq k^+_{m_1 \cdots m_\nu}, \qquad k^-_{m_1 \cdots m_\nu} \equiv k^+_{m_1 \cdots m_\nu} \pmod 2.$$

For the relative density of the classes we find $\Delta^+\overline{\Delta^+}$, $\Delta^-\overline{\Delta^-}$ on the proper and improper part respectively, where

$$\text{(9.7)} \qquad \begin{cases} \Delta^+ = \prod_i s\left(\frac{\varphi_i}{2}\right) \cdot \prod_{i<k} \{c(\varphi_i) - c(\varphi_k)\}, \\ \Delta^- = \prod_i c\left(\frac{\varphi_i}{2}\right) \cdot \prod_{i<k} \{c(\varphi_i) - c(\varphi_k)\}. \end{cases}$$

These functions are double-valued in the manner that by following its continuous variation Δ changes into $-\Delta$ whenever one of the angles φ_i describes the full circle while the others remain fixed. $\xi^+ = \Delta^+\chi^+$ and $\xi^- = \Delta^-\chi^-$ are finite Fourier series with integral coefficients in the modified sense that in the generic term $e(m_1\varphi_1 + \cdots + m_\nu\varphi_\nu)$ the exponents $m_1, \cdots, m_\nu$ are taken from the sequence of "half integers"

$$\cdots, -3/2, -1/2, 1/2, 3/2, \cdots$$

rather than from the sequence of integers

$$\cdots, -2, -1, 0, 1, 2, \cdots.$$

Hence they are linear combinations with integral coefficients of the elementary sums

$$\xi^+(l_1, \cdots, l_\nu) = \sum \pm\, e(l_1\varphi_1 + \cdots + l_\nu\varphi_\nu),$$
$$\xi^-(l_1, \cdots, l_\nu) = \sum \pm\, e(l_1\varphi_1 + \cdots + l_\nu\varphi_\nu).$$

In ξ^+ the sum extends alternatingly over Q_ν while in ξ^- the sum extends alternatingly over the permutations, but directly over the changes of signs. The l's are half-integers satisfying

$$l_1 > l_2 > \cdots > l_\nu > 0.$$

$$\text{(9.8}^+\text{)} \qquad \xi^+(l_1, \cdots, l_\nu) = |\, \epsilon^{l_1} - \epsilon^{-l_1}, \cdots, \epsilon^{l_\nu} - \epsilon^{-l_\nu} \,| = |\, s(l_1\varphi), \cdots, s(l_\nu\varphi) \,|,$$

$$\text{(9.8}^-\text{)} \qquad \xi^-(l_1, \cdots, l_\nu) = |\, \epsilon^{l_1} + \epsilon^{-l_1}, \cdots, \epsilon^{l_\nu} + \epsilon^{-l_\nu} \,| = |\, c(l_1\varphi), \cdots, c(l_\nu\varphi) \,|.$$

Δ^+, Δ^- are the lowest $\dot{\xi}^+$, $\dot{\xi}^-$ with

$$l_1^0 = \nu - 1/2, \cdots, l_\nu^0 = 1/2.$$

In exhibiting the individual terms in lexicographic order we must have

$$\dot{\xi}^+ = c_+ \cdot \xi^+(l_1 \cdots l_\nu) + \cdots,$$
$$\dot{\xi}^- = c_- \cdot \xi^-(l_1 \cdots l_\nu) + \cdots.$$

where

$$(9.9) \qquad c_+ > 0, \qquad c_- \equiv c_+ \pmod 2, \qquad |c_-| \leqq c_+.$$

Denoting by $\mathfrak{M}$, $\mathfrak{M}^+$, $\mathfrak{M}^-$ mean values over the whole group, its proper, and its improper part respectively, one has

$$\mathfrak{M} = \tfrac{1}{2}(\mathfrak{M}^+ + \mathfrak{M}^-).$$

If χ is primitive, its square average $\mathfrak{M} \mid \chi \mid^2$ equals 1; hence the equation

$$\tfrac{1}{2}(c_+^2 + c_-^2) + \cdots = 1$$

which on account of (9.9) leaves two possibilities only: either

$$(9.10) \qquad \chi^+ = \frac{\xi^+(l_1 \cdots l_\nu)}{\xi^+(l_1^0 \cdots l_\nu^0)}, \qquad \chi^- = \frac{\xi^-(l_1 \cdots l_\nu)}{\xi^-(l_1^0 \cdots l_\nu^0)}$$

or

$$(9.11) \qquad \chi^+ = \frac{\xi^+(l_1 \cdots l_\nu)}{\xi^+(l_1^0 \cdots l_\nu^0)}, \qquad \chi^- = -\frac{\xi^-(l_1 \cdots l_\nu)}{\xi^-(l_1^0 \cdots l_\nu^0)}.$$

The highest term in (9.10) is

$$(9.12) \qquad \epsilon_1^{f_1} \cdots \epsilon_\nu^{f_\nu} \qquad (f_i = l_i - l_i^0)$$

for both χ^+ and χ^-, while in (9.11) the leading terms are

$$+\ \epsilon_1^{f_1} \cdots \epsilon_\nu^{f_\nu} \text{ for } \chi^+, \qquad -\ \epsilon_1^{f_1} \cdots \epsilon_\nu^{f_\nu} \text{ for } \chi^-.$$

This is sufficient for the identification with the algebraic construction in Chapter V, §7, which we must adapt to the new shape (9.1) of the fundamental quadratic form. This slight change is only for the better: for a scheme T with $m \leqq \nu$ rows of lengths

$$f_1, \cdots, f_m \qquad (f_{m+1} = \cdots = f_\nu = 0)$$

the only linearly independent component of the generic tensors in $P_0(f_1 \cdots f_\nu)$ of highest weight is now

$$F\begin{pmatrix} 1 & 1 & \cdots\cdots \\ 2 & 2 & \cdots\cdots \\ \cdots & \cdots & \cdots\cdots \\ m & m & \cdots \end{pmatrix},$$

and this weight is (9.12) under both proper and improper transformations. For the associate diagram T' the corresponding component is obtained from the above on extending the first column by

$$m + 1, \cdots, \nu, 0, \nu', (\nu - 1)', \cdots, (m + 1)';$$

it is of weight $\pm \epsilon_1^{f_1} \cdots \epsilon_\nu^{f_\nu}$ according as we take the + or − sign in (9.4). Hence (9.10) belongs to $P_0(f_1 \cdots f_\nu)$ and (9.11) to $P_0'(f_1 \cdots f_\nu)$. By applying Cauchy's lemma in the same form as for the symplectic group, χ^+ and χ^- acquire a unified expression, and in replacing the symbol l by $l + 1/2$ we obtain with the old conventions (8.7),

$$\chi(f_1 \cdots f_\nu) = | p_{l-(\nu-1)} - p_{l-(\nu+1)}, \cdots, p_l - p_{l-2\nu} | \tag{9.13}$$

for $P_0(f_1 \cdots f_\nu)$, and

$$\chi'(f_1 \cdots f_\nu) = | A | \cdot \chi(f_1 \cdots f_\nu) \tag{9.14}$$

for $P_0'(f_1 \cdots f_\nu)$.

We have now reached the point where we may cast off the unitary restriction and the result will hold good in any number field (of characteristic zero). Hence there is no reason why we should not return to the metric ground form (9.2); then our considerations have proved that the algebraically constructed

$$\langle P_0(f_1 \cdots f_\nu)\rangle, \qquad \langle P_0'(f_1 \cdots f_\nu)\rangle$$

yield a complete set of inequivalent continuous irreducible representations of the real orthogonal group.

The treatment of the odd case could have been simplified by using $-E$ as the improper element rather than (9.3) for performing the transition from proper to improper elements. The unified expression (9.13) on proper and improper part is then perfectly plain. However, we followed our course because it serves as a model for the even case $n = 2\nu$. We shall briefly indicate the modifications.

Instead of the normal form (9.4) of the elements of $UO(n)$ we now get

$$x_1 \to \epsilon_1 x_1, \; x_1' \to \epsilon_1' x_1', \; \cdots, \; x_\nu \to \epsilon_\nu x_\nu, \; x_\nu' \to \epsilon_\nu' x_\nu'$$

for the proper and

$$x_1 \to \epsilon_1 x_1, \; x_1' \to \epsilon_1' x_1', \; \cdots, \; x_{\nu-1} \to \epsilon_{\nu-1} x_{\nu-1}, \; x_{\nu-1}' \to \epsilon_{\nu-1}' x_{\nu-1}',$$
$$x_\nu \to x_\nu', \; x_\nu' \to x_\nu$$

for the improper elements. In the latter case we therefore have but $\nu - 1$ angles $\varphi_1, \cdots, \varphi_{\nu-1}$. For the definition of J_n see (9.3). Instead of (9.6) we find that the same relations prevail for the coefficients of

$$\chi_+(\varphi_1, \cdots, \varphi_{\nu-1}, \varphi_\nu = 0) \text{ and } \chi_-(\varphi_1, \cdots, \varphi_{\nu-1}).$$

(9.7) has to be replaced by

$$\begin{cases} \Delta^+ = \prod_{i<k} \{c(\varphi_i) - c(\varphi_k)\} & (i, k = 1, \cdots, \nu); \\ \Delta^- = \prod_i s(\varphi_i) \cdot \prod_{i<k} \{c(\varphi_i) - c(\varphi_k)\} & (i, k = 1, \cdots, \nu - 1). \end{cases} \tag{9.15}$$

The elementary sums $\xi^+(l_1 \cdots l_\nu)$ are

$$| c(l_1\varphi), \cdots, c(l_\nu\varphi) |$$

for $l_\nu > 0$, *but only half of it*:

$$| c(l_1\varphi), \cdots, c(l_{\nu-1}\varphi), 1 |$$

for $l_\nu = 0$. This has the consequence that

$$\int_0^1 \cdots \int_0^1 | \xi^+(l_1 \cdots l_\nu) |^2 \, d\varphi_1 \cdots d\varphi_\nu = \Omega \text{ or } \tfrac{1}{2}\Omega$$

according as $l_\nu > 0$ or $l_\nu = 0$, with $\Omega = 2^\nu \cdot \nu!$, in particular

$$\int_0^1 \cdots \int_0^1 \Delta^+ \overline{\Delta^+} \, d\varphi_1 \cdots d\varphi_\nu = \tfrac{1}{2}\Omega$$

for $\Delta^+ = \xi^+(\nu - 1, \cdots, 0)$. The elementary sums $\xi^-(l_1 \cdots l_{\nu-1})$ are like those for the symplectic group:

$$| s(l_1\varphi), \cdots, s(l_{\nu-1}\varphi) |,$$

and $\Delta^- = \xi^-(\nu - 1, \cdots, 1)$. Using the notation

$$l_i^0 = \nu - i, \qquad f_i = l_i - l_i^0, \tag{9.16}$$

the resulting possibilities are:

$$\chi^+ = \frac{\xi^+(l_1 \cdots l_\nu)}{\xi^+(l_1^0 \cdots l_\nu^0)}, \qquad \chi^- = 0 \qquad \text{with } l_\nu > 0;$$

$$\chi^+ = \frac{\xi^+(l_1 \cdots l_{\nu-1} 0)}{\xi^+(l_1^0 \cdots l_{\nu-1}^0 0)}, \qquad \chi^- = \frac{\xi^-(l_1 \cdots l_{\nu-1})}{\xi^-(l_1^0 \cdots l_{\nu-1}^0)}, \quad \text{and}$$

$$\chi^+ = \frac{\xi^+(l_1 \cdots l_{\nu-1} 0)}{\xi^+(l_1^0 \cdots l_{\nu-1}^0 0)}, \qquad \chi^- = -\frac{\xi^-(l_1 \cdots l_{\nu-1})}{\xi^-(l_1^0 \cdots l_{\nu-1}^0)}$$

for $l_\nu = 0$. They correspond to the tensor spaces

$$\begin{aligned} &P_0(f_1 \cdots f_\nu) = P_0'(f_1 \cdots f_\nu) \qquad (f_\nu > 0), \\ &P_0(f_1 \cdots f_{\nu-1}\, 0), \\ &P_0'(f_1 \cdots f_{\nu-1}\, 0), \end{aligned} \tag{9.17}$$

respectively. Indeed in the even case we have the self-associate diagrams with exactly ν rows, $f_\nu > 0$, for which P_0 and P_0' coincide. No wonder that the corresponding character is zero for the improper elements A; for the representation is equivalent to its associate so that

$$\chi(A) = -\chi(A) \text{ for } | A | = -1.$$

By means of Cauchy's lemma we finally arrive at the same result (9.13), (9.14) as for $n = 2\nu + 1$, both on the proper and improper part, and whether f_ν

is $= 0$ or > 0. The only difference is that in the latter case, in agreement with (9.17),

$$\chi(f_1 \cdots f_\nu) \text{ coincides with } \chi'(f_1 \cdots f_\nu).$$

Indeed we shall prove in a moment that the determinant (9.13) then vanishes for improper orthogonal A.

THEOREM (7.9.A). *The characters of*

$$\langle \mathrm{P}_0(f_1 \cdots f_\nu)\rangle \text{ and } \langle \mathrm{P}_0'(f_1 \cdots f_\nu)\rangle$$

are given by the determinants

$$\chi(f_1 \cdots f_\nu) = | p_{l-(\nu-1)} - p_{l-(\nu+1)}, \cdots, p_l - p_{l-2\nu} |, \tag{9.18}$$
$$\chi'(f_1 \cdots f_\nu) = | A | \cdot \chi(f_1 \cdots f_\nu),$$

the l being defined by (8.7). *For a self-associate diagram, $n = 2\nu$, $f_\nu > 0$, χ and χ' coincide.*

The formulas for the degrees can be as readily supplied as for $GL(n)$ and $Sp(n)$.

THEOREM (7.9.B). *The table*

$$\langle \mathrm{P}_0(f_1 \cdots f_\nu)\rangle, \qquad \langle \mathrm{P}_0'(f_1 \cdots f_\nu)\rangle$$

(with the proviso that $\mathrm{P}_0 = \mathrm{P}_0'$ is taken only once for a self-associate diagram, $n = 2\nu$, $f_\nu > 0$) contains a complete set of inequivalent continuous irreducible representations of the real orthogonal group.

For any orthogonal matrix A, $A^*A = E$, one readily verifies the fact that the characteristic polynomial

$$\varphi(z) = | E - zA |$$

satisfies the functional equation

$$z^n \cdot \varphi(1/z) = \delta \cdot \varphi(z),$$

where δ is the sign $(-1)^n | A |$. In particular, if n is even, $= 2\nu$, and A improper, $| A | = -1$:

$$z^{2\nu} \varphi(1/z) = -\varphi(z),$$

and consequently $\varphi(z)$ has the form

$$\varphi(z) = c_0(1 - z^{2\nu}) + c_1(z - z^{2\nu-1}) + \cdots + c_{\nu-1}(z^{\nu-1} - z^{\nu+1}) \qquad (c_0 = 1),$$

whence springs the recursive relation

$$c_0(p_l - p_{l-2\nu}) + \cdots + c_{\nu-1}(p_{l-\nu+1} - p_{l-\nu-1}) = \begin{cases} 0 \text{ for } l > 0 \\ 1 \text{ for } l = 0. \end{cases}$$

This proves that (9.18) indeed vanishes if $n = 2\nu$, $| A | = -1$, and all l's > 0.

More generally, the polynomial

$$\psi_r(z) = \{1 - \delta z^{2r-n}\} \varphi(z) \qquad (r > \nu)$$

of degree $2r$ satisfies

$$z^{2r}\psi_r(1/z) = -\psi_r(z),$$

and is therefore of the form

$$c_0(1 - z^{2r}) + \cdots + c_{r-1}(z^{r-1} - z^{r+1}) \qquad (c_0 = 1).$$

This leads readily to formulas similar to those established for the symplectic group.[21]

THEOREM (7.9.C). *In order to adapt Theorems (7.8.F) and (7.8.G) to the orthogonal group the following changes are to be made:*

(1) *Put*

$$L^{\pm}_{f_1\cdots f_\nu} = |z^{f_1+(r-1)}, \cdots, z^{f_\nu+(r-\nu)} \,|\, z^{r-\nu-1} \pm z^{r-\nu}, \cdots, 1 \pm z^{2r-n}|$$

for $n = 2\nu + 1$, *and*

$$L^{\pm}_{f_1\cdots f_\nu} = |z^{f_1+(r-1)}, \cdots, z^{f_\nu+(r-\nu)} \,|\, z^{r-\nu-1} \mp z^{r-\nu+1}, \cdots, 1 \mp z^{2r-n}|$$

for $n = 2\nu$. *The L-functions for the types* $P_0(f_1 \cdots f_\nu)$ *and* $P_0'(f_1 \cdots f_\nu)$ *are given by*

$$L = \tfrac{1}{2}(L^+ + L^-), \qquad L' = \tfrac{1}{2}(L^+ - L^-)$$

except for a self-associate diagram ($n = 2\nu, f_\nu > 0$) *where*

$$L(= L') = L^+.$$

(2) *Change the denominator*

$$\prod_{i<k}(1 - z_i z_k) \text{ into } \prod_{i\leq k}(1 - z_i z_k).$$

The last modification is in agreement with the necessity of including the case $i = k$ when forming the (symmetric) table of scalar products $(x^i x^k)$ of a given number of vectors $x^1, \cdots, x^n$.

10. Decomposition and ×-multiplication

Let us illustrate what we have in mind by the symplectic group, which is not quite so simple as $GL(n)$ and not so complicated as $O(n)$. Any invariant subspace of P_f is the substratum of a representation $\mathfrak{D}$ of $Sp(n)$, the character $\chi(\mathfrak{D})$ of which is a polynomial

$$\sum k_{m_1\cdots m_\nu}\epsilon_1^{m_1}\cdots\epsilon_\nu^{m_\nu} \tag{10.1}$$

with non-negative integral coefficients $k_{m_1\cdots m_\nu}$, when expressed in terms of the diagonal elements of $Sp(n)$:

$$x_\alpha \to \epsilon_\alpha x_\alpha, \qquad x'_\alpha \to \epsilon_\alpha^{-1}x'_\alpha \qquad (\alpha = 1, \cdots, \nu).$$

The exponents $m_1, \cdots, m_\nu$ are integral, not necessarily positive, numbers. The corresponding argument was explicitly carried out in Chapter IV, §5, for the full linear group. The representation $\mathfrak{D}$ breaks up into rreducible representations of the type

$$\langle P(f_1 \cdots f_\nu)\rangle = \mathfrak{d}(f_1 \cdots f_\nu).$$

We want to determine how often each of these irreducible constituents occur in $\mathfrak{D}$, under the assumption that we know the coefficients $k_{m_1 \cdots m_\nu}$ characterizing the given representation $\mathfrak{D}$.

A second important problem is the decomposition of the product $\mathfrak{d}' \times \mathfrak{d}$ of two irreducible representations $\mathfrak{d}'$ and $\mathfrak{d}$ into its irreducible constituents $\mathfrak{d}(f_1 \cdots f_\nu)$. We shall solve both problems together by giving a formula for the number of times $\mathfrak{d}(f_1' \cdots f_\nu')$ occurs in the product

$$\mathfrak{D} \times \mathfrak{d}(f_1 \cdots f_\nu) \tag{10.2}$$

where the representation $\mathfrak{D}$ is defined by means of the coefficients $k_{m_1 \cdots m_\nu}$ of its character.

We prefer to think in terms of the unitarily restricted group $Sp(n)$. Instead of the character

$$\chi(\mathfrak{D}) = \sum k_{m_1 \cdots m_\nu} e(m_1\varphi_1 + \cdots + m_\nu\varphi_\nu) \tag{10.3}$$

we might use the "ξ-function" of $\mathfrak{D}$:

$$\xi(\mathfrak{D}) = \Delta \cdot \chi(\mathfrak{D})$$

arising by multiplication with Δ, (8.3). χ is symmetric, ξ antisymmetric, under the octahedral group Q_ν operating on the angles φ_α. The ξ-function of $\mathfrak{d}(f_1 \cdots f_\nu)$ is

$$\xi(l_1 \cdots l_\nu) = | \epsilon^{l_1} - \epsilon^{-l_1}, \cdots, \epsilon^{l_\nu} - \epsilon^{-l_\nu} | \tag{10.4}$$

with

$$f_1 \geqq \cdots \geqq f_\nu \geqq 0, \qquad l_\alpha = f_\alpha + (\nu - \alpha + 1).$$

We retain the definition (10.4) of $\xi(l_1 \cdots l_\nu)$ for arbitrary integers whether or not they satisfy the inequalities

$$l_1 > \cdots > l_\nu > 0.$$

The multiplicities $m\begin{pmatrix} f_1 \cdots f_\nu \\ f_1' \cdots f_\nu' \end{pmatrix}$:

$$\mathfrak{D} \times \mathfrak{d}(f_1 \cdots f_\nu) \sim \sum_{(f')} m\begin{pmatrix} f_1 \cdots f_\nu \\ f_1' \cdots f_\nu' \end{pmatrix} \cdot \mathfrak{d}(f_1' \cdots f_\nu')$$

are to be derived from the character:

$$\chi(\mathfrak{D}) \cdot \chi(f_1 \cdots f_\nu) = \sum_{(f')} m \cdot \chi(f_1' \cdots f_\nu')$$

or, after multiplication with Δ, from:

$$\chi(\mathfrak{D}) \cdot \xi(l_1 \cdots l_\nu) = \sum m \cdot \xi(l_1' \cdots l_\nu').$$

We propose to show that the left side, or the ξ-function of $\mathfrak{D} \times \mathfrak{d}(f_1 \cdots f_\nu)$, equals[22]

$$\sum k_{m_1 \cdots m_\nu} \xi(l_1 + m_1, \cdots, l_\nu + m_\nu), \tag{10.5}$$

in other words, that it is obtained by substituting

$$\xi(l_1 + m_1, \cdots, l_\nu + m_\nu) \text{ for each term } e(m_1\varphi_1 + \cdots + m_\nu\varphi_\nu)$$

in the Fourier expansion (10.3) of $\chi(\mathfrak{D})$. Indeed, let

$$\pm e(l_1\varphi_1' + \cdots + l_\nu\varphi_\nu') \tag{10.6}$$

be any of the $2^\nu \cdot \nu!$ terms making up $\xi(l_1 \cdots l_\nu)$; $\varphi_1', \cdots, \varphi_\nu'$ arises from $\varphi_1, \cdots, \varphi_\nu$ by a substitution of the octahedral group. Since (10.3) is symmetric under that group, we might write

$$\chi(\mathfrak{D}) = \sum k_{m_1 \cdots m_\nu} e(m_1\varphi_1' + \cdots + m_\nu\varphi_\nu'),$$

and therefore the product of $\chi(\mathfrak{D})$ by the term (10.6) equals

$$\pm \sum_{(m)} k_{m_1 \cdots m_\nu} e((m_1 + l_1)\varphi_1' + \cdots + (m_\nu + l_\nu)\varphi_\nu').$$

Alternating summation over Q_ν leads to the formula (10.5).

At first sight it seems to solve our problem without any ado: the irreducible constituent of signature $f_1 + m_1, \cdots, f_\nu + m_\nu$ occurs with the multiplicity $k_{m_1 \cdots m_\nu}$. But this would simplify matters a little bit too much, because the sum (10.5) will contain plenty of terms for which

$$\lambda_1 = l_1 + m_1, \cdots, \lambda_\nu = l_\nu + m_\nu$$

do not keep proper order. The correct conclusion to be drawn from (10.5) is thus: the irreducible representation $\mathfrak{d}(f_1' \cdots f_\nu')$ will occur in the product (10.2)

$$\sum \pm k_{\lambda_1 - l_1, \cdots, \lambda_\nu - l_\nu}$$

times, where the sum extends alternatingly over all sequences $(\lambda_1, \cdots, \lambda_\nu)$ arising from $(l_1', \cdots, l_\nu')$ by the operations of the group Q_ν, and where

$$l_1 = f_1 + \nu, \cdots, l_\nu = f_\nu + 1; \qquad l_1' = f_1' + \nu, \cdots, l_\nu' = f_\nu' + 1.$$

A form easier to keep in mind is obtained by using the symbolic notation

$$k_{m_1 \cdots m_\nu} \sim k_1^{m_1} \cdots k_\nu^{m_\nu}.$$

Then the multiplicity in question is given by the symbolic expression

$$| k^{l_1'} - k^{-l_1'}, \cdots, k^{l_\nu'} - k^{-l_\nu'} | \cdot k_1^{-l_1} \cdots k_\nu^{-l_\nu}. \tag{10.7}$$

This explicit formula includes as the special case $f_1 = \cdots = f_\nu = 0$ the decomposition of $\mathfrak{D}$ itself into irreducible constituents.

In the handy form (10.5) our result depended on nothing but the fact that $\chi(\mathfrak{D})$ is symmetric under the group Q which served to build up the elementary sums. Hence it will hold for the linear and the orthogonal group as well.

THEOREM (7.10.A). *Let $\mathfrak{D}$ be any representation of the (unitarily restricted) group $Sp(n)$ with the character*

$$\chi(\mathfrak{D}) = \sum k_{m_1 \cdots m_\nu} e(m_1\varphi_1 + \cdots + m_\nu\varphi_\nu).$$

The ξ-function of the product of $\mathfrak{D}$ and the irreducible representation $\mathfrak{d}(f_1 \cdots f_\nu)$ is then given by

$$\sum_{(m)} k_{m_1 \cdots m_\nu} \cdot \xi(l_1 + m_1, \cdots, l_\nu + m_\nu).$$

Similarly for the linear and the orthogonal group.

11. The Poincaré polynomial

In an n-dimensional vector space we form the 1-, 2-, 3-, $\cdots$ dimensional elements spanned by 1, 2, 3, $\cdots$ vectors $x, y, z, \cdots$ with the components

$$x_i, \qquad \begin{vmatrix} x_i & x_k \\ y_i & y_k \end{vmatrix} \ (i < k), \qquad \begin{vmatrix} x_i & x_k & x_l \\ y_i & y_k & y_l \\ z_i & z_k & z_l \end{vmatrix} \ (i < k < l), \cdots.$$

A linear substitution

$$A: x_i' = \sum_k a_{ik} x_k$$

induces a linear substitution in the set of p-dimensional elements. Denoting its trace by $\psi_p(A)$, we have the relation

$$\det(zE + A) = z^n + z^{n-1}\psi_1(A) + \cdots + \psi_n(A).$$

Let there be given for a finite or a compact Lie group a representation $s \to A(s)$ with the character $X(s)$. It will be decomposable according to formula (1.1) into irreducible representations. By means of the orthogonality relations we found

$$m = \mathfrak{M}_s\{X(s)\bar{\chi}(s)\}.$$

In particular, the number of times the unit representation $s \to 1$ is contained in the given one is the mean value

$$\mathfrak{M}_s\{X(s)\};$$

this is at the same time the number of linearly independent linear invariants in the representation space P. We wish to determine the number ν_p of invariants

$$\sum f(i_1 \cdots i_p) x_{i_1}^{(1)} \cdots x_{i_p}^{(p)} \tag{11.1}$$

depending linearly on an arbitrary p-dimensional element in P. (When written in the form (11.1) the coefficients $f(i_1 \cdots i_p)$ will be skew-symmetric.) Our remark above shows that the polynomial $P(z)$ with the coefficients ν_p:

$$P(z) = z^n + \nu_1 z^{n-1} + \cdots + \nu_n$$

is

$$= \mathfrak{M}_s \,|\, zE + A(s) \,|.$$

With any Lie group of r parameters there is associated the adjoint representation

$$K(s): x \longrightarrow sxs^{-1}$$

whose r-dimensional vector space consists of the infinitesimal elements x of the group. The polynomial

$$P(z) = z^r + \nu_1 z^{r-1} + \cdots + \nu_r ,$$

whose coefficient ν_p indicates the order of the linear set of all invariants

$$\sum f(i_1 \cdots i_p) \delta x_{i_1}^{(1)} \cdots \delta x_{i_p}^{(p)}$$

depending in linear and antisymmetric fashion on p infinitesimal elements of our group, was called by E. Cartan its "Poincaré polynomial." In the case of a compact group it is given by

$$P(z) = \mathfrak{M}_s \mid zE + K(s) \mid . \tag{11.2}$$

We shall apply this formula in particular to our groups $GL(n)$, $O(n)$ and $Sp(n)$ after having introduced the unitary restriction. In the next chapter we shall see that the coefficients ν_p have at the same time a deep topological significance for the group manifold.

THEOREM (7.11.A). *The Poincaré polynomial of* $GL(n)$ *is*

$$(1 + z)(1 + z^3) \cdots (1 + z^{2n-1}). \tag{11.3}$$

To the diagonal element

$$\{\epsilon\} = \{\epsilon_1 , \cdots , \epsilon_n\}$$

of the unitary group there corresponds in the adjoint representation the following linear transformation

$$K\{\epsilon\}: x'_{ik} = \epsilon_i x_{ik} \epsilon_k^{-1}.$$

Hence the determinant in (11.2) equals

$$\prod_{i,k} (z + \epsilon_i/\epsilon_k) = (1 + z)^n \psi(z)\bar{\psi}(z),$$

where

$$\psi(z) = \prod_{i<k} (\epsilon_i + z\epsilon_k).$$

On using again the difference product

$$\Delta = D(\epsilon_1 , \cdots , \epsilon_n)$$

and the integral of its squared modulus:

$$\int_0^1 \cdots \int_0^1 \Delta\bar{\Delta}\, d\varphi_1 \cdots d\varphi_n = \Omega\ (= n!),$$

one finds

$$(11.4)\qquad P(z) = \frac{(1+z)^n}{\Omega}\int_0^1 \cdots \int_0^1 \psi(z)\bar{\psi}(z)\Delta\bar{\Delta}\, d\varphi_1 \cdots d\varphi_n .$$

Although one would hardly deem it difficult to compute the elementary integral (11.4), nobody so far has succeeded in doing this directly. R. Brauer proceeded as follows.[23] By enumerating the invariants he showed that $P(z)$ is *majorized* by the polynomial (11.3),

$$P_0(z) = z^r + \overset{0}{\nu_1} z^{r-1} + \cdots + \overset{0}{\nu_r} \qquad (r = n^2);$$

i.e.

$$(11.5)\qquad \nu_p \leqq \overset{0}{\nu_p} .$$

The formula (11.4) is used for the value $z = 1$ only, with the result

$$(11.6)\qquad P(1) = 2^n.$$

As this means that the values of $P(z)$ and $P_0(z)$ coincide for $z = 1$:

$$\sum_p \nu_p = \sum_p \overset{0}{\nu_p}$$

one infers from (11.5) the desired equations $\nu_p = \overset{0}{\nu_p}$.

For $z = 1$ one finds in (11.4):

$$\psi(1)\Delta = \prod_{i<k} (\epsilon_i + \epsilon_k)(\epsilon_i - \epsilon_k) = \prod_{i<k} (\epsilon_i^2 - \epsilon_k^2) = \Delta_2 .$$

By the substitution $2\varphi_i \to \varphi_i$ one realizes at once that the integral of $\Delta_2\bar{\Delta}_2$ over $\varphi_1, \cdots, \varphi_n$ from 0 to 1 is the same as for $\Delta\bar{\Delta}$. Hence (11.6) as predicted.

Next we must try actually to enumerate the invariants of the type in question. The element A of $GL(n)$ induces the transformation

$$(11.7)\qquad X' = AXA^{-1}$$

in the adjoint group, $X = \| x_{ik} \|$ denoting a variable matrix. A typical matrix is the product $x\xi$ of a column x with a row ξ: $x_{ik} = x_i\xi_k$. Indeed, under the influence of the linear transformation $x' = Ax$ for the covariant vector x, the contravariant ξ changes into ξA^{-1} and hence

$$x'\xi' = A(x\xi)A^{-1}.$$

A form

$$(11.8)\qquad \omega = \sum \omega(i_1k_1, \cdots, i_pk_p) x^{(1)}_{i_1k_1} \cdots x^{(p)}_{i_pk_p}$$

which depends in linear and antisymmetric fashion on p matrices X and is invariant under the substitutions (11.7) may therefore without any danger of ambiguity be replaced by the invariant

$$(11.9)\qquad \sum \omega(i_1k_1, \cdots, i_pk_p) x^{(1)}_{i_1} \cdots x^{(p)}_{i_p} \xi^{(1)}_{k_1} \cdots \xi^{(p)}_{k_p}$$

depending linearly on p covariant and p contravariant vectors

$$x^{(1)}, \cdots, x^{(p)} \mid \xi^{(1)}, \cdots, \xi^{(p)}. \tag{11.10}$$

This formulation affords an opportunity to bring in the invariant-theoretic main theorem: each such invariant is expressible in terms of products (ξx), and hence is a linear combination of the invariants

$$(\xi^{(1)}x^{(1')})(\xi^{(2)}x^{(2')}) \cdots (\xi^{(p)}x^{(p')}) \tag{11.11}$$

each of which corresponds to a permutation

$$\sigma: 1 \to 1', 2 \to 2', \cdots, p \to p'$$

of p figures. Using the symbol σ for the "monomial" (11.11) as in Chapter V, §5, all our invariants are of the form

$$\omega = \sum a(\sigma)\cdot\sigma. \tag{11.12}$$

They must have the further property that they change into $\pm\omega$ when both sequences (11.10) are submitted to the same permutation ρ, with the $+$ sign for the even, the $-$ for the odd ρ's. As this process changes the monomial σ into $\rho\sigma\rho^{-1}$, one finds

$$\omega = \pm\sum_{\sigma} a(\sigma)\cdot\rho\sigma\rho^{-1} = \pm\sum_{\sigma} a(\rho^{-1}\sigma\rho)\cdot\sigma,$$

and by summing over all ρ one obtains ω as a linear combination of the special invariants

$$\Omega_\sigma = \sum_{\rho} \pm \rho\sigma\rho^{-1} \tag{11.13}$$

or as a combination (11.12) where

$$a(\rho^{-1}\sigma\rho) = \delta_\rho\cdot a(\sigma). \tag{11.14}$$

Decompose σ into distinct cycles:

$$\sigma = (1\,2\cdots h)(1^*\,2^*\cdots k^*)\cdots.$$

The cycle $\rho = (1\,2\cdots h)$ commutes with σ. If h is even then ρ is odd and (11.14) yields $a(\sigma) = 0$: σ contributes no term to (11.12) unless its cycles are of odd lengths $h, k, \cdots$. If $k = h$ $\{\equiv 1 \pmod 2\}$ the odd permutation

$$\rho = (11^*)(22^*)\cdots(hh^*)$$

commutes with σ, and so this case $h = k$ is ruled out. Arranging the *different* odd lengths in their natural order:

$$h < k < \cdots \tag{11.15}$$

one finds that one will have as many special invariants (11.13) as there are partitions of p:

$$p = h + k + \cdots$$

into odd unequal addends (11.15). This would at once result in the inequality (11.5) *provided cycles of length* $\geqq 2n$ *could be forbidden.*

Up to now only the *abstract scheme* of permutations has mattered. The limitation concerning the lengths of cycles comes about by taking into account the *representation* of the permutation σ by the monomial (11.11) with vectors x and ξ *in an n-dimensional space.*

When one multiplies two multilinear forms ω and ω' of the type (11.8) of orders p and q, the one depending on p matrices $X^{(1)}, \cdots, X^{(p)}$, the other on q other matrices $X^{(p+1)}, \cdots, X^{(p+q)}$, one gets a multilinear form of order $p + q$. However, the coefficients of the product will not be antisymmetric like those of ω and ω'. On applying *alternation* to the arguments X:

$$\sum \pm \omega(X^{\alpha_1}, \cdots, X^{\alpha_p})\cdot\omega'(X^{\beta_1} \cdots X^{\beta_q}), \tag{11.16}$$

the sum extending to all "mixtures"

$$\alpha_1, \cdots, \alpha_p; \beta_1, \cdots, \beta_q$$

of the first p indices $1, \cdots, p$ with the last q: $p + 1, \cdots, p + q$, this deficiency is removed. The form (11.16) shall now be denoted by $\omega\cdot\omega'$. In this way Ω_σ, (11.13), appears as a product $\Omega_h\cdot\Omega_k \cdots$ where Ω_h is the alternating sum

$$\sum \pm (\xi^{1'}x^{2'})(\xi^{2'}x^{3'}) \cdots (\xi^{h'}x^{1'})$$

running over all permutations $1' \cdots h'$ of the figures $1 \cdots h$. Hence we are called upon to prove: if p is odd, $= 2q - 1$, and $q > n$, then Ω_p is expressible as a combination of such Ω_σ as correspond to permutations σ of the p figures $1 \cdots p$ that break up into cycles of lengths $< p$.

Take the identity

$$\begin{vmatrix} (\xi^{i_1}x^{k_1}), \cdots, (\xi^{i_1}x^{k_q}) \\ \cdots\cdots\cdots\cdots\cdots\cdots \\ (\xi^{i_q}x^{k_1}), \cdots, (\xi^{i_q}x^{k_q}) \end{vmatrix} = 0$$

with

$$\begin{array}{c|c} i_1 \quad i_2 \quad \cdots \quad i_{q-1} \quad i_q & k_1 \quad k_2 \quad \cdots \quad k_{q-1} \quad k_q \\ \hline = \; 2, \; 4, \; \cdots, \; p-1, \; p & 3, \; 5, \; \cdots, \; p, \; 1 \end{array}$$

and multiply by

$$(\xi^1x^2)(\xi^3x^4) \cdots (\xi^{p-2}x^{p-1}):$$

$$\sum_\tau \delta_\tau(\xi^1x^2)(\xi^2x^{\tau_1}) \cdots (\xi^{p-2}x^{p-1})(\xi^{p-1}x^{\tau_{q-1}})(\xi^px^{\tau_q}) = 0. \tag{11.17}$$

τ is the permutation

$$\begin{pmatrix} 3 & 5 & \cdots & p & 1 \\ \tau_1 & \tau_2 & \cdots & \tau_{q-1} & \tau_q \end{pmatrix}.$$

The corresponding term on the left is the monomial σ where

$$\sigma = \begin{pmatrix} 1 & 2 & 3 & 4 & \cdots & p-2, & p-1, & p \\ 2 & \tau_1 & 4 & \tau_2 & \cdots & p-1, & \tau_{q-1}, & \tau_q \end{pmatrix}.$$

From our equation (11.17):

$$\sum_\tau \delta_\tau \cdot \sigma = 0$$

one derives

$$(11.18) \qquad \sum_\tau \delta_\tau \cdot \Omega_\sigma = 0.$$

$\tau^{-1}\sigma$ = cycle $(1\,2\,3 \cdots p)$. *If σ is also a cycle of length p* then σ and $\tau^{-1}\sigma$ are even; hence τ even, $\delta_\tau = +1$. Let us suppose that this event happens N times; it happens at least once, viz., when τ is the identity. When σ is a single cycle of length p, it is of the following type:

$$(p, 2\kappa_1 - 1, 2\kappa_1, 2\kappa_2 - 1, 2\kappa_2, \cdots, 2\kappa_{q-1} - 1, 2\kappa_{q-1}),$$

and

$$\pi\sigma\pi^{-1} = (1, 2, \cdots, p)$$

with π denoting the permutation

$$\begin{pmatrix} p, & 2\kappa_1 - 1, & 2\kappa_1, & 2\kappa_2 - 1, & 2\kappa_2, & \cdots \\ p, & 1, & 2, & 3, & 4, & \cdots \end{pmatrix}.$$

π leaves p fixed and *exchanges pairs* $(2\kappa - 1, 2\kappa)$. A transposition of two pairs like

$$\begin{pmatrix} 1 & 2 & 3 & 4 \\ 3 & 4 & 1 & 2 \end{pmatrix}$$

is even. Hence π is even, and all the N terms Ω_σ corresponding to the cyclic σ are $= \Omega_p$ *with the same sign*. Consequently (11.18) is turned into an equation

$$N \cdot \Omega_p + \cdots = 0$$

where the dots indicate a number of Ω_σ whose σ break up into several cycles of shorter lengths than p.

In combination with (11.6) this not only proves our theorem, but at the same time establishes the linear independence of the basic invariants to which our construction leads:

THEOREM (7.11.B). *The invariants*

$$\Omega_{h_1} \cdot \Omega_{h_2} \cdots$$

corresponding to all partitions $p = h_1 + h_2 + \cdots$ *into odd unequal numbers* $h_1, h_2, \cdots$ *less than* $2n$:

$$h_1 < h_2 < \cdots ; h_\alpha \equiv 1 \pmod 2), 0 < h_\alpha < 2n,$$

are linearly independent and constitute a basis for the linear invariants of order p *of the adjoint group.*

The integral (11.4) is now evaluated as the polynomial (11.3).

Similar results may be derived for the orthogonal and symplectic group in a similar way. We content ourselves with stating the result:[23]

THEOREM (7.11.C). *The Poincaré polynomials of* $Sp(2\nu)$, $O(2\nu + 1)$ *on the one hand, and of* $O(2\nu)$ *on the other hand, are*

$$(1 + z^3)(1 + z^7) \cdots (1 + z^{4\nu-1})$$

and

$$(1 + z^3)(1 + z^7) \cdots (1 + z^{4\nu-5})(1 + z^{2\nu-1})$$

respectively.

CHAPTER VIII

GENERAL THEORY OF INVARIANTS

A. Algebraic Part

1. Classic invariants and invariants of quantics. Gram's theorem

In the classic theory of invariants[1] one deals with the group $GL(n)$ and considers an arbitrary (covariant) form u of given degree r depending on a contravariant vector ξ; we write it as

$$u = \sum \frac{r!}{r_1! \cdots r_n!} u_{r_1 \cdots r_n} \xi_1^{r_1} \cdots \xi_n^{r_n} \quad (r_1 + \cdots + r_n = r). \tag{1.1}$$

A homogeneous polynomial $J(u)$ of the coefficients u of degree μ is an invariant of weight g if

$$J(u') = \Delta^g \cdot J(u)$$

where $u'_{r_1 \cdots r_n}$ are the coefficients of the form into which u changes by the arbitrary substitution

$$(x'_i = \sum_k a_{ik} x_k) \qquad \xi_i = \sum a_{ki} \xi'_k \tag{1.2}$$

and Δ denotes $\det(a_{ik})$. Instead of one form u serving as argument in J one might have several arbitrary ground forms $u, v, \cdots$ of given degrees $r, r', \cdots$. If $\mu, \mu', \cdots$ are the degrees of $J(u, v, \cdots)$ with respect to $u, v, \cdots$ one must have

$$ng = r\mu + r'\mu' + \cdots \tag{1.3}$$

as one realizes by comparing the degrees of both sides of the equation

$$J(u', v', \cdots) = \Delta^g \cdot J(u, v, \cdots)$$

with respect to the transformation coefficients a_{ik}. For simplicity's sake we shall most of the time talk about one or two ground forms although we have an arbitrary number in mind.

If J depends on a contravariant vector ξ besides u and v, and again

$$J(u', v'; \xi') = \Delta^g \cdot J(u, v; \xi),$$

J is called a *covariant*. J is supposed to be a homogeneous form of some degree m in ξ; then

$$ng + m = r\mu + r'\mu'.$$

g is now capable of negative values. The ground forms u, v themselves are absolute covariants or covariants of weight zero.

Examples. (1) If $u^{(1)}, \cdots, u^{(n)}$ are n forms then the functional determinant or *Jacobian*

$$\begin{vmatrix} \frac{\partial u^{(1)}}{\partial \xi_1}, & \cdots, & \frac{\partial u^{(1)}}{\partial \xi_n} \\ \cdots & \cdots & \cdots \\ \frac{\partial u^{(n)}}{\partial \xi_1}, & \cdots, & \frac{\partial u^{(n)}}{\partial \xi_n} \end{vmatrix}$$

is a covariant of weight 1. Indeed

$$du^{(1)} = \frac{\partial u^{(1)}}{\partial \xi_1} d\xi_1 + \cdots + \frac{\partial u^{(1)}}{\partial \xi_n} d\xi_n,$$
$$\cdots\cdots\cdots\cdots\cdots\cdots$$
$$du^{(n)} = \frac{\partial u^{(n)}}{\partial \xi_1} d\xi_1 + \cdots + \frac{\partial u^{(n)}}{\partial \xi_n} d\xi_n$$

is a system of invariant linear forms of the differentials $d\xi_1, \cdots, d\xi_n$. Under the influence of a substitution

$$d\xi_i = \sum_k a_{ki} d\xi_k'$$

the determinant of n such forms is multiplied by the transformation determinant Δ.

(2) For a single form u the "Hessian"

$$\left| \frac{\partial^2 u}{\partial \xi_i \partial \xi_k} \right|$$

is a covariant of weight 2, for the reason that

$$\sum_{i,k} \frac{\partial^2 u}{\partial \xi_i \partial \xi_k} d\xi_i d\xi_k$$

is an invariant quadratic form of the differentials $d\xi_i$.

The classical notion of a covariant lends itself to an immediate generalization by admitting several contravariant vectors $\xi, \eta, \cdots$ as arguments; we then speak of a multiple covariant. A system of equations

$$K_1(u, v) = 0, \cdots, K_p(u, v) = 0 \tag{1.4}$$

whose left sides are polynomials homogeneous in the coefficients $u_{r_1 \cdots r_n}$ of u as well as in those of v is *of invariant significance* providing each set of values

$$u_{r_1 \cdots r_n}, \qquad v_{r_1' \cdots r_n'}$$

satisfying (1.4) also satisfies the equations

$$K_1(u', v') = 0, \cdots, K_p(u', v') = 0$$

by whatever transformation (1.2) the u', v' arise from u, v.

THEOREM (8.1.A). (Gram's theorem.) *A system of relations* (1.4) *of invariant significance can always be expressed by the vanishing of a number of multiple absolute covariants.*

The proof rests upon a simple formal device we describe as follows. Let $\overset{1}{\eta}, \cdots, \overset{n}{\eta}$ be any n contravariant vectors and replace the argument ξ in the form $u = u(\xi)$ by

$$t_1\overset{1}{\eta} + \cdots + t_n\overset{n}{\eta}$$

with indeterminate t's:

$$(1.5)\quad u(t_1\overset{1}{\eta} + \cdots + t_n\overset{n}{\eta}) = \sum \frac{r!}{r_1! \cdots r_n!} t_1^{r_1} \cdots t_n^{r_n} u^*_{r_1 \cdots r_n}(\overset{1}{\eta}, \cdots, \overset{n}{\eta}).$$

The whole expression is an absolute covariant and so is each coefficient

$$(1.6)\qquad u^*_{r_1 \cdots r_n}(\overset{1}{\eta}, \cdots, \overset{n}{\eta}).$$

The same is true for any homogeneous polynomial depending on them. If one takes the unit vectors

$$\overset{1}{e} = (1, 0, \cdots, 0), \cdots, \qquad \overset{n}{e} = (0, 0, \cdots, 1)$$

for $\overset{1}{\eta}, \cdots, \overset{n}{\eta}$ then the $u^*_{r_1 \cdots r_n}$ change back into the given coefficients $u_{r_1 \cdots r_n}$. If $\overset{1}{\eta}, \cdots, \overset{n}{\eta}$ are linearly independent the equation

$$(1.7)\qquad \xi = t_1\overset{1}{\eta} + \cdots + t_n\overset{n}{\eta}$$

may be looked upon as a linear transformation introducing the new coördinates $t_1, \cdots, t_n$ instead of the components $\xi_1, \cdots, \xi_n$ of ξ.

Hence the assumption of invariant significance allows us to infer from (1.4) the equations

$$(1.8)\qquad K_1(u^*, v^*) = 0, \cdots, \qquad K_p(u^*, v^*) = 0$$

which vice versa lead back to (1.4) by the specialization

$$\overset{1}{\eta} = \overset{1}{e}, \cdots, \overset{n}{\eta} = \overset{n}{e}.$$

The system (1.8) is of the desired form:

$$(1.9)\qquad C_1(u, v; \overset{1}{\eta}, \cdots, \overset{n}{\eta}) = 0, \cdots, C_p(u, v; \overset{1}{\eta}, \cdots, \overset{n}{\eta}) = 0.$$

The absolute covariants C_α are required to vanish for the given values of u, v identically with respect to the arguments $\overset{1}{\eta}, \cdots, \overset{n}{\eta}$.

Very likely the identity

$$C_1(u, v; \overset{1}{\eta}, \cdots, \overset{n}{\eta}) = 0$$

will absorb a certain number of the equations (1.4) and not merely the first one. Then the number of identities (1.9) may be reduced accordingly.

Our multiple covariants C_α depend on n vectors $\overset{i}{\eta}$. This however is by no means a surprising feature of our theorem. For even if the C_α involved any number of argument vectors η one could reduce their number to n by means of

Capelli's general identity. By application of the special identity one is even capable of bringing down that number to $n - 1$ provided one admits covariants of a weight $\geqq 0$.

An arbitrary form depending on two contravariant vectors ξ and η in given degrees r, r' is not a primitive quantity; it is rather the $\times$-product of the primitive quantities of signatures

$$(r, 0, \cdots, 0) \quad \text{and} \quad (r', 0, \cdots, 0).$$

One ought to replace it by the string of independent primitive quantities into which it may be split. For this reason we reject the notion of multiple covariants as not genuine. It is reasonable to break up each of our equations (1.9) according to the several symmetry diagrams. Our critical attitude towards the classical concepts once aroused, we realize that we should not have limited ourselves to ground forms but should rather allow *quantics* of any signature $(f_1, \cdots, f_n)$ to figure as arguments in our invariants.[2] And when considering covariants we ought to take the term in the wide sense introduced in Chapter I, §5, as meaning a quantic J depending on a number of variable quantics $u, v, \cdots$ of pre-assigned signatures

$$(r_1, \cdots, r_n), \qquad (r_1', \cdots, r_n'), \cdots.$$

This scheme comprises contravariant ground forms and even mixed "concomitants" (depending on some contravariant and some covariant vectors) beside covariant forms. Assuming $J(u, v, \cdots)$ to be of the respective degrees $\mu, \mu', \cdots$ one has the relation

$$f = r\mu + r'\mu' + \cdots \tag{1.10}$$

between the degrees

$$f = f_1 + \cdots + f_n; \qquad r = r_1 + \cdots + r_n, \qquad r' = r_1' + \cdots + r_n', \cdots$$

of the quantics $J; u, v, \cdots$, generalizing (1.3), as follows readily from the fact that under the substitution

$$x_i' = ax_i$$

the components of a quantic of degree f are multiplied by a^f. If for an argument u the signature $(r_1, \cdots, r_n)$ has a negative r_n, one can replace the representation according to which the components of u are transformed by the representation of signature $(r_1 + e, \cdots, r_n + e)$, choosing e such that $r_n + e \geqq 0$. Indeed this has merely the effect that the signature $(f_1, \cdots, f_n)$ of the dependent quantity J is changed into $(f_1 + \mu e, \cdots, f_n + \mu e)$. Hence we might assume without any essential loss of generality that $r_n \geqq 0$, or that u ranges over all tensors of rank r and of symmetry $T(r_1 \cdots r_n)$.

THEOREM (8.1.B). (Gram's generalized theorem.) *A system of relations*

$$K_1(u, v, \cdots) = 0, \cdots, K_p(u, v, \cdots) = 0$$

between quantics u, v, ⋯ of prescribed signatures $(r_1, \cdots, r_n), \cdots$, *when of invariant significance, is equivalent to a system which states the vanishing of a number of covariants, i.e. of quantics depending on u, v, ⋯ .*

PROOF. We may assume $r_n \geqq 0$, or that u ranges over all tensors $F(i_1 \cdots i_r)$ of a given rank r and given symmetry $T = T(r_1 \cdots r_n)$. In the invariant form

$$\sum_{i_1, i_2, \cdots} F(i_1 i_2 \cdots) \xi_{i_1} \xi'_{i_2} \cdots \tag{1.11}$$

one substitutes for the contravariant vectors $\xi, \xi', \cdots$ linear combinations of n such vectors $\eta^1, \cdots, \eta^n$:

$$\begin{aligned} \xi &= t_1 \eta^1 + \cdots + t_n \eta^n, \\ \xi' &= t'_1 \eta^1 + \cdots + t'_n \eta^n, \\ &\cdots\cdots\cdots\cdots\cdots\cdots \end{aligned}$$

(1.11) is changed into a multilinear form of the sets of variables $(t_1, \cdots, t_n)$, $(t'_1, \cdots, t'_n), \cdots$:

$$\sum_{i_1, i_2, \cdots} F^*(i_1 i_2 \cdots ; \eta^1 \cdots \eta^n) t_{i_1} t'_{i_2} \cdots .$$

The coefficients

$$F^*(i_1 i_2 \cdots ; \eta^1 \cdots \eta^n)$$

are absolutely invariant functions of the n contravariant arguments $\eta^1, \cdots, \eta^n$. From here on the proof proceeds as before.

2. The symbolic method

The symbolic method is best illustrated by the classical example of an invariant $J(u)$ of degree μ depending on an arbitrary covariant form u of degree r. On specializing u as the rth power of a linear form:

$$u_\xi^r = (u_1 \xi_1 + \cdots + u_n \xi_n)^r, \qquad u_{r_1 \cdots r_n} = u_1^{r_1} \cdots u_n^{r_n}, \tag{2.1}$$

$J(u)$ changes into an invariant $j(u)$ depending on a covariant vector $u = (u_1, \cdots, u_n)$. In this primitive form our method aiming at replacing form invariants by vector invariants is of no great avail because $J(u)$ is not unambiguously recognizable from its symbolic representative $j(u)$. However this deficiency is repaired by first completely polarizing $J(u)$. (Polarization as applied to arguments u is called the Aronhold process.) The polarized form $J(u^{(1)}, \cdots, u^{(\mu)})$ depends linearly on μ arbitrary forms $u^{(1)}, \cdots, u^{(\mu)}$ of degree r; it is moreover symmetric in its μ arguments and leads back to $J(u)$ by the identification

$$u^{(1)} = \cdots = u^{(\mu)} = u.$$

On substituting the rth power of μ linear forms for $u^{(1)}, \cdots, u^{(\mu)}$, the form invariant $J(u^{(1)}, \cdots, u^{(\mu)})$ is changed into an invariant $j(u^{(1)}, \cdots, u^{(\mu)})$ depending on μ covariant vectors $u^{(1)}, \cdots, u^{(\mu)}$; in each of its arguments j is of degree r. j is called the *symbolic expression* of J with the covariant vectors $u^{(1)}, \cdots, u^{(\mu)}$ as μ *equivalent symbols* substituting for the same form u. One wins $J(u^{(1)}, \cdots, u^{(\mu)})$ back from $j(u^{(1)}, \cdots, u^{(\mu)})$ by observing that a linear form in the coefficients $u_{r_1 \cdots r_n}$ of u:

$$\sum \gamma_{r_1 \cdots r_n} u_{r_1 \cdots r_n} \qquad (\gamma_{r_1 \cdots r_n} \text{ constants})$$

is uniquely determined by its value for the specialized u, (2.1), viz. by

$$\sum \gamma_{r_1 \cdots r_n} u_1^{r_1} \cdots u_n^{r_n}.$$

If for the moment we designate the vector arguments in j by $u, v, \cdots$ rather than by $u^{(1)}, u^{(2)}, \cdots$:

$$j(u, v, \cdots) = \sum \lambda(r_1 \cdots r_n ; s_1 \cdots s_n ; \cdots) u_1^{r_1} \cdots u_n^{r_n} v_1^{s_1} \cdots v_n^{s_n} \cdots$$
$$(r_1 + \cdots + r_n = s_1 + \cdots + s_n = \cdots = r),$$

$J(u)$ is produced from $j(u, v, \cdots)$,

$$j(u, v, \cdots) \rightarrow J(u),$$

in this way:

$$J(u) = \sum \lambda(r_1 \cdots r_n ; s_1 \cdots s_n ; \cdots) u_{r_1 \cdots r_n} u_{s_1 \cdots s_n} \cdots.$$

The process works even when $j(u, v, \cdots)$ is not symmetric in its μ equivalent vector arguments $u, v, \cdots$. One has

$$j(v, u, \cdots) \rightarrow J(u)$$

for any permutation $v, u, \cdots$ of the symbols $u, v, \cdots$.

By the first main theorem we know how to deal with vector invariants: they are all expressible in terms of bracket factors $[uv \cdots]$. Thus the symbolic method provides a rule for computing all invariants $J(u)$ of given degree μ, as explicit and finitistic as anybody could desire: one forms all possible products of bracket factors of μ vectors $u^{(1)}, \cdots, u^{(\mu)}$ that are of degree r in each of them and performs the translation $j \rightarrow J$ into non-symbolic form. Every invariant of degree μ is a linear combination with constant coefficients of the J's thus obtained. Great as this accomplishment is, one ought to point out, however, that the method is far from reducing the construction of a finite integrity basis for form invariants to the same for vector invariants. For the number of symbolic vector arguments $u^{(1)}, \cdots, u^{(\mu)}$ we have to introduce is dependent on the degree of $J(u)$, and we must have an unlimited supply of such symbols at our disposal when we are to take into account invariants J of all possible degrees.

The generalization to invariants $J(u, v, \cdots)$ depending on more than one form $u, v, \cdots$ is immediate. When dealing with a covariant $J(u, v; \xi)$ the symbolic expression

$$j(u^{(1)}, \cdots, u^{(\mu)}; v^{(1)}, \cdots, v^{(\mu')}; \xi)$$

will depend on a contravariant vector ξ besides the symbolic covariant vectors

$$u^{(1)}, \cdots, u^{(\mu)}; v^{(1)}, \cdots, v^{(\mu')}$$

and hence will be expressible in terms of Latin bracket factors and products of type $(\xi u) = u_\xi ; v_\xi ; \cdots$.

Examples. (a) Discriminant

$$D = u_{11}u_{22} - u_{12}^2 \tag{2.2}$$

of a binary quadratic form

$$u_{11}\xi_1^2 + 2u_{12}\xi_1\xi_2 + u_{22}\xi_2^2 . \tag{2.3}$$

Polarization changes $2D$ into

$$u_{11}v_{22} + u_{22}v_{11} - 2u_{12}v_{12}$$

which by specialization becomes

$$u_1^2v_2^2 + u_2^2v_1^2 - 2u_1u_2v_1v_2 = (u_1v_2 - u_2v_1)^2.$$

Consequently

$$\tfrac{1}{2}[uv]^2$$

is the symbolic expression of D.

(b) The Jacobian of three ternary forms which we write symbolically as

$$u_\xi^r, \qquad v_\xi^{r'}, \qquad w_\xi^{r''}$$

is easily found to be

$$rr'r''u_\xi^{r-1}v_\xi^{r'-1}w_\xi^{r''-1}[uvw].$$

No polarization is needed because the Jacobian is linear in the coefficients of the forms.

(c) The Hessian of the ternary form

$$f = u_\xi^r (= v_\xi^r = w_\xi^r)$$

is easily rendered into the symbolic expression

$$\tfrac{1}{6}r^3(r - 1)^3(u_\xi v_\xi w_\xi)^{r-2}[uvw]^2.$$

In order to secure a natural generality we now admit again as arguments $u, v, \cdots$ in our invariants J variable quantics of any pre-assigned signatures $(r_1, \cdots, r_n), \cdots$. How are we to adapt the symbolic method to this general case? We saw before that to assume $r_n \geqq 0$ means no serious restriction. u then ranges over all tensors $F(i_1 \cdots i_r)$ of rank $r = r_1 + \cdots + r_n$ and of symmetry $T(r_1 \cdots r_n)$. Let $\mathbf{e}$ be the generating idempotent of that symmetry class, a numerical multiple of the Young symmetrizer $\mathbf{c}$. We replace the tensor F by $\mathbf{e}F$ so that F now varies over *all* tensors of rank r:

$$J^*(F, \cdots) = J(\mathbf{e}F, \cdots).$$

Because $\mathbf{e}F = F$ for tensors F in $P(r_1 \cdots r_n)$, we have thus succeeded in constructing an invariant $J^*(F, \cdots)$ defined for *arbitrary* tensors $F, \cdots$ which coincides with the given $J(F, \cdots)$ in the domain:

$$F \text{ in } P(r_1 \cdots r_n), \cdots$$

for which J was defined. After having removed any symmetry restriction we can now perform the Aronhold process on J^* and then specialize F as the product of r variable covariant vectors $u, v, \cdots, w$:

$$F(i_1 i_2 \cdots i_r) = u_{i_1} v_{i_2} \cdots w_{i_r}.$$

The original invariant J is reproducible by a simple process from the vector invariant j thus obtained.

When we pass from invariants to the study of covariant quantics $Q(u, v, \cdots)$ it is natural to assume that only the differences $f_i - f_k$ in the signature $(f_1, \cdots, f_n)$ of Q are given; they characterize the behavior of Q under *unimodular* transformations A. But for a factor Δ^e ($e = f_n$) in its transformation law, Q behaves as a tensor $Q(i_1 \cdots i_f)$ of symmetry

$$T(f_1 - e, \cdots, f_{n-1} - e, f_n - e) \qquad \{f = \sum f_i - ne\}$$

and

$$\sum Q(i_1 \cdots i_f)\xi_{i_1} \cdots \zeta_{i_f}$$

is then an invariant of weight e depending on f contravariant vectors $\xi, \cdots, \zeta$; in its dependence on $u, v, \cdots$ it lends itself readily to the symbolic treatment.

3. The binary quadratic

THEOREM (8.3.A). *Every invariant of the binary quadratic is expressible in terms of the discriminant, every covariant (simple or multiple) in terms of the discriminant, the (polarized) form itself, and bracket factors.*

We take up this simple case as an illustration of how, despite our critical remarks, the symbolic method, when combined with arguments of a different order, may sometimes be used for the actual determination of an integrity basis.

The symbolic expression of an invariant $J(u)$ of degree μ for the quadratic (2.3) will be an invariant $j(a, b, c, \cdots)$ depending on μ equivalent symbolic binary vectors $a, b, c, \cdots$ and of degree 2 in each of them. j is expressible by bracket factors of the type

$$[ab] = a_1 b_2 - a_2 b_1.$$

A product of such factors in which each argument occurs exactly twice breaks up into a number of closed chains like

$$[ab][bc][cd] \cdots [ha] \to K \tag{3.1}$$

where $a, b, c, d, \cdots, h$ are distinct. If the length of the chain is odd we obtain by reversing the order of arguments in each bracket:

$$[ah] \cdots [cb][ba] \to -K.$$

As the right side differs from (3.1) merely by the labels attached to the equivalent symbols one finds in this case for the actual invariant K symbolically expressed by (3.1):

$$K = -K, \quad \text{hence} \quad K = 0.$$

Let the length now be even $= 2l$. The lowest case of length 2 is settled by the relation

$$[ab][ba] = -[ab]^2 \to -2D \qquad (D = \text{discriminant}).$$

In the higher cases we make use of the identity

$$(3.2) \qquad [ab][cd] + [ca][bd] + [bc][ad] = 0.$$

Substituting

$$-[ca][bd] - [bc][ad]$$

for the product $[ab][cd]$ in (3.1) we get

$$K = K_l = -[ca][bd][bc][de] \cdots - [bc]^2[ad][de] \cdots$$
$$= -[ac][cb][bd][de] \cdots - [bc]^2 \cdot [ad][de] \cdots$$

or in passing to the non-symbolic entities:

$$K_l = -K_l - 2D \cdot K_{l-1} \quad \text{or} \quad K_l = -D \cdot K_{l-1}.$$

By induction with respect to l we thus find

$$K = 0 \quad \text{or} \quad 2(-D)^l$$

according as the length of the chain is odd or even. This settles the question of invariants.

In taking up covariants we replace the contravariant vector $\xi = (\xi_1, \xi_2)$ by the covariant vector

$$\xi^{\cdot} = (\xi_2, -\xi_1)$$

according to (4.5.1), thus changing the signature from $(0, -1)$ to $(1, 0)$. Besides closed chains one will now encounter chains joining two Greek symbols by Latin links:

$$[\xi^{\cdot} a][ab] \cdots [gh][h\eta^{\cdot}].$$

The shortest chain, besides the bracket factor $[\xi^{\cdot}\eta^{\cdot}] = [\eta\xi]$ itself, is

$$-[\xi^{\cdot} a][a\eta^{\cdot}] = a_\xi \cdot a_\eta,$$

and this is the symmetric bilinear form corresponding to the given quadratic; the longer ones, as far as they are not zero, arise from it by multiplication with powers of the discriminant. The treatment is the same as before.

The symbolic method and its applications have found so widespread circulation through the current textbooks of invariant theory that we shall content ourselves with this one example.

4. Irrational methods

The symbolic method, as a matter of fact, is far from being the only road along which one may successfully approach the determination of a finite integrity basis of invariants in concrete simple cases. Sometimes the use of appropriate irrationalities gives surprisingly quick returns. As an example we prove:

THEOREM (8.4.A). *Every invariant of a quadratic*

$$\sum g_{ik}\xi_i\xi_k \qquad (g_{ki} = g_{ik}) \tag{4.1}$$

in n variables is expressible in terms of the discriminant

$$D = \det(g_{ik}).$$

Let c be the value of the given invariant J for the unit form

$$\xi_1^2 + \cdots + \xi_n^2. \tag{4.2}$$

If it is of weight h, J will equal

$$c \cdot \Delta^h$$

for the form (4.1) proceeding from (4.2) by the linear substitution

$$\xi_i \to \sum a_{ki}\xi_k$$

with the determinant $\Delta = \det(a_{ik})$. Since

$$D = \Delta^2$$

one finds

$$J^2(g) = c^2 \cdot D^h. \tag{4.3}$$

Let now $g_{ik} = g_{ki}$ be arbitrarily given values with a non-vanishing determinant D. *After suitable quadratic extensions of the underlying field* the form with these coefficients is transformable into the unit form. Hence (4.3) holds good for any such values g_{ik} and is therefore an identity. Taking into account that D is an *irreducible* polynomial of the $\frac{1}{2}n(n+1)$ variables g_{ik} one derives from (4.3) that $J(g)$ itself must be a constant multiple of a power of D, or that h is even and

$$J(g) = c \cdot D^{h/2}.$$

In order to prove the irreducibility of the symmetric determinant

$$D_n = \begin{vmatrix} g_{11}, & \cdots, & g_{1n} \\ \cdots & \cdots & \cdots \\ g_{n1}, & \cdots, & g_{nn} \end{vmatrix}$$

we make the inductive hypothesis that D_{n-1} is irreducible. D_n is a linear function of the variable g_{nn} :

$$D_n = D_{n-1} \cdot g_{nn} + D_n(g_{nn} = 0).$$

Were D_n resolvable into two factors, one of them would be of degree 1 and the other of degree 0 in g_{nn} :

$$D_n = (Bg_{nn} + B')\cdot C$$

(B, B', C independent of g_{nn}); or

$$D_{n-1} = BC, \qquad D_n(g_{nn} = 0) = B'C.$$

Because D_{n-1} is irreducible, either B or C must be a constant independent of all the g_{ik}. Since the second case does not lead to a real decomposition of D_n we may assume $B = 1$ and then get

$$D_n(g_{nn} = 0) = B'\cdot D_{n-1}. \tag{4.4}$$

The simple example

$$\| g_{ik} \| = E_{n-2} \dot{+} \begin{Vmatrix} a & b \\ b & 0 \end{Vmatrix}$$

with

$$D_n(g_{nn} = 0) = -b^2, \qquad D_{n-1} = a$$

shows that $D_n(g_{nn} = 0)$ is not divisible by D_{n-1}, contrary to (4.4).

The binary cubic

$$f = u_0\xi_1^3 + u_1\xi_1^2\xi_2 + u_2\xi_1\xi_2^2 + u_3\xi_2^3$$

can easily be treated by a similar "irrational" procedure. On putting $u_0 = 1$ and $\xi = \xi_1/\xi_2$ one solves the cubic equation:

$$\xi^3 + u_1\xi^2 + u_2\xi + u_3 = (\xi - \alpha_1)(\xi - \alpha_2)(\xi - \alpha_3).$$

An invariant J will thus become a symmetric polynomial of the roots α_1, α_2, α_3. From the fact that the only projectively invariant relation between three points α_1, α_2, α_3 on a straight line is coincidence, one realizes easily that the invariant must be a constant multiple of a power of the discriminant

$$D = 27(\alpha_1 - \alpha_2)^2(\alpha_2 - \alpha_3)^2(\alpha_3 - \alpha_1)^2.$$

The only absolute projective invariant depending on four points ξ, α_1, α_2, α_3 is their cross ratio

$$(\xi - \alpha_1)(\alpha_3 - \alpha_2)/(\xi - \alpha_2)(\alpha_3 - \alpha_1) = \eta_1/\eta_2 = \eta$$

which however takes on 6 values

$$\eta, \quad 1/\eta, \quad 1 - \eta, \quad 1/(1 - \eta), \quad \eta/(\eta - 1), \quad (\eta - 1)/\eta$$

under permutation of the three roots α_1, α_2, α_3. Without much ado this leads to a complete list of the covariants. Beside the form f itself one obtains two other forms which are readily verified to be the Hessian $-4H$ of f, and the Jacobian t of f and H. Among the four covariants D, f, H, t there is one relation (syzygy)

$$4H^3 = t^2 + D\cdot f^2.$$

In the textbooks on invariants one finds these results derived by the symbolic method.[3]

5. Side remarks

Before plunging into deep waters we skim the surface with one or two observations about general invariants. We take up the situation as described in Chapter I, §5: given a group γ and a number of variable quantities $x, y, \cdots$ the types of which are defined by (irreducible) representations of γ, we investigate *relative invariants* J of such arguments. As a polynomial of the components of $x, y, \cdots, J(x, y, \cdots)$ might be reducible; we factorize it into prime factors

$$J = c \cdot J_1^{e_1} \cdots J_h^{e_h}.$$

The question arises as to whether the prime polynomials $J_1, \cdots, J_h$ will be relative invariants. This is true under a certain restriction concerning the nature of γ:

THEOREM (8.5.A). *Under the assumption that the group γ has no subgroup of finite index (except itself), each prime factor of a relative invariant is a relative invariant.*

After the transformations

$$x \to x', \qquad y \to y', \cdots$$

induced by the group element s we have an equation

$$J(x', y', \cdots) = \lambda \cdot J(x, y, \cdots)$$

with $\lambda = \lambda(s)$ as a multiplier, or

$$J_1^{e_1}(x'y' \cdots) J_2^{e_2}(x'y' \cdots) \cdots = \lambda \cdot J_1^{e_1}(xy \cdots) J_2^{e_2}(xy \cdots) \cdots.$$

Hence $J_1' = J_1(x', y', \cdots)$ must be a product of the prime factors on the right side; the same for $J_2', \cdots$. In each case the number of factors (which is at least one) must be exactly one because otherwise the product on the left side would consist of more than $e_1 + e_2 + \cdots$ prime factors. Consequently we shall have equations like

$$J_i' = \lambda_i \cdot J_{\alpha_i}$$

where $\alpha_1, \cdots, \alpha_h$ is a permutation of $1, \cdots, h$. Those elements s for which this permutation is the identity evidently constitute a subgroup of γ of an index $\leqq h!$.

For the full linear group and the classical case of form invariants $J(u, v, \cdots)$ it is more convenient to consider the transform $J(u', v', \cdots)$ as a polynomial of $u, v, \cdots$ *and of the transformation coefficients* a_{ik}. The multiplier $\lambda(A)$ is $= \Delta^g$. Operating in the domain just described we should find an equation like

$$J_1(u', v', \cdots) = \Delta^{g_1} \cdot J_\alpha(u, v, \cdots).$$

The specialization $a_{ik} = \delta_{ik}$ shows at once that $\alpha = 1$. Here we need not enter upon the structure of the group:

THEOREM (8.5.B). *In case of the full linear group each prime factor of an invariant depending on quantics $u, v, \cdots$ is again an invariant.*

As an application take the discriminant D of the binary cubic. Because there exists no invariant of lower degree it is of necessity irreducible. This fact has to be used in carrying out the sketch given above for the proof that every covariant is expressible in terms of the four basic covariants D, f, H, t.

Our second remark aims in an entirely different direction. Let us suppose that the invariants to be considered depend among others on a number of quantities $x, x', \cdots$ all of the same type as described by the representation $A(s)$ of degree n.

THEOREM (8.5.C). (Pascal's theorem.) *Invariants depending on m quantities $x, x', \cdots$ of the same type $A(s)$ of degree n can be expressed by means of polarization and linear combination in terms of invariants depending on not more than n such quantities. On making use of the relative invariant $[xx' \cdots]$ whose multiplier $= |A(s)|$ one may even reduce the number of arguments to $n - 1$.*

The proposition is an immediate consequence of Capelli's identities. For instance, in studying simultaneous invariants of a number of binary cubics one can limit oneself to four or even three such cubics.

6. Hilbert's theorem on polynomial ideals

As pointed out in Chapter II, §2, Hilbert founded the proof of the invariant-theoretical main theorems on a general proposition concerning polynomial ideals that is one of the simplest and most important in the whole of algebra. Let us consider a ring R *in which every ideal has a finite ideal basis.* A field k or the ring of ordinary integers may serve as examples. Hilbert's theorem states that this property is not lost under adjunction of an indeterminate.

THEOREM (8.6.A). *If every ideal in the ring R has a finite ideal basis then the same is true for the ring $R[x]$.*

In this modern generalized form the proposition at once suggests the steps one has to take in proving it. I refer the reader to van der Waerden's *Moderne Algebra.*[4] Repeated application carries the theorem over to the ring $R[x_1 \cdots x_n]$ of polynomials in R of any number of indeterminates $x_1, \cdots, x_n$. If one specializes R either as a field k or as the ring of all common integers one obtains the two propositions as formulated by Hilbert himself:[4]

THEOREM (8.6.B). *If k is any field, then every ideal in the ring $k[x_1 \cdots x_n]$ has a finite ideal basis. The same holds good if k be replaced by the ring of common integers.*

Take any set $\mathfrak{S} = \{\alpha\}$ of numbers α in a ring R in which every ideal has a finite ideal basis. All numbers of the form

$$\lambda_1\alpha_1 + \lambda_2\alpha_2 + \cdots \tag{6.1}$$

where $\alpha_1, \alpha_2, \cdots$ is any finite sequence of elements in $\mathfrak{s}$ while the λ's are arbitrary elements of R, constitute an ideal $(\mathfrak{s})$, the smallest ideal containing $\mathfrak{s}$. After determining an ideal basis for $(\mathfrak{s})$ and expressing each of its elements in the form (6.1) one gets a finite set of numbers $\alpha_1, \alpha_2, \cdots, \alpha_h$ in $\mathfrak{s}$ such that every number in $\mathfrak{s}$ is of the form

$$\lambda_1\alpha_1 + \lambda_2\alpha_2 + \cdots + \lambda_h\alpha_h \qquad (\lambda_i \text{ in } R).$$

(However, $\mathfrak{s}$ not being an ideal, one is not at all sure that conversely every number of this form will belong to $\mathfrak{s}$.)

7. Proof of the first main theorem for $GL(n)$

For simplicity's sake we consider invariants $J(u)$ depending on one variable form (1.1) of given degree r. In the ring $k[u]$ of all polynomials depending on the variable coefficients $u_{r_1\cdots r_n}$ we consider the set $\mathfrak{J}$ *consisting of all non-constant invariants* $J(u)$ (it is quite important to exclude the constants). According to the concluding observation of the preceding section we pick out a finite number of non-constant invariants

$$J_1(u), \cdots, J_h(u)$$

such that every non-constant invariant $J(u)$ appears as a linear combination

$$J(u) = L_1(u)J_1(u) + \cdots + L_h(u)J_h(u) \tag{7.1}$$

with polynomial coefficients $L_i(u)$. Suppose that $J; J_1, \cdots, J_h$ are homogeneous of degrees $\mu; \mu_1, \cdots, \mu_h$. The equation (7.1) will not break down if one cancels in $L_i(u)$ all terms that are not of degree $\mu - \mu_i$, so that we may assume the $L_i(u)$ to be homogeneous of the "right" degrees $\mu - \mu_i$. (For $\mu_i > \mu$, the coefficient $L_i(u)$ will then be zero.)

This first step is of universal significance and not at all limited to the classical case. We shall now endeavor to show that *the invariants $J_i(u)$ just determined constitute an integrity basis for all invariants.* This second step will be peculiar to the group $GL(n)$. We make use of the same device which served to prove Gram's theorem. We substitute the absolute covariants

$$\overset{*}{u}_{r_1\cdots r_n}(\overset{1}{\eta}, \cdots, \overset{n}{\eta}) \tag{1.6}$$

as defined in (1.5) for $u_{r_1\cdots r_n}$ in the relation (7.1). For an invariant $J(u)$ of weight g we have

$$J(u^*) = \mathrm{H}^g \cdot J(u)$$

where H is the determinant $[\overset{1}{\eta} \cdots \overset{n}{\eta}]$. Hence

$$\mathrm{H}^g \cdot J(u) = L_1(u^*) \cdot \mathrm{H}^{g_1} J_1(u) + \cdots + L_h(u^*) \cdot \mathrm{H}^{g_h} J_h(u). \tag{7.2}$$

H is a covariant of weight -1; the factors $\mathrm{H}^{g_i} L_i(u^*)$ are therefore covariants of weight $-g_i$. Cayley's Ω-process

$$\begin{vmatrix} \frac{\partial}{\partial \eta_1^1}, & \cdots, & \frac{\partial}{\partial \eta_n^1} \\ \cdots & \cdots & \cdots \\ \frac{\partial}{\partial \eta_1^n}, & \cdots, & \frac{\partial}{\partial \eta_n^n} \end{vmatrix}$$

changes a covariant of weight g^* depending on the n contravariant vectors $\overset{1}{\eta}, \cdots, \overset{n}{\eta}$ into a covariant of weight $g^* + 1$. Consequently if we perform the Ω-process g times upon (7.2) a relation

$$c_g \cdot J(u) = L_1^*(u) \cdot J_1(u) + \cdots + L_h^*(u) \cdot J_h(u) \tag{7.3}$$

will ensue, where c_g is the constant

$$c_g = \Omega^g(\mathrm{H}^g)$$

and

$$L_i^*(u) = \Omega^g\{\mathrm{H}^{g_i} L_i(u^*)\}$$

are covariants of weight $g - g_i$, or, as they no longer involve the variables $\overset{1}{\eta}, \cdots, \overset{n}{\eta}$, rather *invariants* of that weight. We shall soon make sure that c_g is $\neq 0$. Taking this important point for granted we divide (7.3) by c_g and thus succeed in normalizing the coefficients $L_i(u)$ in (7.1) so as to make them *invariant*.

The statement that any invariant $J(u)$ of degree μ is expressible in terms of $J_i(u)$ is now proved by induction with respect to μ. The statement is trivial for $\mu = 0$, where $J(u)$ is a constant. Since each of the invariants $J_i(u)$ is at least of degree 1, the *invariant* coefficients $L_i(u)$ in (7.1) we have just obtained by the Ω-process are of *lower* degree than μ; and if they are expressible in terms of the $J_i(u)$, so also is $J(u)$.

In order to show that $c_g \neq 0$ we observe that H^g is a form in the variables $\overset{i}{\eta}_k$ with integral coefficients, and that Ω^g is precisely the same form in the differential operators $\dfrac{\partial}{\partial \overset{i}{\eta}_k}$. But if

$$f(x_1, \cdots, x_r) = \sum a(i_1 \cdots i_r) x_1^{i_1} \cdots x_r^{i_r} \qquad (i_1 + \cdots + i_r = s)$$

is any form of degree s, and

$$f\left(\frac{\partial}{\partial x_1}, \cdots, \frac{\partial}{\partial x_r}\right) = \sum a(i_1 \cdots i_r) \frac{\partial^s}{\partial x_1^{i_1} \cdots \partial x_r^{i_r}},$$

then a simple calculation shows that

$$f\left(\frac{\partial}{\partial x_1}, \cdots, \frac{\partial}{\partial x_r}\right) f(x_1, \cdots, x_r) = \sum i_1! \cdots i_r!\, a^2(i_1 \cdots i_r).$$

It follows that c_g is in fact a *positive integer*.

When we have to deal with invariants J depending on some quantics $u, v, \cdots$ we may suppose without any loss of generality that the signatures $(r_1, \cdots, r_n)$

of these quantics satisfy the condition $r_n \geqq 0$. It is then obvious from the proof of Gram's generalized theorem how to introduce the analogue of the covariants (1.6). We summarize:

THEOREM (8.7.A). *The relative invariants $J(u, v, \cdots)$ for the full linear group $GL(n)$, i.e. the absolute invariants for $SL(n)$, depending on a number of quantics $u, v, \cdots$, possess a finite integrity basis.*

As was mentioned in Chapter II, §2 the second main theorem is contained in the following general algebraic statement:

THEOREM (8.7.B). *All relations holding among given polynomials are algebraic consequences of a finite number among them.*

Indeed

$$J_1(z_1, \cdots, z_l) = J_1(z), \cdots, J_p(z),$$

being the given polynomials of any number of variables $z_1, \cdots, z_l$, a relation is a polynomial $R(t_1, \cdots, t_p)$ of p independent variables t_i such that

$$R(J_1(z), \cdots, J_p(z)) = 0.$$

The relations obviously form an ideal within the polynomial ring, and this ideal has a finite ideal basis.[5]

The field k in which we operated throughout might be any field of characteristic 0.

8. The adjunction argument

Consider the group characteristic for the $(n-1)$-dimensional affine geometry:

$$(8.1)\qquad \begin{cases} \xi_1 = \xi_1' \\ \xi_i = a_i\xi_1' + \sum_k a_{ki}\xi_k' \end{cases} \qquad (i, k = 2, \cdots, n)$$

where the only restriction imposed upon the a_i, a_{ik} is

$$\det_{i,k=2,\cdots,n}(a_{ik}) = 1.$$

The $\xi_1, \xi_2, \cdots, \xi_n$ are homogeneous point coördinates. Affine geometry results from projective geometry with the full unimodular group $SL(n)$ by adjoining the "plane at infinity"

$$\sum_{i=1}^{n} e_i\xi_i = 1\cdot\xi_1 + 0\cdot\xi_2 + \cdots + 0\cdot\xi_n = 0$$

as an absolute entity. An affine property of a geometric structure is a projective relation to the absolute plane. Hence this procedure for the formation of affine invariants of ground forms $u, v, \cdots$ is suggested: We adjoin an arbitrary linear form

$$(8.2)\qquad l: \sum_i l_i\xi_i$$

to the ground forms and construct projective invariants $J(u, v, \cdots; l)$ for the system $u, v, \cdots ; l$. Then $J(u, v, \cdots ; e)$ is obviously an absolute invariant under the affine group (8.1). The question is whether *all* affine invariants are obtained in this way. As far as invariants go, we should then be justified in treating the affine space as the projective space with an absolute plane.[6]

The affirmative answer to our question is readily arrived at by combining the symbolic method with the construction of an integrity basis of typical vector invariants for the affine group as accomplished in Chapter II, §7. The given affine invariant $J(u, v, \cdots)$ is first replaced by its symbolic expression, an affine invariant $j(u, u', \cdots ; v, v', \cdots ; \cdots)$ depending on a lot of covariant vectors $u, u', \cdots, v, v', \cdots, \cdots$. j is expressible in terms of bracket factors

$$[xy \cdots z]_n \quad \text{and} \quad [x \cdots y]_{n-1} = [ex \cdots y]_n .$$

We introduce the arbitrary linear form (8.2) and replace in this expression of j every factor

$$[x \cdots y]_{n-1} \quad \text{by} \quad [lx \cdots y],$$

thus changing the affine invariant $j(uu' \cdots, vv' \cdots, \cdots)$ into a projective invariant* $j^*(l; uu' \cdots, vv' \cdots, \cdots)$. We then revert to the non-symbolic expression, which is possible because j^* the same as j has the right degree in each of the arguments $uu' \cdots, vv' \cdots, \cdots$ (equal to the degrees of the ground forms $u, v, \cdots$). We thus obtain a projective invariant $J^*(l; u, v, \cdots)$ for which

$$J^*(e; u, v, \cdots) = J(u, v, \cdots).$$

In Chapter II, §7, we generalized the affine group to the case where the absolute consists of a linear manifold of $n - 2$ or $n - 3$ or $\cdots$ rather than of $n - 1$ dimensions. Let us study the case $n - 2$. An $(n - 2)$-dimensional linear manifold M_{n-2} is the intersection of two planes l and l':

$$\sum_i l_i \xi_i = 0, \qquad \sum_i l'_i \xi_i = 0$$

and has the anti-symmetric coördinates

$$l_{ik} = l_i l'_k - l'_i l_k . \tag{8.3}$$

A one-dimensional line $\{\xi, \eta\}$ joining two points ξ, η intersects M_{n-2} if

$$\sum_{i,k} l_{ik} \xi_i \eta_k = 0.$$

For an arbitrary anti-symmetric (covariant) tensor l_{ik} of rank 2 this equation defines a *line complex*. The "special" line complexes M_{n-2} as described by (8.3) satisfy the quadratic relations

$$l_{14} l_{23} + l_{24} l_{31} + l_{34} l_{12} = 0 \tag{8.4}$$

* j^* might consist of several terms of different degrees in l, each term being an invariant homogeneous in l as well as in $uu' \cdots, vv' \cdots, \cdots$.

(1, 2, 3, 4 standing for any four of the n indices 1, 2, $\cdots$, n). For the group we have in mind the absolute is such a special line complex M_{n-2} with

$$l = e = (1, 0, 0, \cdots, 0), \qquad l' = e' = (0, 1, 0, \cdots, 0)$$

or

$$l_{12} = -l_{21} = 1, \qquad \text{all other } l_{ik} = 0. \tag{8.5}$$

The symbolic expression j of invariants $J(uv \cdots)$ for this group will involve the bracket factor $[x \cdots y]_{n-2}$ besides the total bracket $[xy \cdots z]_n$. We replace the former by the alternating sum

$$\tfrac{1}{2} \sum \pm l_{i_1 i_2} x_{i_3} \cdots y_{i_n}$$

extending to all permutations $i_1 i_2 i_3 \cdots i_n$ of $1, \cdots n$, which is projectively invariant and turns back into $[x \cdots y]_{n-2}$ by the specialization (8.5). The result will be that an invariant $J(uv \cdots)$ of a number of ground forms $uv \cdots$ in the affine space "of rank 2" is derived by the specialization (8.5) from a projective invariant *involving an arbitrary anti-symmetric tensor l_{ik} of rank* 2 (or a line complex) besides the given system of ground forms. What we wish to emphasize in this case is the fact that one must introduce an *arbitrary* line complex as a new argument into the invariants; one must not be content with an arbitrary *special* line complex, although its characteristic conditions (8.4) are projectively invariant.

When the ground forms involve Latin as well as Greek variables the list of typical basic vector invariants is to be extended by the Greek bracket factor and $[\xi\eta]_2$, which we replace by $\sum l_{ik}\xi_i\eta_k$. As this replacement shows, the essential result holds good.

The same adjunction argument is applicable to the important case of the *orthogonal group*. Through relativity theory one has become thoroughly familiar with treating orthogonal vector invariants as affine vector invariants after adjoining the metric ground form

$$\sum \gamma_{ik} x_i x_k \qquad (\gamma_{ik} = \gamma_{ki}) \tag{8.6}$$

which becomes the unit form,

$$\gamma_{ik} = \delta_{ik}, \tag{8.7}$$

in the Cartesian coördinate systems. The above considerations afford a justification of that procedure when we remark that any (even or odd) orthogonal vector invariant is expressible in terms of bracket factors $[xy \cdots z]$ and scalar products (xy). The first is invariant under the full linear group, and so is the scalar product when written as

$$(xy) = \sum \gamma_{ik} x_i y_k \tag{8.8}$$

and considered as depending on an arbitrary (contravariant!) quadratic (8.6) in addition to the two covariant vectors x and y. The form invariants for the orthogonal group arise from the form invariants under the full linear group when one first adjoins a contravariant quadratic (8.6) to the system of ground forms and then specializes by (8.7). When one wants to avoid the introduction of a contravariant quadratic (8.6) one might resort to a *covariant* quadratic

$$\sum g_{ik}\xi_i\xi_k \qquad (g_{ik} = g_{ki}) \tag{8.9}$$

instead, and then replace the scalar product (xy) by the projective invariant

$$-\begin{vmatrix} g_{11} & \cdots & g_{1n} & x_1 \\ \cdots & \cdots & \cdots & \cdots \\ g_{n1} & \cdots & g_{nn} & x_n \\ y_1 & \cdots & y_n & 0 \end{vmatrix}. \tag{8.10}$$

[If $\| \gamma_{ik} \|$ is the inverse of $\| g_{ik} \|$, then (8.10) is $\det(g_{ik})$ times (8.6).]

Similarly for the symplectic group. In each case a finite integrity basis is ascertained by applying Theorem (8.7.A). There is no difficulty in replacing the ground forms by any *quantics* $u, v, \cdots$ with respect to the full linear group. I summarize:

THEOREM (8.8.A). *For the group of step transformations, the orthogonal, and the symplectic group, every invariant* $J(u, v, \cdots)$ *can be written as an invariant under the group* $SL(n)$ *after adjoining a suitable "absolute entity" to the arguments* $u, v, \cdots$ *provided* $u, v, \cdots$ *are quantics defined under the group* $SL(n)$.

I do not wish to detract from this triumphant attainment of the symbolic method. However, with regard to the first main theorem, the state of affairs as laid down in our proposition is still unsatisfactory, in so far as quantics for $SL(n)$ are not the primitive and not the most general quantities for the more limited groups we are dealing with. I do not see that much can be done about it in the affine case. As a matter of fact the representations of the affine group are not fully reducible and that complicates the survey of possible quantities under that group almost beyond repair. For the orthogonal and symplectic groups the outlook is much brighter. Indeed a "quantic" will here range over the tensors F in one of the subspaces designated as $P_0(T)$, T being a permissible diagram. With $\mathbf{e}$ being the idempotent generator of that symmetry class T, we first replace the argument F in our invariant $J(F, \cdots)$ by $\mathbf{e}F$:

$$J^*(F, \cdots) = J(\mathbf{e}F, \cdots)$$

so that F now ranges over *all* tensors in P_r^0 (removal of the symmetry restriction). Any tensor F of rank r may be uniquely decomposed according to Theorem (5.6.A), and the first part F^0 of vanishing traces proceeds from F by a certain linear operator $\mathbf{t}$:

$$F^0 = \mathbf{t}F, \qquad F^0(i_1 \cdots i_r) = \sum_{(k)} t(i_1 \cdots i_r\,;\, k_1 \cdots k_r)F(k_1 \cdots k_r)$$

which is invariant under the orthogonal group or commutes with all operators in $\mathfrak{A}_r$; it therefore lies in the algebra described in Chapter V, §5, as ω_r^n. **t** is evidently idempotent. The second step consists in replacing the argument F in J^* ranging over all tensors of vanishing traces by $\mathbf{t}F$:

$$J^*(\mathbf{t}F, \cdots) = J^{**}(F, \cdots).$$

J^{**} is now an orthogonal invariant defined for *all* tensors F of rank r whatsoever (removal of the trace condition). The adjunction argument then leads to a projective invariant $J^{***}(g_{ik}\,; F, \cdots)$ in which the quadratic (8.9) is added as an argument to the tensors $F, \cdots$, and such that

$$J^{***}(g_{ik}\,; F, \cdots) = J(F, \cdots) \qquad \text{for } g_{ik} = \delta_{ik} \text{ and } F \text{ in } \mathrm{P}_0(r_1 \cdots r_n), \cdots.$$

THEOREM (8.8.B). *A finite integrity basis for the orthogonal invariants* $J(uv \cdots)$ *depending on some orthogonal quantics* $u, v, \cdots$ *is ascertained by constructing the integrity basis for projective invariants* $J(g_{ik}\,; F, \cdots)$ *in which each argument* $u, \cdots$ *is replaced by a free tensor* F *of the corresponding rank and the quadratic* (8.9) *is added to the arguments. The same is true mutatis mutandis for the symplectic group.*

B. Differential and Integral Methods

9. Group germ and Lie algebras

A neighborhood of the unit element I on a locally Euclidean continuous group is a topological image of Euclidean space in which composition applies only to elements sufficiently near to the center I. Let us call such a manifold a *group germ* provided composition as far as defined has the same properties as for a continuous group.[7] Whether every group germ is capable of blossoming forth into a full group (on which it forms a certain neighborhood of the unit element) is a difficult question beyond the pale of our present knowledge. The elements of the would-be-group could be introduced as arbitrary finite chains $a_1a_2 \cdots a_h$ of elements a_i of the group germ. The intricate point is to decide under what conditions two such chains are to be identified.

However if there *is* a group γ with the given group germ γ_0 we can ascertain exactly to what extent it is determined by the germ. (If γ should consist of several disconnected pieces we take into regard only the piece containing the point I, which is a group in itself, called the proper part of γ; the other pieces, its cosets, are out of the game.) We simply face a particular case of the question how far a manifold is known in its whole extent when it is locally known at every point.[8] The *universal covering manifold* provides the answer to this sort of question; we shall prove in a moment that it is a continuous group with the same group germ γ_0. This simply connected group we now call γ. Every group γ' with the germ γ_0 has γ as its universal covering manifold. In other words, every extension of the germ γ_0 into a full group γ' is obtained from the most "complete" one γ by a process of projection: one takes an arbitrary discrete group $\{S\}$ of continuous automorphisms $S:p \to p'$ on γ without fixed points, and identifies points on γ equivalent under $\{S\}$. Points p, p' are equivalent under $\{S\}$ which arise from each other by a transformation S of the group $\{S\}: p' = Sp$. Discreteness means that no set of mutually equivalent points ever has a condensation point on γ. Absence of fixed points means that for no transformation S of $\{S\}$ different from the identity there is a point p such that $Sp = p$. What we have to show is this: Let γ, γ' be two groups with a common germ γ_0 and let γ be simply connected. Then there is a definite continuous homomorphic mapping $p \to p'$ of γ upon γ' which is the identity within the germ γ_0. The function $p' = \varphi(p)$ is constructed by the familiar process of continuation, first used by Weierstrass in the domain of analytic functions. If $\varphi(p) = p'$ is known for a point p_0 we define it in the neighborhood of p_0 by

$$\varphi(p_0 a) = p_0' \cdot a$$

where a is in the germ γ_0. In this way by continuation along a path leading from $\mathfrak{l}$ to p we come to a definite image $\varphi(p) = p'$ of the end point p. In general p' would depend on the path and not only on the end point. However, if γ is simply connected, continuation along different paths from $\mathfrak{l}$ to p necessarily leads to the same $\varphi(p)$. On considering the point $p' = \varphi(p)$ to be the "trace" of p, γ is molded into an (unbounded, unramified) covering manifold over γ' and the group of covering transformations S of γ over γ' is constructed in the familiar way.

All this is nothing else than the ever-recurring topological device by which one links events in the small and in the large; only one point is peculiar to group theory, namely that *the universal covering manifold $\tilde{\gamma}$ of a continuous group γ is a continuous group* again. A point $\tilde{p}$ of $\tilde{\gamma}$ is defined by means of a point p of γ and a path leading on γ from the center $\mathfrak{l}$ to p. Two paths leading from $\mathfrak{l}$ to p define the same point $\tilde{p}$ on $\tilde{\gamma}$ if and only if they can be continuously deformed into each other without moving the end points $\mathfrak{l}$ and p (or if both paths lead to the same end point on every unramified, unbounded manifold γ^* over γ provided one starts at the same point $\mathfrak{l}^*$ on γ^*). A neighborhood of $\tilde{p}$ is formed by hitching onto the path defining $\tilde{p}$ all paths starting from p and lying in a given (Euclidean) neighborhood of p. A path is described by a continuous function $p(\lambda)$ whose argument λ ranges over the interval $0 \leqq \lambda \leqq 1$ of real numbers while its value $p(\lambda)$ varies on γ; the path joins $\mathfrak{l}$ and p if $p(0) = \mathfrak{l}$, $p(1) = p$. If $\tilde{p}$ over p is given by the path $p(\lambda)$ and $\tilde{q}$ over q by the path $q(\mu)$ we may define $\tilde{p}\tilde{q}$ by the path consisting of $p(\lambda)$ $\{0 \leqq \lambda \leqq 1\}$ followed by the path

$$p \cdot q(\lambda - 1) \qquad \{1 \leqq \lambda \leqq 2\}$$

arising from $q(\mu)$ by the left translation p. The resulting path is obtained from

$$p(\lambda) \cdot q(\mu) \qquad \{0 \leqq \lambda \leqq 1, \qquad 0 \leqq \mu \leqq 1\}$$

by letting the point (λ, μ) in the $\lambda\mu$-square describe the broken line $(0,0) \to (1, 0) \to (1, 1)$. The result does not change on replacing it by the line $(0, 0) \to (0, 1) \to (1, 1)$ since the one may be deformed into the other within the square. Another suitable choice would be the diagonal

$$\lambda = \tau, \mu = \tau \qquad \{0 \leqq \tau \leqq 1\}.$$

The first two ways show that $\tilde{p}\tilde{q}$ is continuous in $\tilde{q}$ uniformly with respect to $\tilde{p}$ and continuous in $\tilde{p}$ uniformly with respect to $\tilde{q}$ whence follows continuity in the pair $(\tilde{p}, \tilde{q})$. The diagonal definition proves the element $\tilde{p}^{-1}$ as defined by $p^{-1}(\lambda)$ to be the inverse of $\tilde{p}$.

By the same argument we applied in constructing the function $p' = \varphi(p)$ one concludes that any continuous realization, in particular any continuous representation, of the group germ γ_0 may be extended in a unique fashion to the whole simply connected group γ. However the resulting representation will in general not be single-valued on any "less complete" group γ' engendered by γ_0.

Granted certain differentiability conditions the group germ is reducible to, and reproducible from, its *infinitesimal elements* $a, b, \cdots$, which not only form a linear set, the tangent plane of the group γ at the origin I, but even a kind of algebra.[9] Besides addition and multiplication by numbers satisfying the rules common to all linear sets, we have a multiplication $[ab]$ (corresponding to the commutator $sts^{-1}t^{-1}$ of two group elements s, t) satisfying the distributive law in both factors:

$$(9.1)\qquad \begin{array}{rl|rl} [a + a', b] = [ab] + [a'b] & & [a, b + b'] = [ab] + [ab'] \\ [\lambda a, b] = \lambda[ab] & & [a, \lambda b] = \lambda[ab] \end{array}$$

(λ any number). Instead of the associative law, however, we have anti-commutativity

$$(9.2)\qquad [ba] = -[ab]$$

and the Jacobi rule

$$(9.3)\qquad [a[bc]] + [b[ca]] + [c[ab]] = 0.$$

The numbers are taken from the continuum K. In homage to Sophus Lie such an algebra is nowadays called a *Lie algebra*. The Lie algebra consisting of the infinitesimal elements of a group or a group germ may be called its *nucleus*. Every Lie algebra $\mathfrak{a}$ in K engenders and uniquely determines a group germ with $\mathfrak{a}$ as its nucleus. It remains doubtful however whether the group germ is extensible to a full group.[9a] Of this last point Lie himself had not been sufficiently aware, his interests being entirely concentrated on questions in the small. One ought to emphasize that apart from its significance for continuous groups the notion of a Lie algebra is applicable to *any* number field k; it is as purely algebraic and as worthy of an independent algebraic investigation as the notion of associative algebras. For a group of linear transformations or matrices the "commutator product" of two infinitesimal elements A, B turns out to be

$$[AB] = AB - BA. \tag{9.4}$$

Hence $a \to A$ is a representation by matrices of the Lie algebra over which a varies if $a \to A$, $b \to B$ entail

$$a + b \to A + B, \qquad \lambda a \to \lambda A, \qquad [ab] \to AB - BA$$
$$\{\lambda \text{ any number}\}.$$

Again this is a purely algebraic concept. Without the laws (9.1–3) any hope of faithfully representing the abstract Lie algebra would obviously be nipped in the bud.

The formula (9.4) is the only point in the whole of Lie's theory that actually matters for our investigations. We therefore give the proof. By integral-like repetition of the infinitesimal linear transformation A one gets a one-parameter group $X(s)$ of transformations; one has to integrate the differential equation

$$dX/ds = AX$$

with the initial value $X(0) = E$. This is done by means of the exponential function $\exp(sA)$ which is defined for matrices exactly as for numbers:

$$\exp(A) = \lim_{\nu \to \infty} \left(E + \frac{A}{\nu}\right)^{\nu} = \sum_{\nu=0}^{\infty} \frac{1}{\nu!} A^{\nu}.$$

One has

$$X(s + t) = X(s) \cdot X(t).$$

With

$$X(s) = \exp(sA), \qquad Y(t) = \exp(tB)$$

we form the commutator

$$Z(s, t) = X(s)Y(t)X^{-1}(s)Y^{-1}(t) = X(s)Y(t)X(-s)Y(-t)$$

for which we want to show the limit relation

$$(Z(s, t) - E)/st \to AB - BA \qquad \text{with } (s, t) \to (0, 0).$$

For any function $f(s, t)$ vanishing for $s = 0$ identically in t and for $t = 0$ identically in s one has

$$f(s, t) = \int_0^s \int_0^t \frac{\partial^2 f}{\partial s \partial t}\, dt ds,$$

provided the derivative under the integral exists and is continuous, and hence

$$f(s, t)/st \to \left(\frac{\partial^2 f}{\partial s \partial t}\right)_{0,0} \qquad \text{with} \qquad (s, t) \to (0, 0).$$

In order to apply this to $Z(s, t) - E$ we compute

$$\frac{\partial^2 Z}{\partial s \partial t} = X'(s)Y'(t)X(-s)Y(-t) - X(s)Y'(t)X'(-s)Y(-t)$$
$$- X'(s)Y(t)X(-s)Y'(-t) + X(s)Y(t)X'(-s)Y'(-t)$$

which for $s = 0$, $t = 0$ passes into

$$AB - BA - AB + AB = AB - BA.$$

γ' is an invariant subgroup of γ if for any t in γ' and any s in γ the element sts^{-1} or the commutator $sts^{-1}t^{-1}$ lies in γ'. Accordingly for Lie algebras: the linear subspace $\mathfrak{a}'$ of $\mathfrak{a}$ is an invariant subalgebra of $\mathfrak{a}$ if for any x in $\mathfrak{a}'$ and any a in $\mathfrak{a}$ the product $[ax]$ lies in $\mathfrak{a}'$. The term "ideal in $\mathfrak{a}$" would also be an adequate description of that situation. All elements of the form $[ab]$ (a, b in $\mathfrak{a}$) and their linear combinations obviously form such an invariant subalgebra $\mathfrak{a}'$ in $\mathfrak{a}$, corresponding to the commutator group in group theory; Lie called it the *derived algebra.*

10. Differential equations for invariants. Absolute and relative invariants

If $f(x, x', \cdots)$ is a polynomial depending in a homogeneous manner on the components of each of the vectors $x, x', \cdots$ and if

$$a = (A, A', \cdots) \tag{10.1}$$

is a set of linear operators operating on the respective vector spaces, one can form the differential

$$d_a f = \sum_i \frac{\partial f}{\partial x_i} dx_i + \sum_j \frac{\partial f}{\partial x'_j} \partial x'_j + \cdots$$

where one puts

$$dx = Ax, \; dx' = A'x', \cdots.$$

f is invariant with respect to the set a of infinitesimal transformations if $d_a f = 0$. Let $b = (B, B', \cdots)$ be another set of such transformations. With any two numbers λ and μ one can form

$$\lambda a + \mu b = (\lambda A + \mu B, \quad \lambda A' + \mu B', \cdots).$$

Moreover one readily finds

$$d_b(d_a f) - d_a(d_b f) = d_{[ab]} f$$

where $[ab]$ is the set

$$(AB - BA, \; A'B' - B'A', \cdots).$$

Hence if

$$d_a f = 0, \qquad d_b f = 0$$

one must have at the same time

$$d_{\lambda a + \mu b} f = 0, \qquad d_{[ab]} f = 0.$$

This is perhaps the quickest way to establish the three fundamental operations of an infinitesimal group or a "Lie algebra". With a ranging over an abstract Lie algebra $\mathfrak{a}$, and on interpreting the matrices $A, A', \cdots$ in (10.1) as so many representations of $\mathfrak{a}$, the differential equations

$$d_a f = 0 \tag{10.2}$$

characterize the invariants f. This again is a purely algebraic concept of which we already made use in Chapter II for the orthogonal group.

A relative invariant is characterized by equations

$$d_a f = \kappa_a \cdot f. \tag{10.3}$$

The number κ_a, the infinitesimal multiplier, is a linear function of a, and because

$$d_a f = \kappa_a \cdot f, \qquad d_b f = \kappa_b \cdot f$$

imply

$$d_b(d_a f) - d_a(d_b f) = \kappa_a \kappa_b f - \kappa_b \kappa_a f = 0,$$

$\kappa_{[ab]}$ vanishes, or *the infinitesimal multiplier* κ_a *disappears for all elements* a *of the derived Lie algebra.*

Examples. (1) $SL(n)$. The nucleus $^s\mathfrak{l}(n)$ consists of the n-rowed matrices of trace 0. The equations

$$\begin{pmatrix} 1 & 0 \\ 0 & -1 \end{pmatrix}\begin{pmatrix} 0 & 1 \\ 0 & 0 \end{pmatrix} - \begin{pmatrix} 0 & 1 \\ 0 & 0 \end{pmatrix}\begin{pmatrix} 1 & 0 \\ 0 & -1 \end{pmatrix} = \begin{pmatrix} 0 & 2 \\ 0 & 0 \end{pmatrix},$$

$$\begin{pmatrix} 0 & 1 \\ 0 & 0 \end{pmatrix}\begin{pmatrix} 0 & 0 \\ 1 & 0 \end{pmatrix} - \begin{pmatrix} 0 & 0 \\ 1 & 0 \end{pmatrix}\begin{pmatrix} 0 & 1 \\ 0 & 0 \end{pmatrix} = \begin{pmatrix} 1 & 0 \\ 0 & -1 \end{pmatrix}$$

show that for $n \geqq 2$ the derivative $^s\mathfrak{l}'$ coincides with the whole Lie algebra $^s\mathfrak{l}$. Hence *there exist only absolute invariants.*

(2) $O(n)$. Its nucleus $\mathfrak{o}(n)$ consists of the skew-symmetric matrices. Denoting by $S_{ik}\{i \neq k\}$ the matrix having a $+1$ at the place (ik) of the matrix, a -1 at (ki) and zeros elsewhere, one readily verifies

$$S_{12}S_{23} - S_{23}S_{12} = S_{31}.$$

Hence for $n \geqq 3$ the derivative $\mathfrak{o}'(n)$ again is the whole $\mathfrak{o}(n)$, and there exist no relative invariants.

(3) The same is found to be true for $Sp(n)$ with its nucleus $\mathfrak{sp}(n)$.

(4) Concerning $GL(n)$ whose nucleus $\mathfrak{l}$ consists of all n-rowed matrices $A = \| a_{ik} \|$ whatsoever, example (1) shows that the derivative $\mathfrak{l}'$ is $^s\mathfrak{l}$. Hence κ_A must be a multiple of the *trace*:

$$\kappa_A = g(a_{11} + \cdots + a_{nn}). \tag{10.4}$$

These results concerning the Lie algebras hold good in any number field k. In case $k = K$ they have their consequences for the continuous groups. In studying our groups for arbitrary *complex* values of the components a_{ik} of the generic matrix one ought to observe that their real and imaginary parts are to be considered as separate parameters and the Lie algebra as one in K rather than in $K^{\dagger}$ unless we limit ourselves to *analytic* representations where differentiation with respect to the complex parameters is possible. Hence in example (4) one has to replace (10.4) by the statement that κ_A is a linear combination of the real and the imaginary part of the trace or by an equation

$$\kappa_A = g(a_{11} + \cdots + a_{nn}) + g'(\bar{a}_{11} + \cdots + \bar{a}_{nn}) \tag{10.5}$$

with two constants g, g'. A further remark is to the effect that from the infinitesimal elements we can draw conclusions upon the proper part of the group only. With this in mind we may advance the statement that the groups of all real or unitary or complex unimodular transformations, the group of all real or complex proper orthogonal transformations and the group of all real or unitary or complex symplectic transformations have no relative invariants. For in each case the group indicated consists of one piece. For the whole orthogonal group comprising both the proper and the improper part one encounters the distinction between even and odd invariants. As to the group of all real linear transformations A with a positive determinant $|A|$ or of all unitary transformations we find that the multipliers of relative invariants are necessarily of the form $|A|^g$, while for the group of all non-singular complex linear transformations

$$|A|^g|\bar{A}|^{g'}$$

is their universal form; g (and g') are arbitrary real or complex constants. If one requires the representations and hence the multipliers to be single-valued on the whole group then none but integral values of g (and g') are admissible for the unitary and the total complex group. Topology thus casts its shadow on the algebraic scene.

(5) We add one further example: the group of all real transformations of the form

$$\begin{Vmatrix} A_1 & 0 \\ * & A_2 \end{Vmatrix}$$

with positive determinants $|A_1|$, $|A_2|$ (step transformations). The multiplier of a relative invariant is necessarily of the form

$$|A_1|^{g_1}|A_2|^{g_2}$$

with two constant exponents g_1 and g_2.

The method by which we obtained these results presupposes that we deal with representations satisfying the differentiability conditions which enable us to pass to the infinitesimal elements (Lie representations). This is a serious drawback of the infinitesimal approach from the standpoint of continuous groups. However if we study the Lie algebras as such, our results contain a complete algebraic answer to a straightforward algebraic question, leaving nothing to be desired.

11. The unitarian trick

Determining the representations of the nuclei of the classical groups

$$GL(n), \qquad SL(n), \qquad O(n), \qquad Sp(n)$$

is an algebraic problem to the independent solution of which E. Cartan made the decisive contributions. It is more difficult than the simplest case of representations of degree 1, as considered in the previous section, leads one to expect. However we are able to foretell the results, for the field K or $K^\dagger$ at least, from our treatment of the representations of the unitarily restricted groups by the integration method.

Take the group $SL(n)$. We shall soon find out that the group ${}^sU(n)$ of all unimodular unitary transformations is simply connected. Hence any irreducible representation of its nucleus ${}^s\mathfrak{u}(n)$ leads to a single-valued continuous representation of ${}^sU(n)$ which must be equivalent to one of the representations

$$(11.1) \qquad \langle P(f_1 \cdots f_n)\rangle \quad \text{with} \quad f_1 \geqq f_2 \geqq \cdots \geqq f_n = 0$$

we constructed in Chapter IV [cf. Theorem (7.5.C)].

We now compare ${}^s\mathfrak{u}(n)$ with the Lie algebras ${}^s\mathfrak{l}(n)$ and ${}^s\mathfrak{r}(n)$ of all complex and of all real matrices of vanishing trace. ${}^s\mathfrak{u}(n)$ and ${}^s\mathfrak{r}(n)$ both arise from ${}^s\mathfrak{l}(n)$ by reality restrictions. But these are immaterial for any linear problems. We face a special case of the following general situation.

When we are given a Lie algebra $\mathfrak{a}$ with the basis $e_1, \cdots, e_r$ in a field k we may extend the field k to a larger field K. $\mathfrak{a}$ is then changed into a Lie algebra $\mathfrak{a}_K$ in K whose elements are all linear combinations

$$(11.2) \qquad a = \lambda_1 e_1 + \cdots + \lambda_r e_r$$

with coefficients λ_i varying in K. (In our case $k = \mathrm{K}$, $K = \mathrm{K}^\dagger$.) Let

$$\mathfrak{a} = (e_1, \cdots, e_r), \qquad \mathfrak{a}' = (e_1', \cdots, e_r')$$

be two Lie algebras in k which coincide after the extension: $\mathfrak{a}_K = \mathfrak{a}_K'$. This situation arises if the e_i are expressible in terms of the e_i:

$$(11.3) \qquad e_i' = \sum_k \gamma_{ik} e_k$$

with coefficients γ_{ik} in K and $\det(\gamma_{ik}) \neq 0$. We shall then have

$$\lambda_1' e_1' + \cdots + \lambda_r' e_r' = \lambda_1 e_1 + \cdots + \lambda_r e_r$$

when we tie the parameters λ_i, λ_i' together by the substitution

$$\sum_i \gamma_{ik} \lambda_i' = \lambda_k .$$

If the structure of $\mathfrak{a}$ is described by the multiplication table for the basis:

$$[e_i e_k] = \sum_{l=1}^{r} \alpha_{ik,l} e_l \qquad \{\alpha_{ik,l} \text{ in } k\}$$

then the transformation (11.3) must bring it about that the resulting coefficients α' in

$$[e'_i e'_k] = \sum_l \alpha'_{ik,l} e'_l$$

also lie in k. Any representation

$$e_i \to A_i \qquad (\lambda_1 e_1 + \cdots + \lambda_r e_r \to \lambda_1 A_1 + \cdots + \lambda_r A_r)$$

of $\mathfrak{a}$ in K is associated with a representation of $\mathfrak{a}'$ in K:

$$e'_i \to A'_i \qquad (A'_i = \sum_k \gamma_{ik} A_k)$$

and vice versa, such that both coincide after the extension of k into K.

In our case we conclude that every representation of ${}^{s}\mathfrak{r}(n)$ is fully reducible and that there are no other irreducible representations than those described by (11.1). From this result follows by integration:

THEOREM (8.11.A). *Every Lie representation of the group of all real unimodular transformations is fully reducible. There are no other irreducible such representations than the rational ones formerly described as "quantics".*

More intricate is the question of deriving all representations in K of the Lie algebra $\mathfrak{a}_K$ from those of $\mathfrak{a}$ if $\mathfrak{a}_K$ is considered as a Lie algebra *in* k. Let us keep closer to the case which interests us here by assuming K to be a quadratic field over k with the determining quadratic equation

$$x^2 - \vartheta = 0 \qquad (\vartheta \text{ in } k;\ \Theta = \sqrt{\vartheta}).$$

Every number of K,

$$\lambda = \alpha + \beta\Theta \qquad (\alpha, \beta \text{ in } k),$$

has its "k-components" α, β and its conjugate $\bar{\lambda} = \alpha - \beta\Theta$. When taken as a Lie algebra in k, $\mathfrak{a}_K$ is of order $2r$ with the basic elements e_i, Θe_i. In a given K-representation the representing matrix of the element (11.2) is a linear combination of the k-components of the coefficients λ_i or a linear combination of λ_i and $\bar{\lambda}_i$:

$$(\lambda_1 A_1 + \cdots + \lambda_r A_r) + (\bar{\lambda}_1 A'_1 + \cdots + \bar{\lambda}_r A'_r).$$

On setting up the condition

$$[ab] \to AB - BA = [AB]$$

one finds that both

$$\mathfrak{A}: e_i \to A_i \quad \text{and} \quad \mathfrak{A}': e_i \to A'_i \tag{11.4}$$

must constitute K-representations of $\mathfrak{a} = \mathfrak{a}_k$ commutable with each other:

$$[A_i A'_j] = 0.$$

Let us suppose that every K-representation of $\mathfrak{a}$ breaks up into absolutely irreducible constituents. Then one sees that the pair of commutable representations (11.4) must break up into blocks of the form

$$\mathfrak{A}: e_i \to (A_i \times E'), \qquad \mathfrak{A}': e_i \to (E \times A_i')$$

where

$$a \to \lambda_1 A_1 + \cdots + \lambda_r A_r = S(a)$$

and

$$a \to \lambda_1 A_1' + \cdots + \lambda_r A_r' = S'(a)$$

are absolutely irreducible. E, E' are the unit matrices of correct degrees. Hence our representation of $\mathfrak{a}_K$ breaks up into parts of the form

$$(S(a) \times E') + (E \times S'(\bar{a})) \tag{11.5}$$

$$(a = \lambda_1 e_1 + \cdots + \lambda_r e_r, \qquad \bar{a} = \bar{\lambda}_1 e_1 + \cdots + \bar{\lambda}_r e_r).$$

The enveloping (associative) algebra of the linear set of matrices $S(a)$ is the complete matric algebra; the same for $S'(a)$. The enveloping algebra of (11.5) contains every matrix of either of the forms

$$(S(a) \times E') \text{ and } (E \times S'(a))$$

and hence of the form $S(a) \times S'(a')$. As the direct product of two complete matric algebras it is therefore itself a complete matric algebra, and (11.5) is absolutely irreducible.

Returning to our special case and realizing by the differential equation

$$d(x_i y_k) = dx_i \cdot y_k + x_i \cdot dy_k$$

that (11.5) describes the Kronecker multiplication in terms of the infinitesimal substitutions we can advance[10] the

THEOREM (8.11.B). *Every Lie representation of the complex unimodular group is fully reducible. The irreducible parts are of the form*

$$R(A) \times R'(\bar{A}) \tag{11.6}$$

where R and R' are two irreducible representations of the type (11.1).

The same arguments as lead to Theorems (8.11.A) and (8.11.B) go through for the symplectic group because the unitarily restricted $Sp(n)$ is also simply connected. However it must of necessity break down for $GL(n)$. A representation of the nucleus $\mathfrak{l}(n)$ for $n = 1$ is just any individual matrix, and one knows matrices like

$$\begin{Vmatrix} 1 & 1 \\ 0 & 1 \end{Vmatrix}$$

which are not fully reducible. This could not happen if the universal covering manifold of $U(n)$ were compact. As a matter of fact, $U(n)$ is not simply connected, and its universal covering manifold consists of infinitely many sheets.

Nor is the group of real proper orthogonal transformations simply connected, but its universal covering manifold consists of two sheets only. Hence full reducibility will prevail here again although there is to be expected a whole set of representations of the nucleus $\mathfrak{o}(n)$ leading to double-valued representations of the group. It is almost certain what the characters of these double-valued representations will be: the old formulas of the type (7.9.10) will hold for them also, but in the case $n = 2\nu + 1$ one will have to admit sets of integers $(l_1, \cdots, l_\nu)$ besides sets of half-integers, while for $n = 2\nu$ the reverse situation prevails.[11] The equation (11.6) will give all the single-valued irreducible representations of the complex orthogonal group if R and R' are either two single-valued or two double-valued representations. Moreover the fact that the orthogonal group consists of two pieces has to be taken into proper account. The results for the real orthogonal group remain valid for the real transformations leaving invariant a non-degenerate quadratic form of any index of inertia.

The differentiability assumptions indicated by Lie's name in Theorems (8.11.A) and (8.11.B) can be removed in replacing each of the basic infinitesimal transformations of the group by the one parameter group it engenders. I. Schur[10] carried this out for $SL(n)$, E. Mohr [12] for $Sp(n)$. In the case of $O(n)$ the double connectivity causes some difficulties which prevented R. Brauer, in his treatment of $O(n)$ along such lines, from completely eliminating the topologic spectre.[13] Another way of accomplishing the same ends is by appealing to general theorems concerning the Lie nature of linear groups.[14]

12. The Connectivity of the Classical Groups

THEOREM (8.12.A). *The group* $^sU(n)$ *of all unimodular unitary transformations is simply connected.*

I give the sketch of a proof which I owe to an oral communication of Dr. Witold Hurewicz; but here, as in all other topological considerations, I shall not insist on complete details.[15]

$^sU(n)$ contains $^sU(n-1)$ as the subgroup of its matrices of the form

$$U_{n-1} = \begin{Vmatrix} 1 & 0 & \cdots & 0 \\ 0 & * & \cdots & * \\ \cdots & \cdots & \cdots & \cdots \\ 0 & * & \cdots & * \end{Vmatrix}.$$

Two elements A and A' of $^sU(n)$ are left-equivalent modulo this subgroup, $A' = A \cdot U_{n-1}$, i.e. they belong to the same coset of $^sU(n-1)$, if and only if they coincide in their first column

$$(a_1, a_2, \cdots, a_n),$$

which is a vector of length 1:

$$|a_1|^2 + |a_2|^2 + \cdots + |a_n|^2 = 1. \tag{12.1}$$

As long as $n \geqq 2$, any such vector a occurs as the first column in some element of ${}^{s}U(n)$; indeed, one can determine a unitary unimodular vector basis $e_1, \cdots, e_n$ of which a is the first member: $e_1 = a$. Hence the manifold of the cosets is topologically equivalent to the $(2n - 1)$-dimensional sphere (12.1) in a Euclidean space with the real coördinates $\mathfrak{R}a_i, \mathfrak{J}a_i$. The sphere is simply connected. A closed curve (cycle) C on ${}^{s}U(n)$ is at the same time a cycle in the manifold of cosets. As such we may shrink it to the unit point $(1, 0, \cdots, 0)$ and thus deform C into a cycle on the subgroup ${}^{s}U(n - 1)$. The induction with respect to n thus started can be carried down to $n = 1$. As ${}^{s}U(1)$ consists of the element I only, we have then succeeded in shrinking C step by step to the unit point on ${}^{s}U(n)$.

It is exactly at this last instance that the argument fails for the complete group $U(n)$. $U(1)$ is the circle of all complex numbers of modulus 1, and a circle is of infinite connectivity: its universal covering manifold is a spiral with an infinity of coils. Every cycle C of $U(n)$ is deformable into $(\sim)$ a multiple of the cycle C_0 which the element

$$A = E_{n-1} \dot{+} e(\varphi)$$

describes with φ running from 0 to 1. The real function Φ defined on $U(n)$ by $\det A = e(\Phi)$ (which is multivalued but unramified) shows at once that no multiple of C_0 is ~ 0. The universal covering manifold of $U(n)$ consists of infinitely many "coils"; its group of covering transformations is the infinite discrete cyclic group.

Hurewicz's argument applies to the proper real orthogonal group $O^{+}(n)$, proving that any cycle on $O^{+}(n)$ is $\sim$ a multiple of the cycle C_0 described by

$$E_{n-2} \dot{+} \begin{Vmatrix} \cos\varphi, & -\sin\varphi \\ \sin\varphi, & \cos\varphi \end{Vmatrix} \qquad (0 \leqq \varphi \leqq 2\pi).$$

For $n = 2$ no such multiple is ~ 0. On $O^{+}(3)$, however, and a fortiori on $O^{+}(n)$ for $n \geqq 3$, $2C_0$ is ~ 0. One proves this either by the quaternion representation of the rotations in 3-space mapping the 3-dimensional sphere in 4-space on $O^{+}(3)$ so that antipodic points are identified, or by the following picture: Take two solid straight circular cones of aperture α with common vertex and touching each other along a generator, the one fixed in space, the other rolling on the first. The roller describes a closed motion which is $2C_0$ for $\alpha = 90°$ and approaches rest as $\alpha \rightarrow 180°$. By continuous variation of the parameter α one thus deforms $2C_0$ into the point I. So far it would still be possible for C_0 itself to become ~ 0 for some higher n. That this is not the case is most easily proved by explicit construction of the simplest of the double-valued representations, the so-called spin representation, first discovered and infinitesimally described for any n by E. Cartan. Dirac found that the spinors for $n = 4$ (4-dimensional space-time-world) account in quantum theory for the spin of the electron; hence the name. The existence of spinors is such an important feature of the orthogonal group that we shall briefly give their algebraic description in the next section.[16]

The unitary symplectic group $USp(n)$, $n = 2\nu$, affords the best occasion for applying Hurewicz's argument; for here the induction leads straight down to $n = 0$. The first two columns

$$a = (a_1, \cdots, a_n), \qquad b = (b_1, \cdots, b_n)$$

of an element A of $USp(n)$ characterize the coset to which it belongs modulo $USp(n-2)$. The conditions to be satisfied are

$$[ab] = 1; \tag{12.2}$$

$$(aa)_H = 1, \qquad (bb)_H = 1, \qquad (ab)_H = 0.$$

On introducing $\tilde{a}$ by (6.2.14) one sees that the equation (12.2), or

$$(\tilde{a}b)_H = 1,$$

is in agreement with

$$(\tilde{a}\tilde{a})_H = 1, \qquad (bb)_H = 1$$

only if $b = \tilde{a}$. Indeed,

$$(\tilde{a} - b, \tilde{a} - b)_H = (\tilde{a}\tilde{a})_H + (bb)_H - (\tilde{a}b)_H - (b\tilde{a})_H = 1 + 1 - 1 - 1 = 0.$$

Consequently we obtain as necessary and sufficient conditions

$$(aa)_H = |a_1|^2 + \cdots + |a_n|^2 = 1, \qquad b = \tilde{a}.$$

The cosets form a manifold topologically equivalent to the $(2n - 1) = (4\nu - 1)$-dimensional sphere.

Incidentally Hurewicz's procedure is by far the simplest way to compute the number of dimensions of each of our groups.

THEOREM (8.12.B). *$UO^+(n)$ has a universal covering manifold of two sheets when $n \geqq 3$, $USp(n)$ is simply connected for every $n = 2\nu$.*

13. Spinors

We choose the same starting point as Dirac in his classical paper on the spinning electron by asking whether it is not possible to interpret the square sum of n variables x_i as the square of a linear form:[17]

$$x_1^2 + \cdots + x_n^2 = (p_1x_1 + \cdots + p_nx_n)^2. \tag{13.1}$$

The coefficients p_i must then be quantities satisfying the relations

$$p_i^2 = 1, \qquad p_ip_k + p_kp_i = 0 \qquad (i \neq k). \tag{13.2}$$

n quantities p_i of this kind define a certain non-commutative associative abstract algebra $\mathfrak{p}$; as its basic elements we may take the 2^n monomials

$$p_1^{\alpha_1} \cdots p_n^{\alpha_n} = p_{\alpha_1 \cdots \alpha_n},$$

where the exponents α_i are integers mod 2. The rule of multiplication is

$$p_{\alpha_1 \cdots \alpha_n} \cdot p_{\beta_1 \cdots \beta_n} = (-1)^{\delta} \cdot p_{\gamma_1 \cdots \gamma_n},$$

where

$$\gamma_i = \alpha_i + \beta_i, \qquad \delta = \sum_{i>k} \alpha_i \beta_k .$$

We first consider the *even* case $n = 2\nu$ and use the notation

(13.3) $$p_1, q_1, \cdots, p_\nu, q_\nu \quad \text{besides} \quad p_1, \cdots, p_n .$$

A representation

$$p_\alpha \to P_\alpha, \qquad q_\alpha \to Q_\alpha \qquad (\alpha = 1, \cdots, \nu)$$

of degree 2^ν of our algebra is obtained by the following equations (familiar to the quantum theorist from the process of superquantization with Fermi statistics):

$$P_\alpha = 1' \times \cdots \times 1' \times \left\| \begin{matrix} 0 & 1 \\ 1 & 0 \end{matrix} \right\| \times 1 \times \cdots \times 1,$$

$$Q_\alpha = 1' \times \cdots \times 1' \times \left\| \begin{matrix} 0 & i \\ -i & 0 \end{matrix} \right\| \times 1 \times \cdots \times 1.$$

The number of factors is ν. The two matrices

$$P = \left\| \begin{matrix} 0 & 1 \\ 1 & 0 \end{matrix} \right\|, \qquad Q = \left\| \begin{matrix} 0 & i \\ -i & 0 \end{matrix} \right\| \qquad (i = \sqrt{-1})$$

appear at the αth place ; 1, 1′ mean

$$\left\| \begin{matrix} 1 & 0 \\ 0 & 1 \end{matrix} \right\|, \qquad \left\| \begin{matrix} 1 & 0 \\ 0 & -1 \end{matrix} \right\|$$

respectively. The two rows and columns of our matrices may be marked by + and −. Hence the variables $x_{\sigma_1 \cdots \sigma_\nu}$ in our representation space bear combinations of signs $\sigma_\alpha = \pm$ as their indices. $iP_\alpha Q_\alpha = R_\alpha$ equals

$$1 \times \cdots \times 1 \times 1' \times 1 \times \cdots \times 1$$

with 1′ at the αth place. Thus

$$\tfrac{1}{2}(E \pm R_\alpha), \qquad \tfrac{1}{2}R_1 \cdots R_{\alpha-1}(P_\alpha \pm iQ_\alpha)$$

are of the form

$$1 \times \cdots \times 1 \times V \times 1 \times \cdots \times 1$$

where V at the αth place is one of the four matrices

$$\left\| \begin{matrix} 1 & 0 \\ 0 & 0 \end{matrix} \right\|, \quad \left\| \begin{matrix} 0 & 0 \\ 0 & 1 \end{matrix} \right\|, \quad \left\| \begin{matrix} 0 & 1 \\ 0 & 0 \end{matrix} \right\|, \quad \left\| \begin{matrix} 0 & 0 \\ 1 & 0 \end{matrix} \right\|.$$

Consequently our representation of the algebra $\mathfrak{p}$ yields the *complete matric algebra* in 2^ν dimensions. Because of coincidence of orders, $2^n = (2^\nu)^2$, the correspondence is a one-to-one isomorphism.

We now perform an arbitrary orthogonal transformation o:

$$x_i = \sum_k o(ik)x'_k, \qquad p'_i = \sum_k o(ki)p_k.$$

The p'_i fulfill the same conditions (13.2) as the p_i; hence

$$p_i \rightarrow P'_i = \sum_k o(ki)P_k \tag{13.4}$$

is another irreducible representation of the algebra $\mathfrak{p}$. We know by Theorem (3.3.E) that there is only one such representation of the complete matric algebra in the sense of equivalence. In other words there exists a non-singular matrix $S(o)$ uniquely determined but for a numerical factor such that

$$P'_i = S(o)P_iS^{-1}(o) \qquad (i = 1, \cdots, n). \tag{13.5}$$

For two orthogonal transformations o, o' one has

$$S(oo') = \kappa\, S(o)S(o') \tag{13.6}$$

with the numerical factor κ depending on o and o'. Indeed from

$$\sum_l o'(li)P_l = S(o')P_iS^{-1}(o')$$

one deduces by forming (13.5):

$$\sum_l o'(li)P'_l = S(o)S(o')P_iS^{-1}(o')S^{-1}(o),$$

and according to (13.4) the left side is

$$\sum_k oo'(ki)P_k$$

which, on the other hand, equals

$$S(oo')P_iS^{-1}(oo').$$

In brief, $S(o)$ is a "projective" representation of $O(n)$.

So far all goes well in any field k to which one has adjoined $\sqrt{-1}$. Now follows the attempt to normalize the arbitrary gauge factor in $S(o)$ so as to turn the projective into an ordinary "affine" representation. The transposed matrices P_i^* again satisfy the conditions (13.2), and hence there is a matrix C such that

$$P_i^* = CP_iC^{-1}. \tag{13.7}$$

Without much ado we may write down C explicitly:

$$C = \begin{Vmatrix} 0 & 1 \\ 1 & 0 \end{Vmatrix} \times \begin{Vmatrix} 0 & i \\ -i & 0 \end{Vmatrix} \times \begin{Vmatrix} 0 & 1 \\ 1 & 0 \end{Vmatrix} \times \begin{Vmatrix} 0 & i \\ -i & 0 \end{Vmatrix} \times \cdots.$$

The equation (13.5) implies

$$\sum_k o(ki)P_k^* = \hat{S}(o)P_i^*\hat{S}^{-1}(o),$$

and on account of (13.7),

$$C^{-1}\hat{S}(o)C$$

is another solution of the equations (13.5) for $S(o)$. Hence

$$\hat{S}(o) = \rho \cdot CS(o)C^{-1}.$$

In replacing $S(o)$ by $\lambda S(o)$, $\hat{S}(o)$ is multiplied by $1/\lambda$ and ρ replaced by ρ/λ^2. By choosing $\lambda = \sqrt{\rho}$ we normalize the gauge factor in such a way that $S(o)$ satisfies the condition

$$S(o) = CS(o)C^{-1}$$

The sign in $\pm S(o)$ still remains indeterminate. Because $X = S(o)S(o')$ and $S(oo')$ both satisfy the normalizing condition

$$X = CXC^{-1},$$

the factor κ in (13.6) is ± 1. Instead of the projective we have thus obtained an ordinary though double-valued representation $\pm S(o)$ of degree 2^ν, called the *spin representation.*

The normalization requires the possibility of *extracting square roots.* The constructions in Euclidean geometry with ruler and compass are algebraically equivalent to the four species and the extraction of square roots. A field in which every quadratic equation $x^2 - \rho = 0$ is solvable may therefore be called a *Euclidean field.* Our result is then that *in every Euclidean field we can construct the spin representation*; the Euclidean nature of the field is essential. The orthogonal transformations are the automorphisms of Euclidean vector space. Only with the spinors do we strike that level in the theory of its representations on which Euclid himself, flourishing ruler and compass, so deftly moves in the realm of geometric figures. In some way Euclid's geometry must be deeply connected with the existence of the spin representation.

When we operate in the field $K^\dagger$, one readily verifies that the diagonal transformation o of $UO(n)$:

$$p_\alpha \pm iq_\alpha \to e(\pm \varphi_\alpha)\cdot(p_\alpha \pm iq_\alpha)$$

is represented by the following diagonal $S(o)$:

$$x_{\sigma_1\cdots\sigma_\nu} \to e(\tfrac{1}{2}(\sigma_1\varphi_1 + \cdots + \sigma_\nu\varphi_\nu))\cdot x_{\sigma_1\cdots\sigma_\nu}.$$

This shows that $S(o)$ changes sign if we follow its continuous variation while o races over the cycle

$$C_0: \quad \varphi_0 = \cdots = \varphi_{\nu-1} = 0;\ 0 \leqq \varphi_\nu \leqq 1.$$

Hence $O^+(n)$ is certainly not simply connected. The character of the spin representation turns out to be the sum

$$\sum e(\tfrac{1}{2}(\pm \varphi_1 \pm \varphi_2 \pm \cdots \pm \varphi_\nu))$$

extending over all combinations of signs, a formula which confirms our conjecture about the general nature of characters of the double-valued representations.

The modifications needed for $n = 2\nu + 1$ are not very serious. To (13.3) we add one further $p = p_n$ represented by

$$P_n = 1' \times 1' \times \cdots \times 1'.$$

The product of all the P's in the order $P_1Q_1 \cdots P_\nu Q_\nu P_n$ then equals $i^\nu E$. Supplementing (13.2) by the one further relation

$$i^{-\nu} p_1 \cdots p_n = 1$$

reduces the order $2^n = 2\cdot(2^\nu)^2$ of the algebra $\mathfrak{p}$ to $(2^\nu)^2$, and our representation then maps it by a one-to-one isomorphism on the complete matric algebra of degree 2^ν. Since (13.4) entails

$$P_1' \cdots P_n' = \det(o(ik))\cdot P_1 \cdots P_n$$

we are able to associate with any orthogonal o a matrix $S(o)$ in 2^ν dimensions such that

$$\pm P_i' = S(o)P_iS^{-1}(o)$$

the + or − sign prevailing according as o is proper or improper.

For further details, including the case of an indefinite metric ground-form so important for physics, see the paper by R. Brauer and the author, quoted above.

14. Finite integrity basis for invariants of compact groups

We now come to the essential point in Part B of the present chapter to which all its other considerations are subservient.

THEOREM (8.14.A). *The (absolute) invariants $J(x, y, \cdots)$ corresponding to a given set of representations of a finite or a compact Lie group have a finite integrity basis.*

By means of Hilbert's ideal theorem we first select a finite number among the non-constant invariants

$$J_1(xy\cdots), \cdots, J_h(xy\cdots) \tag{14.1}$$

such that all such invariants $J(xy\cdots)$ are linear combinations of them with polynomial coefficients:

$$J(xy\cdots) = \sum_{i=1}^{h} L_i(xy\cdots)J_i(xy\cdots). \tag{14.2}$$

The next step should be to devise a linear process $\omega(L)$ which carries each polynomial L into an invariant and moreover satisfies the conditions

$$\omega(1) = 1, \qquad \omega(L\cdot J) = \omega(L)\cdot J$$

for every invariant J. Once in possession of such a process we derive from (14.2):

$$J = \sum_i \omega(L_i) \cdot J_i.$$

The $\omega(L_i)$ being invariants, induction with respect to the degree will prove that every invariant is expressible in terms of the basis (14.1).

A process ω of the desired nature is given by averaging over the group manifold:

$$\omega(L) = \mathfrak{M}_s(sL).$$

It works for any finite group (provided the number field is not of a prime characteristic dividing the order of the group) and for any compact continuous Lie group. It works on all compact groups accessible to Haar's measure; finally, it works on any group whatsoever if one limits oneself to von Neumann's almost periodic representations. The most important example, besides the finite groups, is the real orthogonal group $O(n)$ in the real field K and its two-sheeted universal covering manifold.

In the case of the classical groups in K or $K^\dagger$ one applies the procedure after having introduced the unitary restriction. By means of the considerations in §11, one can get rid again of that restriction or replace it by some other reality condition, with the result:

THEOREM (8.14.B). *The invariants of the classical groups corresponding to any Lie representations in* K *have a finite integrity basis.*

As a matter of fact, we have proved this before by a detailed algebraic investigation which rested on the explicit determination of all Lie representations, on the Cayley Ω-process and the adjunction argument. But the advantages of the integration method, its directness and its generality with respect to the nature of the group, are obvious. Its great drawback is that it deals with the one field K only. Moreover, the group of *step transformations* which we were able to cover in some way by the adjunction argument is definitely out of its reach. For wherever the integration method applies, directly or indirectly through the unitarian trick, it establishes both full reducibility and a finite integrity basis for invariants. But the first statement is simply not true for the group just mentioned. We know of no single instance where the first main theorem about invariants fails, but a proof holding for all groups whatsoever is likewise unknown.[18]

15. The first main theorem for finite groups

An elementary proof for finite groups not depending on Hilbert's general theorem on polynomial ideals was given by E. Noether.[19] Here is her straightforward construction, as it applies to a given group of linear substitutions

$$(15.1)\quad A^{(\alpha)}: x^{(\alpha)} = A^{(\alpha)}x, \quad x_i^{(\alpha)} = \sum_k a_{ik}^{(\alpha)} x_k \quad (i, k = 1, \cdots, n; \alpha = 1, \cdots, h).$$

If $J(x)$ is any invariant one writes

$$J(x) = \frac{1}{h} \sum_{\alpha=1}^{h} J(x^{(\alpha)}).$$

Consider the right side as a function of $n \cdot h$ independent variables $x_i^{(\alpha)}$. As such it is a symmetric function of the n vectors

$$\mathfrak{x}_1 = (x_1^{(1)}, \cdots, x_1^{(h)}),$$
$$\cdots\cdots\cdots\cdots\cdots\cdots$$
$$\mathfrak{x}_n = (x_n^{(1)}, \cdots, x_n^{(h)})$$

in an h-dimensional vector space in the sense of Chapter II, §3, and thus expressible in terms of the polarized elementary symmetric functions. With the arguments $\mathfrak{x}_1, \cdots, \mathfrak{x}_n$ put in in all possible combinations, we can define these functions as the coefficients $G_{r_1 \cdots r_n}$ in the product

$$\prod_{\alpha=1}^{h} \{u + u_1 x_1^{(\alpha)} + \cdots + u_n x_n^{(\alpha)}\} = \sum u^r u_1^{r_1} \cdots u_n^{r_n} G_{r_1 \cdots r_n}(x_i^{(\alpha)})$$
$$(r + r_1 + \cdots + r_n = h)$$

involving the indeterminates $u, u_1, \cdots, u_n$ besides the $x_i^{(\alpha)}$. The functions one derives from

$$G_{r_1 \cdots r_n}(x_i^{(\alpha)})$$

by the substitution (15.1) constitute an integrity basis for our finite group. They are all of degree $\leqq h$ and their number is

$$(h + 1) \cdots (h + n)/1 \cdot 2 \cdots n.$$

One could not demand anything more explicit.

16. Invariant differentials and Betti numbers of a compact Lie group

One of the most beautiful applications of the integration method is E. Cartan's theory of invariant differentials on a compact Lie group.[20] In concluding this chapter we shall give a general outline of its leading ideas.

On a differentiable manifold with coördinates $x_1, \cdots, x_n$ we may consider a scalar field $f(x)$, or a differential linear form

$$\sum_i f_i(x)\, dx_i \tag{16.1}$$

depending on a line element dx, or a differential linear form of rank 2,

$$\sum_{i,k} f_{ik}(x)\, dx_i \delta x_k,$$

which we suppose to be skew-symmetric so that it depends on the two-dimensional element with the components

$$dx_i\, \delta x_k - \delta x_i\, dx_k$$

which is spanned by the two line elements dx, δx. Thus we can go on forming "*differentials*" of rank $p = 0, 1, 2, \cdots, n$. The differential forms indicate how their coefficients f should be transformed under transition to other coördinates. There is a process of *derivation* that is independent of the coördinate system and quite familiar to anybody who ever studied Maxwell's theory of the electromagnetic field: the derivative of the scalar f is the gradient

$$f_i = \partial f/\partial x_i ,$$

the derivative of the differential (16.1) of rank 1 is given by

$$f_{ik} = \partial f_k/\partial x_i - \partial f_i/\partial x_k ,$$

and in general, the derivative ω' of ω with the skew-symmetric components $f_{i_1 \cdots i_p}$ has the components

$$\frac{\partial f_{i_2 \cdots i_{p+1}}}{\partial x_{i_1}} - + \cdots$$

(alternating sum of $p + 1$ terms). Derivation raises the rank by 1. A differential ω_p of rank p can be integrated over a p-chain C_p. Stokes's general theorem states that the integral of the derivative ω_p' over a $(p + 1)$-chain C_{p+1} equals the integral of ω_p over the p-cycle C'_{p+1} bounding C_{p+1}.

A differential ω whose derivative vanishes is called *exact.* What we mean by $\omega \sim 0$ (ω homologous zero) may be explained in two ways:[21] either differentially as indicating that ω is the derivative of a differential of next lower rank, or integrally as demanding that the integral of ω over any cycle vanishes. Every differential ~ 0 is exact; one readily proves this in both ways. "In the small" both notions, exact and ~ 0, coincide, but not in the large. The study of exact differentials and their homologies is a dual or contragredient counterpart of that of cycles where the derivative takes the place of the boundary. Some recent advances in the foundations of topology were made by stressing the dual aspect and "topologizing" these operations with differentials. Anyhow, the *Betti number* B_p may be explained as the number of exact differentials of rank p that are linearly independent in the sense of homology.[22]

We now assume that our manifold is a compact Lie group. The differential ω is called *invariant* if it stays unaltered under left and right translations:

$$x \to sx \quad \text{and} \quad x \to xs.$$

Since left translations carry the origin I into any other point it is sufficient to know an invariant differential at the origin, where it is a skew-symmetric multilinear form

$$\sum_{(i)} a_{i_1 \cdots i_p} \delta_1 x_{i_1} \cdots \delta_p x_{i_p} \tag{16.2}$$

with constant coefficients a, depending on p infinitesimal group elements $\delta_1 x$, $\cdots$, $\delta_p x$. The requirement to be invariant also under right translations means that (16.2) is invariant under the adjoint group

$$\delta x \to s^{-1} \cdot \delta x \cdot s.$$

Hence the invariant differentials are exactly those invariants of the adjoint group which we considered in Chapter VII, §11. The numbers of the linearly independent ones among them for the several ranks p are the coefficients of the Poincaré polynomial computable by (7.11.2).

Three facts establish a close relationship between arbitrary exact differentials under homology and invariant differentials under equality:

LEMMA (8.16.A). (1) *Every invariant differential is exact.*

(2) *Every exact differential is homologous to an invariant one.*

(3) *An invariant differential* ~ 0 *is necessarily* $= 0$.

(1) is simply proved by explicit computation of the derivative of the invariant differential ω. (2) can be obtained either in integral (a) or in differential (b) fashion. (a) Integrate $s\omega$, the differential arising from ω by the left translation s, over a given p-cycle C. One obviously has

$$\int_C s\omega = \int_{s^{-1}C} \omega.$$

$s^{-1}C$ arises by continuous deformation from C in passing from $\mathfrak{l}$ to s^{-1} along a continuous path. When integrated over any two cycles C, C' which are deformable into each other, an exact differential ω yields the same integral:

$$\int_C \omega = \int_{C'} \omega.$$

In particular

$$\int_C s\omega = \int_{s^{-1}C} \omega = \int_C \omega.$$

Consequently the mean value

$$\psi = \mathfrak{M}_s(s\omega)$$

is $\sim \omega$. The differential ψ is left-invariant. By performing right translations on ψ in the same fashion one obtains

$$\mathfrak{M}_{s'}\mathfrak{M}_s(s\omega s') \sim \omega;$$

the differential on the left is two-sided invariant. (b) For any *infinitesimal* s one can easily ascertain an infinitesimal differential φ of rank $p - 1$ such as to make $s\omega - \omega = \varphi'$.

(3) If the invariant ω is a derivative, $\omega = \varphi'$, then $\omega = s\omega s'$ is the derivative of $s\varphi s'$ and thus of the invariant

$$\psi = \mathfrak{M}_{s'}\mathfrak{M}_s(s\varphi s').$$

According to point (1), the derivative of an invariant differential ψ is zero.

From the lemma it follows that the Betti number, i.e. the number of *exact* differentials linearly independent in the sense of *homology*, equals the number of *invariant* differentials linearly independent in the sense of *equality*. Hence:

THEOREM (8.16.B). *The coefficients of the Poincaré polynomial of a compact Lie group give its Betti numbers.*

From this relationship one can deduce quite a surprising amount of information about the Betti numbers $B_1, B_2, \cdots$ of a compact Lie group of r parameters. Since the number of linearly independent skew-symmetric multilinear forms (16.2), whether invariant under the adjoint group or not, equals the binomial coefficient $\binom{r}{p}$, one has

$$B_p \leqq \binom{r}{p}.$$

On the other hand, let

$$\sum_i a_i'\delta x_i, \cdots, \sum_i a_i^{(p)}\delta x_i \tag{16.3}$$

be any p of the invariant forms of rank 1. The equation

$$a_{i_1 \cdots i_p} = \begin{vmatrix} a_{i_1}', & \cdots, a_{i_p}' \\ \cdots & \cdots \\ a_{i_1}^{(p)}, & \cdots, a_{i_p}^{(p)} \end{vmatrix} \tag{16.4}$$

then defines an *invariant* form of rank p. While each of the forms (16.3) varies in a linear manifold of $B_1 = \beta$ dimensions, these special forms (16.4) of rank p range over a manifold of $\binom{\beta}{p}$ dimensions. Thus

$$B_p \geqq \binom{\beta}{p}.$$

Further inequalities of the same type may be obtained by considering those among the linear invariants of rank $p = p_1 + p_2$ which arise by multiplication (as defined in Chapter VII, §11) from any pair of invariant forms of ranks p_1 and p_2.

As to the unitarily restricted classical groups, their Betti numbers[23] are determined by the explicit formulas for their Poincaré polynomials, Theorems (7.11.A) and (7.11.C).

CHAPTER IX

MATRIC ALGEBRAS RESUMED

1. Automorphisms

In order to round off our investigations we resume in this chapter the subject of Chapter III, Part A, the study of fully reducible matric algebras.[1] The method used there can be applied to three important questions concerning them: automorphisms, cross multiplication, and extension of the underlying field k.

Take a simple algebra $\mathfrak{a}$ in k and its irreducible representation $a \to A(a)$. Let $a \to a'$ be an automorphism of $\mathfrak{a}$. Then $a \to A(a')$ defines another irreducible representation of $\mathfrak{a}$. But we have observed [Theorem (3.3.E)] that there is only one such representation in the sense of equivalence. Hence there exists a non-singular matrix H such that

$$A(a') = H \cdot A(a) \cdot H^{-1}$$

for all elements a of $\mathfrak{a}$:

LEMMA (9.1.A). *Any automorphism $A \to A'$ of an irreducible matric algebra $\mathfrak{A}$ is generated by a constant non-singular matrix H:*

$$A' = HAH^{-1}.$$

Consider any matric algebra $\mathfrak{A} = \{A\}$. If a constant non-singular matrix H transforms every element A of $\mathfrak{A}$ into an element

$$A' = HAH^{-1} \tag{1.1}$$

of $\mathfrak{A}$, then $A \to A'$ is an automorphism of $\mathfrak{A}$. At the same time the equation

$$B' = HBH^{-1}$$

turns every commutator B of $\mathfrak{A}$ into a commutator and thus defines an automorphism in the commutator algebra $\mathfrak{B}$ as well. We are thus led to study *simultaneous automorphisms* of $\mathfrak{A}$ and $\mathfrak{B}$. Let $\mathfrak{A}$ be a fully reducible matric algebra, and $\mathfrak{B}$ its commutator algebra. $\mathfrak{B}$ is likewise fully reducible, the relationship of $\mathfrak{A}$ and $\mathfrak{B}$ is mutual, and their structure is described by Theorem (3.5.B). The intersection $\mathfrak{Z}$ of $\mathfrak{A}$ and $\mathfrak{B}$ is called the *centrum*; it consists of every element of $\mathfrak{A}$ that commutes with all elements of $\mathfrak{A}$. If the centrum contains only the numerical multiples of the unit element E, then $\mathfrak{A}$ is called *normal* (and so is $\mathfrak{B}$). Our second formulation shows that this property may be ascribed to the abstract algebra. If a fixed non-singular matrix H generates two simultaneous automorphisms $A \to A'$, $B \to B'$ in $\mathfrak{A}$ and $\mathfrak{B}$,

$$A' = HAH^{-1}, \qquad B' = HBH^{-1}, \tag{1.2}$$

the two automorphisms obviously coincide within the centrum $\mathfrak{Z}$. Our aim is the inverse proposition:

THEOREM (9.1.B). *Any two simultaneous automorphisms*

$$A \to A', \qquad B \to B' \tag{1.3}$$

of the fully reducible matric algebra $\mathfrak{A}$ *and its commutator algebra* $\mathfrak{B}$ *which coincide within the centrum* $\mathfrak{Z}$ *are generated by a common non-singular matrix* H:

$$A' = HAH^{-1}, \qquad B' = HBH^{-1}. \tag{1.2}$$

In our ascent in Chapter III, Part A, the first point where we reached full reciprocity between $\mathfrak{A}$ and $\mathfrak{B}$ was at the end of §4:

$$\mathfrak{A} = s(\mathfrak{d})_t, \qquad \mathfrak{B} = t(\mathfrak{d}')_s. \tag{1.4}$$

We then form the linear closure $\mathfrak{C} = \mathfrak{A}\mathfrak{B}$ of all products

$$C = AB = BA \qquad (A \text{ in } \mathfrak{A},\ B \text{ in } \mathfrak{B}).$$

From the description (3.4.10) of the schemes of the generic matrices A and B one derives at once that

$$\mathfrak{C} = \mathfrak{D}_{st},$$

where $\mathfrak{D} = (\mathfrak{d})(\mathfrak{d}')$ is the linear closure of all the transformations in $\mathfrak{d}$,

$$(a)(b)': x' = axb,$$

which correspond to arbitrary elements a, b of $\mathfrak{d}$. $\mathfrak{D}$ is certainly irreducible as it contains the irreducible $(\mathfrak{d})$. A commutator of $\mathfrak{D}$ commutes in particular with each (a): $x \to ax$ and is hence of the form $x \to xj$, where j is the image of the unit of $\mathfrak{d}$. Because of its commutability with all $(b)'$: $x \to xb$, the given commutator is similarly found to be $x \to jx$. Consequently $jx = xj$, or j lies in the centrum $\mathfrak{z}$ of $\mathfrak{d}$. $\mathfrak{z}$ is a commutative division algebra or a field of finite order δ over k. On extending k to $\mathfrak{z}$ we may look upon $\mathfrak{d}$ as a division algebra of order $m = d/\delta$ over the field $\mathfrak{z}$. The centrum of $\mathfrak{D}$ in its concrete form is $m \cdot (\mathfrak{z})$, and by applying Theorem (3.4.B) to the irreducible $\mathfrak{D}$ it follows that

$$\mathfrak{D} = (\mathfrak{z})_m. \tag{1.5}$$

The centrum $\mathfrak{Z}$ of $\mathfrak{A}$ and $\mathfrak{B}$ and their product $\mathfrak{C} = \mathfrak{A}\mathfrak{B}$ are thus given by

$$\mathfrak{Z} = g \cdot (\mathfrak{z}), \qquad \mathfrak{C} = (\mathfrak{z})_g \qquad [g = mst].$$

The situation for the pair of commuting algebras $\mathfrak{Z}$, $\mathfrak{C}$ is considerably simpler than for $\mathfrak{A}$ and $\mathfrak{B}$; compare (1.4).

Let us first study the particular case where $\mathfrak{d}$ (or $\mathfrak{A}$ and $\mathfrak{B}$) are *normal.* Then formula (1.5) shows $\mathfrak{D}$ to be the complete matric algebra $\mathfrak{M}_d$:

LEMMA (9.1.C). *Let* $\mathfrak{d}$ *be a normal division algebra. The linear closure of all the transformations in* $\mathfrak{d}$,

$$x' = axb,$$

which correspond to arbitrary elements a, b *of* $\mathfrak{d}$ *is the complete matric algebra* $\mathfrak{M}_d$.

$\mathfrak{C}$ is now the complete matric algebra

$$\mathfrak{C} = \mathfrak{M}_g$$

of degree $g = dst$. This result is in keeping with the relation (3.4.12). Let

$$A_\lambda \quad (\lambda = 1, \cdots, dt^2), \qquad B_\mu \quad (\mu = 1, \cdots, ds^2)$$

be linear bases of $\mathfrak{A}$ and $\mathfrak{B}$ respectively. We have found that the matrices $A_\lambda B_\mu$ are linearly independent and constitute a basis for $\mathfrak{C} = \mathfrak{M}_g$. With our given automorphisms $A \to A'$, $B \to B'$ of $\mathfrak{A}$ and $\mathfrak{B}$ we construct a corresponding automorphism of $\mathfrak{M}_g$ by

$$\sum \zeta_{\lambda\mu} A_\lambda B_\mu \to \sum \zeta_{\lambda\mu} A'_\lambda B'_\mu \qquad (\zeta_{\lambda\mu} \text{ any numbers})$$

which associates $A'B'$ with AB. By applying Lemma (9.1.A) to the complete matric algebra $\mathfrak{M}_g$ we find a constant non-singular matrix H generating this automorphism,

$$A'B' = H \cdot AB \cdot H^{-1},$$

and thus satisfying in particular (B or $A = E$) the relations (1.2) for all A in $\mathfrak{A}$ and B in $\mathfrak{B}$.

By the simple device of forming the products AB we have thus succeeded in sharpening Lemma (9.1.A) considerably, although it enters into our argument only for the special case of the full matric algebra. When we combine the identical automorphism $B' = B$ of $\mathfrak{B}$ with a given automorphism $A \to A'$ of $\mathfrak{A}$, our H commutes with every B and *hence lies in* $\mathfrak{A}$. We express this beautiful result in abstract terms:[2]

THEOREM (9.1.D). *Any automorphism of a normal simple algebra is inner.*

The next step consists in removing the assumption that $\mathfrak{A}$ and $\mathfrak{B}$ are normal.

$$A_\lambda \qquad\qquad (\lambda = 1, \cdots, mt^2)$$

is a basis of $\mathfrak{A}$ over $\mathfrak{Z}$ if every A in $\mathfrak{A}$ is uniquely expressible as $\sum Z_\lambda A_\lambda$ with "coefficients" Z_λ in $\mathfrak{Z}$. A_λ, B_μ being bases of $\mathfrak{A}$ and $\mathfrak{B}$ over $\mathfrak{Z}$, each matrix of $\mathfrak{C}$ is uniquely expressible in the form

$$\sum Z_{\lambda\mu} A_\lambda B_\mu \qquad\qquad (Z_{\lambda\mu} \text{ in } \mathfrak{Z}).$$

If (1.3) are given automorphisms in $\mathfrak{A}$ and $\mathfrak{B}$ *which coincide within* $\mathfrak{Z}$: $Z \to Z'$, then

$$\sum Z_{\lambda\mu} A_\lambda B_\mu \to \sum Z'_{\lambda\mu} A'_\lambda B'_\mu$$

defines a corresponding automorphism in $\mathfrak{C}$ by which $AB \to A'B'$. Applying the lemma (9.1.A) to the irreducible matric algebra $\mathfrak{C} = (\mathfrak{z})_g$ we obtain the desired proposition (9.1.B) *for the case where* $\mathfrak{A}$ *and* $\mathfrak{B}$ *consist of one block only.*

In order to pass to several blocks we must generalize Lemma (9.1.A) so as to cover the direct sum $\mathfrak{a} = \sum \mathfrak{a}_u$ of v simple algebras $\mathfrak{a}_u$, in concrete form

$$(1.6) \qquad \mathfrak{A} = \mathfrak{A}_1 + \cdots + \mathfrak{A}_v, \qquad A = A_1 \dot{+} \cdots \dot{+} A_v,$$

where each component A_u varies independently over the irreducible matric algebra $\mathfrak{A}_u$. By Theorem (3.5.C) any non-degenerate representation of $\mathfrak{a}$ is a sum

$$\sum_{u=1}^{v} s_u \mathfrak{A}_u . \tag{1.7}$$

The representation (1.6), $a \to A(a)$, with each $s_u = 1$, is faithful. Let $a \to a'$ be an automorphism of $\mathfrak{a}$ and consider the representation $a \to A(a')$. It must be equivalent to some (1.7); its faithfulness would be violated if one of the s_u equaled 0, its degree would be too high if one of the s_u were > 1. Consequently all the $s_u = 1$, and the new representation is equivalent to the old one, or there is a non-singular matrix H such that

$$A(a') = H \cdot A(a) \cdot H^{-1}.$$

LEMMA (9.1.E). *Any automorphism $A \to A'$ of the direct sum $\mathfrak{A}$ of irreducible matric algebras,*

$$\mathfrak{A} = \mathfrak{A}_1 + \cdots + \mathfrak{A}_v , \qquad A = A_1 \dotplus \cdots \dotplus A_v ,$$

is generated by a constant non-singular matrix H:

$$A' \doteq HAH^{-1}.$$

We finally return to an arbitrary fully reducible matric algebra $\mathfrak{A}$ and its commutator algebra $\mathfrak{B}$ as studied in Theorem (3.5.B). Again we form the linear closure $\mathfrak{C}$ of all matrices

$$C = AB = BA \qquad (A \text{ in } \mathfrak{A},\ B \text{ in } \mathfrak{B}),$$

and using notations that hardly need to be explained we find

$$\mathfrak{Z} = \sum g_u(\mathfrak{z}_u), \qquad \mathfrak{C} = \sum (\mathfrak{z}_u)_{g_u} .$$

By application of our last lemma to $\mathfrak{C}$ (rather than to $\mathfrak{A}$) and to the automorphism $AB \to A'B'$, we construct the matrix H whose existence was claimed by Theorem (9.1.B).

Concerning the question of *uniqueness*, we observe that H may be replaced by HZ_0 where Z_0 is a non-singular matrix in the centrum $\mathfrak{Z}$.

On associating the identical automorphism of $\mathfrak{B}$ with any automorphism of $\mathfrak{A}$ which leaves the central elements fixed, one gets in particular:

THEOREM (9.1.F). *Any automorphism of the direct sum $\mathfrak{a}$ of v simple algebras $\mathfrak{a}_u$ which does not touch the elements of the centrum $\mathfrak{z}$ of $\mathfrak{a}$ is inner.*

2. A lemma on multiplication

If $\mathfrak{A} = \{A\}$, $\mathfrak{B} = \{B\}$ are two matric algebras whose elements A, B are transformations in an n- and ν-dimensional vector space P_n, P_ν respectively, then each

$$A \times B \qquad (A \text{ in } \mathfrak{A},\ B \text{ in } \mathfrak{B}) \tag{2.1}$$

operates on the product space $P_n P_\nu$ of $n\nu$ dimensions. The reader may be warned that now the Kronecker product $A \times B$ and not, as in the last section, the ordinary product AB is studied, and that A and B are not commutators of each other. The linear closure of all the matrices (2.1) is an algebra $[\mathfrak{A} \times \mathfrak{B}]$ which we call the algebra product of $\mathfrak{A}$ and $\mathfrak{B}$. The process can be defined in terms of the abstract algebras $\mathfrak{a}$ and $\mathfrak{b}$. If $\mathfrak{a}$ of order m is referred to a basis a_i $(i = 1, \cdots, m)$ and $\mathfrak{b}$ of order μ to a basis b_ι $(\iota = 1, \cdots, \mu)$ with the multiplication rules

$$a_i a_k = \sum_l \alpha^l_{ik} a_l, \qquad b_\iota b_\kappa = \sum_\lambda \beta^\lambda_{\iota\kappa} b_\lambda$$

then $\mathfrak{c} = [\mathfrak{a} \times \mathfrak{b}]$ has a basis $c_{i\iota}$ with the multiplication table

$$c_{i\iota} c_{k\kappa} = \sum_{l,\lambda} \alpha^l_{ik} \beta^\lambda_{\iota\kappa} c_{l\lambda} .$$

Transition to another basis in $\mathfrak{a}$ and in $\mathfrak{b}$ merely causes a change of basis in $[\mathfrak{a} \times \mathfrak{b}]$.

Notice that

$$\mathfrak{A}_v = [\mathfrak{A} \times \mathfrak{M}_v];$$

hence

$$[\mathfrak{A} \times \mathfrak{B}_v] = [\mathfrak{A} \times \mathfrak{B}]_v .$$

Let K be a field over k. Any k-linear set $\mathfrak{A}$ of matrices A in k can be extended to K by forming the linear closure $\mathfrak{A}_K$ of $\mathfrak{A}$ in K. If $A_1, \cdots, A_m$ is a basis of $\mathfrak{A}$, then

$$A = \xi_1 A_1 + \cdots + \xi_m A_m$$

ranges over $\mathfrak{A}$ or $\mathfrak{A}_K$ when the coefficients ξ_i vary in k or K respectively. If $\mathfrak{A}$ is an algebra in k, so likewise is $\mathfrak{A}_K$ in K. Again the process may be described in abstract terms: if $a_1, \cdots, a_m$ is a basis of the k-linear set $\mathfrak{a}$ of order m, then $\mathfrak{a}_K$ consists of all formal sums $\sum_{i=1}^m \xi_i a_i$ with arbitrary coefficients ξ_i in K. Change of basis in $\mathfrak{a}$ amounts to a particular change of basis in $\mathfrak{a}_K$, namely to one in which the transformation coefficients lie in the subfield k.

With $\mathfrak{B}$ being the commutator algebra in k of the algebra $\mathfrak{A}$ in k, $\mathfrak{B}_K$ is the commutator algebra of $\mathfrak{A}_K$ in K. Indeed, if $A_1, \cdots, A_m$ is a basis of $\mathfrak{A}$, then a matrix B in K belongs to $\mathfrak{B}_K$ if it satisfies the m equations

$$BA_i = A_i B.$$

These are homogeneous linear equations for B with coefficients in k. Consequently the solutions B have a basis $B_1, \cdots, B_\mu$ consisting of matrices in k.

$$\eta_1 B_1 + \cdots + \eta_\mu B_\mu$$

yields all the commutators in k or K with η_i ranging over all numbers in k or K respectively. Moreover, the centrum of $\mathfrak{A}_K$, i.e., the intersection of $\mathfrak{A}_K$ and

$\mathfrak{B}_K$, is the extension $\mathfrak{Z}_K$ to K of the centrum $\mathfrak{Z}$ of $\mathfrak{A}$ and $\mathfrak{B}$. Indeed, $A_1, \cdots, A_m$ being a basis of $\mathfrak{A}$, the elements

$$Z = \xi_1 A_1 + \cdots + \xi_m A_m$$

of $\mathfrak{Z}_K$ are obtained as solutions of the equations

$$\sum_{k=1}^{m} \xi_k(A_i A_k - A_k A_i) = 0$$

by numbers ξ_k in K. They have a k-basis, because the coefficients of the equations lie in k. In particular, if $\mathfrak{A}$ is normal, so is $\mathfrak{A}_K$.

Let $\mathfrak{d}$ now be an abstract normal division algebra of order δ and $\mathfrak{B} = \{B\}$ an irreducible matric algebra of degree d. The operations

$$(a) \times B \qquad \{a \text{ in } \mathfrak{d}, B \text{ in } \mathfrak{B}\}$$

operate on the δd-dimensional vector space $\mathfrak{d}\mathrm{P}_d$. We maintain that $\mathfrak{d}\mathrm{P}_d$ under this set $(\mathfrak{d}) \times \mathfrak{B}$ of operators splits into a number u of irreducible invariant subspaces in each of which $(a) \times B$ induces the same transformation; or

LEMMA (9.2.A).

$$(\mathfrak{d}) \times \mathfrak{B} \sim u \cdot \mathfrak{H}, \tag{2.2}$$

where $\mathfrak{H}$ is an irreducible matric set.

The vectors $\mathfrak{x}$ of $\mathfrak{d}\mathrm{P}_d$ are expressed in terms of a basis $\mathfrak{r}_1, \cdots, \mathfrak{r}_d$ of P_d as:

$$\mathfrak{x} = x_1\mathfrak{r}_1 + \cdots + x_d\mathfrak{r}_d \qquad (x_i \text{ in } \mathfrak{d}). \tag{2.3}$$

Considering $\mathfrak{d}$ as a "quasi-field" in which the coefficients x_i vary freely, we define for any a in $\mathfrak{d}$:

$$a\mathfrak{x} = (ax_1)\mathfrak{r}_1 + \cdots + (ax_d)\mathfrak{r}_d, \qquad \mathfrak{x}a = (x_1a)\mathfrak{r}_1 + \cdots + (x_da)\mathfrak{r}_d.$$

An invariant subspace Σ of $\mathfrak{d}\mathrm{P}_d$ is certainly a subset of vectors $\mathfrak{x}$, (2.3), closed with respect to addition and front multiplication ($\mathfrak{x} \to a\mathfrak{x}$); for the latter operation is what we formerly denoted by $(a) \times E_d$. Hence Σ has a $\mathfrak{d}$-basis $\mathfrak{l}_1, \cdots, \mathfrak{l}_n$ in terms of which every $\mathfrak{x}$ in Σ is uniquely expressible as

$$\mathfrak{x} = y_1\mathfrak{l}_1 + \cdots + y_n\mathfrak{l}_n \qquad (y_i \text{ in } \mathfrak{d}),$$

and the dimensionality δn of Σ is a multiple of δ. We assume Σ to be irreducible.

If b is a given quantity in $\mathfrak{d}$, the space Σb containing all vectors $\mathfrak{x}b$ ($\mathfrak{x}$ in Σ) is invariant with respect to $(a) \times B$, as well as Σ, and $(a) \times B$ induces therein the same transformation as in Σ. By making use of a basis $c_1 = 1, \cdots, c_\delta$ of $\mathfrak{d}$ we apply Lemma (3.2.B) to the row of irreducible invariant subspaces

$$\Sigma_1 = \Sigma c_1 = \Sigma, \qquad \Sigma_2 = \Sigma c_2, \cdots, \qquad \Sigma_\delta = \Sigma c_\delta$$

and thus succeed in picking out a number among them which by a proper arrangement may be denoted by $\Sigma_1, \cdots, \Sigma_u$ such that 1) $\Sigma_1, \cdots, \Sigma_u$ are linearly independent, and 2) each Σ_ι ($\iota = 1, \cdots, \delta$) is contained in the sum

$$\Sigma_1 + \cdots + \Sigma_u = (\Sigma).$$

The latter fact shows that (Σ) is also invariant with respect to back multiplications: $(\Sigma)c_\iota$ is contained in (Σ) for $\iota = 1, \cdots, \delta$. Consequently (Σ) is invariant with respect to all transformations of the type

$$(a)(b)' \times B, \qquad a \text{ and } b \text{ varying over a basis of } \mathfrak{d},\ B \text{ in } \mathfrak{B},$$

and thus, according to Lemma (9.1.C), with respect to $\mathfrak{M}_\delta \times \mathfrak{B}$. Lemma (3.2.A) then proves (Σ) to be the total space $\mathfrak{d}P_d$, and this remark finishes our demonstration, at the same time yielding the equation

$$d = nu:$$

u is a divisor of d.

3. Products of simple algebras

The serious work is done with the proof of Lemma (9.2.A); it remains to evaluate and interpret the result in its bearings upon multiplication of simple algebras (§3) and adjunction (§4).

From (2.2) there follow the equations

$$[(\mathfrak{d}) \times \mathfrak{B}] \sim u[\mathfrak{H}], \qquad [(\mathfrak{d})_v \times \mathfrak{B}] \sim u[\mathfrak{H}]_v\,.$$

According to Lemma (3.2.A), the algebra $[\mathfrak{H}]_v$ is irreducible since $[\mathfrak{H}]$ is; and in view of Wedderburn's theorem we may put our result into the equivalence

$$[\mathfrak{A} \times \mathfrak{B}] \sim u \cdot \mathfrak{L}, \qquad \mathfrak{L} \text{ irreducible,} \tag{3.1}$$

holding for any two irreducible matric algebras $\mathfrak{A}$ and $\mathfrak{B}$ the first of which is normal. Our result implies the abstract statement:

THEOREM (9.3.A). *The algebra product of two simple algebras one of which is normal, is a simple algebra again.*

From this we could, by means of Theorem (3.3.E), infer our concrete proposition that $[\mathfrak{A} \times \mathfrak{B}]$ is a certain multiple u of an irreducible matric algebra $\mathfrak{L}$. Our proof here, however, aimed directly at this concrete statement and yielded the further result that u is a divisor of the degree d of $\mathfrak{B}$.

Wedderburn's theorem makes transition from division algebras $\mathfrak{e}$ to simple algebras $\mathfrak{B}$ so easy that it is perhaps convenient to specialize our result (2.2) to the case $\mathfrak{B} = (\mathfrak{e})$, $\mathfrak{e}$ a division algebra of order d, rather than to generalize it to (3.1). Hence let us write down the (special $\times$ special)-equation

$$(\mathfrak{d}) \times (\mathfrak{e}) \sim u \cdot \mathfrak{H}. \tag{3.2}$$

This leads back to the (general $\times$ general)-result (3.1) in the form

$$[(\mathfrak{d})_v \times (\mathfrak{e})_w] \sim u \cdot [\mathfrak{H}]_{vw}\,. \tag{3.3}$$

Transition from $(\mathfrak{d})$ and $(\mathfrak{e})$ to $\mathfrak{A} = (\mathfrak{d})_v$ and $\mathfrak{B} = (\mathfrak{e})_w$ leaves the multiplicity u unchanged, while replacing $[\mathfrak{H}]$ by the likewise irreducible $[\mathfrak{H}]_{vw}$.

Concerning the (special $\times$ special)-case (3.2) I feel bound to make two additional remarks.

First remark. An invariant subspace Σ of $\mathfrak{de}$ has a basis $\mathfrak{l}_1, \cdots, \mathfrak{l}_n$ relative to the quasi-field of coefficients in $\mathfrak{d}$. However, we may exchange the rôles of $\mathfrak{d}$ and $\mathfrak{e}$ and look upon $\mathfrak{d}$ as a vector space and on $\mathfrak{e}$ as a quasifield of multipliers or coefficients. Σ will then have an $\mathfrak{e}$-basis $\mathfrak{k}_1, \cdots, \mathfrak{k}_\nu$ in terms of which

$$z_1\mathfrak{k}_1 + \cdots + z_\nu\mathfrak{k}_\nu$$

describes Σ with the coefficients z_i ranging over $\mathfrak{e}$. The dimensionality of Σ is

$$n\delta = \nu d, \quad \text{therefore} \quad d:\delta = n:\nu.$$

As $d = nu$, we obtain the further relation $\delta = \nu u$ and thus realize that *u is a common divisor of d and δ.* The same relationship prevails in the (general $\times$ general)-case. For in passing from $(\mathfrak{d})$ to $\mathfrak{A} = (\mathfrak{d})_v$ and from $(\mathfrak{e})$ to $\mathfrak{B} = (\mathfrak{e})_w$, the degrees δ and d change into δv and dw respectively, while u is unchanged, equation (3.3). With this additional information on hand we give the concrete counterpart of Theorem (9.3.A) as follows:

THEOREM (9.3.B). *The algebra product of two irreducible matric algebras $\mathfrak{A}$ and $\mathfrak{B}$ one of which is normal, decomposes into a number u of equal irreducible components $\mathfrak{L}$ according to the equivalence*

$$[\mathfrak{A} \times \mathfrak{B}] \sim u\cdot\mathfrak{L}.$$

The multiplicity u is a common divisor of the degrees of both factors.

The u equal parts into which the generic matrix $A \times B$ of $\mathfrak{A} \times \mathfrak{B}$ decomposes will occasionally be denoted by $\Pi(A, B)$. In the special case $\mathfrak{A} = (\mathfrak{d})$ we simply write $\Pi(a, B)$ instead of $\Pi((a), B)$ and similarly when $\mathfrak{B}$ is specialized into $(\mathfrak{e})$.

Second remark. In the (special $\times$ special)-relation (3.2) or in

$$[(\mathfrak{d}) \times (\mathfrak{e})] \sim u\cdot[\mathfrak{H}]$$

we apply Wedderburn's theorem to the irreducible $[\mathfrak{H}]$ and infer from it that

$$[\mathfrak{H}] = (\mathfrak{h})_v,$$

where the abstract division algebra $\mathfrak{h}$, called the Brauer product,[3] is uniquely determined by the factors $\mathfrak{d}$ and $\mathfrak{e}$. Comparison of degrees and orders in the ensuing equivalence

$$[(\mathfrak{d}) \times (\mathfrak{e})] \sim u\cdot(\mathfrak{h})_v$$

leads to the relations

$$\delta d = uvh, \qquad \delta d = v^2h,$$

h being the degree of $(\mathfrak{h})$ = order of $\mathfrak{h}$. Hence $v = u$ and

$$d = nu, \qquad \delta = \nu u, \qquad h = n\nu.$$

THEOREM (9.3.C). *The algebra product of the regular representations* $(\mathfrak{d})$ *and* $(\mathfrak{e})$ *of two division algebras* $\mathfrak{d}$ *and* $\mathfrak{e}$ *of orders* δ *and* d *decomposes according to*

$$[(\mathfrak{d}) \times (\mathfrak{e})] \sim u \cdot (\mathfrak{h})_u$$

provided $\mathfrak{d}$ *is normal. Putting*

$$d = nu, \qquad \delta = \nu u, \qquad (n \text{ and } \nu \text{ integers})$$

the order of the division algebra $\mathfrak{h}$ *equals* $n\nu$.

4. Adjunction

We have not as yet evaluated to the full the idea involved in our backbone proof that an invariant subspace Σ of $\mathfrak{d}P_d$ can be referred to a $\mathfrak{d}$-basis $\mathfrak{l}_1, \cdots, \mathfrak{l}_n$. Let us now consider its implications for the case which is the other way around: $\mathfrak{A} \times (\mathfrak{e})$, $\mathfrak{A}$ being a normal irreducible matric algebra, $\mathfrak{e}$ an arbitrary division algebra. Let Σ be an invariant subspace of the vector space $P\mathfrak{e}$ upon which the operators $A \times (a)$ of $\mathfrak{A} \times (\mathfrak{e})$ work (A are operators in the δ-dimensional space P, $\mathfrak{e}$ is of order d). Σ has an $\mathfrak{e}$-basis $\mathfrak{l}_1, \cdots, \mathfrak{l}_\nu$ such that each $\mathfrak{x}$ of Σ is uniquely expressible as

$$\mathfrak{x} = y_1\mathfrak{l}_1 + \cdots + y_\nu\mathfrak{l}_\nu \qquad (y_\iota \text{ in } \mathfrak{e}). \tag{4.1}$$

We now look upon $P\mathfrak{e} = P_\mathfrak{e}$ as the vector space P under extension of its field of multiplicators k into the quasi-field $\mathfrak{e}$. The elements of $P_\mathfrak{e}$ are rows $\mathfrak{x}$ of δ quantities $(x_1, \cdots, x_\delta)$ in $\mathfrak{e}$. Addition is defined in the obvious manner, multiplication by a quantity a of $\mathfrak{e}$ as:

$$a(x_1, \cdots, x_\delta) = (ax_1, \cdots, ax_\delta).$$

A linear subspace $\Sigma_\mathfrak{e}$ of $P_\mathfrak{e}$ is a subset closed with respect to addition and multiplication by any a in $\mathfrak{e}$. The subspace $\Sigma_\mathfrak{e}$ has a basis $\mathfrak{l}_1, \cdots, \mathfrak{l}_\nu$ as indicated by equation (4.1).

Each A is a linear substitution with ordinary numbers $\alpha_{\iota\kappa}$ in k:

$$x'_\iota = \sum_\kappa x_\kappa \alpha_{\iota\kappa} \qquad (\iota, \kappa = 1, \cdots, \delta)$$

and hence commutes with all the multiplications $\mathfrak{x} \to a\mathfrak{x}$. If $\Sigma_\mathfrak{e}$ is invariant with respect to the transformations A of $\mathfrak{A}$, then each $A: \mathfrak{x} \to \mathfrak{x}'$ carries the basic vectors $\mathfrak{l}_1, \cdots, \mathfrak{l}_\nu$ into linear combinations of themselves:

$$\mathfrak{l}'_\iota = \sum_\kappa a_{\kappa\iota}\mathfrak{l}_\kappa \qquad (\iota, \kappa = 1, \cdots, \nu).$$

Commuting as it does with the multiplications, A then carries (4.1) into

$$\mathfrak{x}' = y_1\mathfrak{l}'_1 + \cdots + y_\nu\mathfrak{l}'_\nu = y'_1\mathfrak{l}_1 + \cdots + y'_\nu\mathfrak{l}_\nu$$

where

$$y'_\iota = \sum_\kappa y_\kappa a_{\iota\kappa} \qquad (\iota, \kappa = 1, \cdots, \nu).$$

The correspondence $A \rightarrow \| a_{\iota\kappa} \|$ constitutes a *representation* of $\mathfrak{A}$ in $\mathfrak{e}$ (by linear transformations in which the coefficients stand *behind* the variables).

THEOREM (9.4.A). *Under extension of k into a quasi-field $\mathfrak{e}$ over k, a given normal matric algebra $\mathfrak{A}$, irreducible in k, breaks up into u equal irreducible representations of $\mathfrak{A}$ in $\mathfrak{e}$.*

This mode of visualizing the situation is related to our former viewpoint in the following manner. Adopting the coördinate system here used, $\Pi(A, \mathfrak{l})$ arises from our representation $A \rightarrow \| a_{\iota\kappa} \|$ in replacing each $a_{\iota\kappa}$ by the matrix $(a_{\iota\kappa})'$ (back multiplication!), whereas $\Pi(E_\delta, a)$ is simply $E_\nu \times (a)$.

Of particular import is the case when $\mathfrak{e}$ is a commutative field K. We then deduce from our theorem that a normal irreducible matric algebra $\mathfrak{A}$ in k splits into u equal irreducible matric algebras in K after extending the reference field k to a finite field K over k:

$$\mathfrak{A}_K \sim u \cdot \mathfrak{L}.$$

The matric algebra $\mathfrak{L}$ in K is not only irreducible in K but at the same time normal. For we observed in §2 that normality is preserved under extension of the field. Let us consider again the special case $\mathfrak{A} = (\mathfrak{d})$. We then must have an equivalence like*

$$(\mathfrak{d}_K) \sim u \cdot (\mathfrak{D})_v$$

where $\mathfrak{D}$ is a normal division algebra in K. D being the degree = order of $(\mathfrak{D})$, comparison of degrees and orders leads to the relations

$$\delta = uvD, \qquad \delta = v^2 D,$$

hence

$$u = v \quad \text{and} \quad \delta = u^2 D.$$

THEOREM (9.4.B). *Under extension of the field k into a finite field K over k, a normal division algebra $\mathfrak{d}$ in k breaks up according to the equation*

$$(\mathfrak{d}_K) \sim u \cdot (\mathfrak{D})_u$$

where $\mathfrak{D}$ is a normal division algebra in K.

Is it always possible by a suitable algebraic adjunction to effect reduction ($u > 1$), as long as $\mathfrak{d}$ is not yet the underlying field k? Yes, indeed. We choose any element b_0 in $\mathfrak{d}$ which is not a numerical multiple of the unit e, and adjoin to k a root θ of the characteristic equation $\varphi(z) = 0$ of the substitution

$$(b_0)': x \rightarrow xb_0 .$$

The transformation $(b_0)' - \theta E$ is then singular, $\neq 0$, and commutes with all matrices (a) of $(\mathfrak{d})$. Hence, by Schur's lemma, $(\mathfrak{d})$ must reduce in $k(\theta)$.

* Cornered by conflicting claims I here violate the convention forbidding the use of German capitals for abstract algebras. May the God who watches over the right use of mathematical symbols, in manuscript, print, and on blackboard, forgive me this and my many other sins!

By repeating this process one obtains an algebraic extension K of k such that

$$(\mathfrak{d}_K) \sim u \cdot \mathfrak{M}_u .$$

THEOREM (9.4.C). *There exist finite algebraic fields K over k, so-called splitting fields,*[4] *in which a given normal division algebra* $\mathfrak{d}$ *in k reduces according to*

$$(\mathfrak{d}_K) \sim u \cdot \mathfrak{M}_u .$$

Hence the order of any normal division algebra is a square number u^2.

For any normal irreducible matric algebra $\mathfrak{A} \sim (\mathfrak{d})_t$ in k one obtains in a splitting field of $\mathfrak{d}$:

$$\mathfrak{A}_K \sim u \cdot \mathfrak{M}_v$$

where $v = ut$ is a multiple of u. The order of the simple algebra $\mathfrak{a}$ is the square number v^2; the degree of $\mathfrak{A}$ is uv.

In the investigations of §§3 and 4 it was essential to suppose one of our factors to be normal. The unruly things happen in the commutative fields: the algebra product of two fields breaks up into *inequivalent* parts, and even to secure its full reducibility at all, assumptions concerning separability are needed.[5] An analogous situation prevailed for the automorphisms: what ambiguity there is in the generating matrix H of Theorem (9.1.B) resulted from the central fields $\mathfrak{z}_u$. The whole superstructure of algebras over commutative fields is of a relatively simple nature when compared with the fields themselves.

CHAPTER X

SUPPLEMENTS

A. Supplement to Chapter II, §§9–13, and Chapter VI §1, Concerning Infinitesimal Vector Invariants

1. An identity for infinitesimal orthogonal invariants

In Chapters II and VI the problem of vector invariants was approached for all groups that came under consideration by the same method that combines the formal apparatus of Capelli's identities with such non-formal arguments as since Euclid are in common use to prove the fundamental propositions about congruent figures. In a natural manner this method settled the question for the orthogonal group *in the field* K *of all real numbers* (II, 9). The topological fact that the proper orthogonal group is a connected manifold shows that in this field the group may be replaced by the set of its infinitesimal elements. Actually I employed in II, §§10–11, 13, an algebraic equivalent for this topological argument, thus carrying by a somewhat devious procedure the results over, first to "formal," then to "infinitesimal" invariants. Although no such difficulties arise in the case of the symplectic group, it seems preferable from an algebraic standpoint, to settle the question of infinitesimal invariants for the orthogonal and the symplectic groups by means of another formal identity of Capelli's type and without appealing to any "non-formal" arguments. I discovered this alternate procedure after the publication of the first edition of this book.

Let $f(x, y, \cdots, z)$ be a polynomial homogeneous of a certain degree with respect to the components of each (covariant) argument vector. Let e be the number of argument vectors and r the degree of f with respect to x. In order to express invariance of f under infinitesimal orthogonal transformations we introduce the matrix $X = x\xi$ with the elements $X(ik) = x_i\xi_k$ composed of a covariant vector x (single column) and a contravariant one ξ (single row), and moreover the matrix $X' = X - X^*$ with the elements $X'(ik) = x_i\xi_k - x_k\xi_i$. On setting $\xi_i = \dfrac{\partial}{\partial x_i}$ the numbers $X(ik)$ and $X'(ik)$ turn into differential operators $d_x(ik)$ and $d'_x(ik)$. Infinitesimal invariance of f requires that the total differential

$$df = \sum_i \frac{\partial f}{\partial x_i}\,dx_i + \sum_i \frac{\partial f}{\partial y_i}\,dy_i + \cdots$$

vanishes by virtue of the substitution $dx = Sx,\ dy = Sy, \cdots,\ S$ being an arbitrary skew-symmetric matrix; it is thus expressed by the equations

$$\left(x_i \frac{\partial f}{\partial x_k} - x_k \frac{\partial f}{\partial x_i}\right) + \left(y_i \frac{\partial f}{\partial y_k} - y_k \frac{\partial f}{\partial y_i}\right) + \cdots = 0$$

or

(1.1) $$\textstyle\sum_v d'_v(ik)f = 0 \qquad (i, k = 1, \cdots, n),$$

the sum $\sum_v$ extending over the e argument vectors of f.

The relation we wish to develop is a differential counterpart of the trivial algebraic identity

(1.2) $$\begin{vmatrix} (xy) & (x\eta) \\ (\xi y) & (\xi\eta) \end{vmatrix} = \operatorname{tr}(XY').$$

Put $\xi = \frac{\partial}{\partial x}, \eta = \frac{\partial}{\partial y}\cdot$ Then $(x\eta)$ and (ξy) become the polar operators D_{xy} and D_{yx} respectively and $(\xi\eta)$ the Laplacian operator

$$\Theta_{(xy)} \equiv \textstyle\sum_i \dfrac{\partial^2}{\partial x_i \partial y_i}\cdot$$

Handling the operators D and d as if multiplication by x_i and differentiation $\frac{\partial}{\partial x_i}$ were commutative one would thus obtain from (1.2) the equation

(1.3) $$(xy)\Theta_{(xy)}f - D_{xy}(D_{yx}f) = \operatorname{tr}(d_x, d'_y f).$$

The assumption is of course incorrect; instead one has

$$\frac{\partial}{\partial x_i}(x_i f) - x_i \frac{\partial}{\partial x_i} f = f$$

(Schrödinger's well known quantum mechanical rule for the commutation of coordinate and momentum). Consequently a similar correction must be applied as in our derivation of Capelli's identity (II, 4). Two cases (i) and (ii) are to be distinguished according to whether y is not or is identical with the argument x.

(i) $$\operatorname{tr}(d_x, d'_y f) = \sum_{i,k} x_i \frac{\partial}{\partial x_k}\left(y_i \frac{\partial}{\partial y_k} - y_k \frac{\partial}{\partial y_i}\right) f$$
$$= \sum_{i,k} x_i y_i \frac{\partial^2 f}{\partial x_k \partial y_k} - \sum_{i,k} x_i y_k \frac{\partial^2 f}{\partial x_k \partial y_i}\cdot$$
$$D_{xy}(D_{yx}f) = \sum_{i,k} x_i \frac{\partial}{\partial y_i}\left(y_k \frac{\partial f}{\partial x_k}\right)\cdot$$
$$= \sum_{i,k} x_i y_k \frac{\partial^2 f}{\partial y_i \partial x_k} + \sum_{i,k} x_i \delta_{ik} \frac{\partial f}{\partial x_k}\cdot$$

The last sum equals $\sum_i x_i \partial f/\partial x_i = rf$. Thus

$$\operatorname{tr}(d_x, d'_y f) + D_{xy}(D_{yx}f) = (xy)\ \Theta_{(xy)}f + rf,$$

or (1.3) is to be replaced by

$$(1.4)\qquad (xy)\cdot\Theta_{(xy)}f - D_{xy}(D_{yx}f) + rf = \mathrm{tr}(d_x, d'_y f).$$

$$\text{(ii)}\qquad \mathrm{tr}(d_x, d'_x) = \sum x_i \frac{\partial}{\partial x_k}\left(x_i\frac{\partial}{\partial x_k} - x_k\frac{\partial}{\partial x_i}\right) =$$

$$\sum_{i,k} x_i^2\frac{\partial^2}{\partial x_k^2} + \sum_{i,k} x_i\delta_{ik}\frac{\partial}{\partial x_k} - n\sum_i x_i\frac{\partial}{\partial x_i} - \sum_{i,k} x_i x_k\frac{\partial^2}{\partial x_i\partial x_k}.$$

Since

$$\sum_i x_i\frac{\partial f}{\partial x_i} = rf, \qquad \sum_{i,k} x_i x_k \frac{\partial^2 f}{\partial x_i \partial x_k} = r(r-1)f$$

we find

$$(1.5)\qquad (xx)\cdot\Theta_{(xx)}f - (r+n-2)rf = \mathrm{tr}(d_x, d'_x f).$$

For an infinitesimal (inf.) invariant f the equation (1.1) holds and hence summation of (1.5) and the relations (1.4) corresponding to the $e-1$ arguments $y, \cdots, z$ different from x yields

FORMULA (10.1.A):

$$\sum_y (xy)\cdot\Theta_{(xy)}f - \sum_y' D_{xy}(D_{yx}f) = r(r+n-1-e)f$$

where the sums $\sum_y, \sum_y'$ *range over all e arguments y of f including or excluding the fixed argument x.*

2. First Main Theorem for the orthogonal group

By means of this equation and the Capelli identities we now prove:

THEOREM (10.2.A). *Every inf. orthogonal invariant f can be built up in an integral rational fashion from bracket factors and scalar products.*

With f also $D_{yx}f$, $\Theta_{(xy)}f$ are inf. orthogonal invariants. The first observation permits us, by means of Capelli's general and special identities, to reduce the proof to the case where the number of arguments $e \leq n-1$. Here the statement asserts that the invariant f can be built up from scalar products alone. Apply induction with respect to r and observe that $D_{yx}f$ for $y \neq x$, and $\Theta_{(xy)}f$ for $y \neq x$ as well as $y = x$, are of degrees $<r$ with respect to x. Since the numerical factor by which f is multiplied on the right side of Formula (10.1.A) cannot vanish for $e \leq n-1$, $r \geq 1$, the theorem is thus proved for all possible degrees $r = 0, 1, 2, \cdots$ provided it holds for $r = 0$ But $r = 0$ means that f depends on the $e-1$ argument vectors $y, \cdots, z$ only, and hence induction with respect to e concludes the demonstration. *It clearly goes through in any field k of characteristic zero.*

Since f is built up from bracket factors and scalar products we see that *infinitesimal* invariance implies invariance with respect to all proper orthogonal transformations in k. The converse is far more obvious: A form f invariant under all proper orthogonal transformations is in particular invariant with respect to $A = \dfrac{E - S}{E + S}$ if $S = ||s_{ik}||$ is a non-exceptional numerical skew-symmetric matrix; consequently it is a formal invariant in the sense of II, §11. Whereupon the first trivial part of the proof of Theorem (2.13.A) on p. 69 shows that f is a fortiori an infinitesimal invariant.

Since the product of two bracket factors is expressible in terms of scalar products every inf. invariant f is a sum of terms of the form Φ or $[u \cdots v] \cdot \Phi$ where Φ stands for any aggregate of scalar products and $u, \cdots, v$ are any n of the e arguments of f. Thus f breaks up into an *even* invariant, Φ, and an *odd* invariant which is a sum of terms of the form $[u \cdots v] \cdot \Phi$. The even part remains unaltered and the odd one changes sign under the influence of the substitution J_n, (II, 9.3). Forms that are invariant under the full orthogonal group $O(n)$ are even inf. invariants and hence can be built up from scalar products alone.

3. The same for the symplectic group

It is easy to see with what modifications our construction carries over to the symplectic group. The notations of Chapter VI will be used: the components of a vector x are x_α and $x'_\alpha = x_{\alpha'}$; Greek indices run from 1 to ν, italic indices from 1 to $n = 2\nu$. The matrix X has the same significance $x\xi$ as above, but X' is now defined by $X' = X + IX^*I$ or

$$X'(\alpha\beta) = x_\alpha\xi_\beta - x'_\beta\xi'_\alpha, \qquad X'(\alpha\beta') = x_\alpha\xi'_\beta + x_\beta\xi'_\alpha,$$
$$X'(\alpha'\beta) = x'_\alpha\xi_\beta + x'_\beta\xi_\alpha, \qquad X'(\alpha'\beta') = x'_\alpha\xi'_\beta - x_\beta\xi_\alpha.$$

The algebraic identity (1.2) is replaced by

$$\begin{vmatrix} [xy] & (x\eta) \\ (\xi y) & [\xi\eta] \end{vmatrix} = -\operatorname{tr}(XY').$$

Substitution of $\dfrac{\partial}{\partial x}$ for ξ changes $X'(ik)$ into a differential operator $d'_x(ik)$, and infinitesimal invariance of f under the symplectic group is again expressed by equation (1.1). The analogue of the Laplacian operator is

$$\Theta_{[xy]}f = \sum_\alpha \left(\frac{\partial^2}{\partial x_\alpha \partial y'_\alpha} - \frac{\partial^2}{\partial y_\alpha \partial x'_\alpha}\right) f$$

and instead of (10.1.A) we obtain for inf. symplectic invariants the

FORMULA (10.3.A):

$$\sum_y [xy] \cdot \Theta_{[xy]}f - \sum'_{y} D_{xy}(D_{yx}f) = r(r + n + 1 - e)f.$$

We wish to prove:

THEOREM (10.3.B). *Every inf. symplectic invariant may be built up from skew products* $[xy]$.

It follows from the previous relation as long as the number of arguments $e \leq n + 1$. But Capelli's general identity enables us to increase the number of arguments indefinitely once the stage of n arguments is reached. (There is an overlapping in as much as the increase of e from n to $n + 1$ may be effected either by that identity or by (10.3.A). Up to $n + 1$ arguments there are no algebraic relations between the skew products. No explicit use is made of the expression (6, 1.8) of the bracket factor in terms of skew products.)

B. Supplement to Chapter V, §3, and Chapter VI, §§2 and 3, Concerning the Symplectic and Orthogonal Ideals

4. A proposition on full reduction

In V, §2, the full reducibility of $\Pi_f(O)$ in a real field is inferred from the fact that $\Pi_f(A)$ is orthogonal if the matrix A is. While this is a perfectly natural procedure, it becomes less satisfactory when in §3, A is replaced by $\dfrac{E - S}{E + S}$ and the elements s_{ik} $(i < k)$ of the skew-symmetric matrix S are treated as indeterminates. In the symplectic case the method can be worked only by resorting to the artifice of unitary restriction (cf. VI, §2). Instead we propound here a general principle that holds in any real field k and in spite of its elementary character seems to have for a long time escaped the algebraists' notice:

Theorem (10.4.A). *Let $\mathfrak{A}$ be a set of linear transformations A of a vector space* P *such that the transpose A^* is a member of $\mathfrak{A}$ whenever A is, and let* P_a *be a subspace of* P *that is invariant with respect to $\mathfrak{A}$. Then the perpendicular subspace* P_b *is likewise invariant. In this precise sense reduction for $\mathfrak{A}$ implies full reduction.*

The scalar product of two vectors (columns) a and b is defined by

$$b^*a = b_1a_1 + \cdots + b_na_n.$$

It should be noticed that because k is real P_a and P_b have no vector in common except 0, and hence the total space P breaks up into $\mathrm{P}_a + \mathrm{P}_b$. Our theorem will be used only for the case where k is the field κ of all rational numbers.

Proof. Let a, b be arbitrary vectors in P_a, P_b respectively and the matrix A be a member of $\mathfrak{A}$. Since A^* is also in $\mathfrak{A}$ and P_a invariant with respect to $\mathfrak{A}$, the vector A^*a lies in P_a, hence $b^*(A^*a) = 0$, or after transposition

$$a^*Ab = 0.$$

This means that Ab is perpendicular to every vector a of P_a and therefore lies in P_b: consequently A carries every vector b of P_b into a vector of P_b.

This theorem is closely related to the investigations of E. Fischer about invariants quoted in footnote 18 of Chapter VIII.

5. The symplectic ideal

Because of the complications caused in the orthogonal case by the distinction between the proper and the full group we propose to apply Theorem (10.4.A) first to the symplectic ideal combining it with a more consistent use of indeterminates. An infinitesimal operation $dx = Sx$ of the symplectic group has a matrix

$$S = \left\| \begin{matrix} s(\alpha\beta), & s(\alpha\beta') \\ s(\alpha'\beta), & s(\alpha'\beta') \end{matrix} \right\|$$

for which

$$(5.1) \qquad s(\alpha'\beta') = -s(\beta\alpha); \qquad s(\alpha\beta') = s(\beta\alpha'), \qquad s(\alpha'\beta) = s(\beta'\alpha).$$

As its $\frac{1}{2}n(n+1)$ parameters s_j we use

$$\text{all } s(\alpha\beta); \quad s(\alpha\beta') \quad \text{for} \quad \alpha \leq \beta, \quad s(\alpha'\beta) \quad \text{for} \quad \alpha \leq \beta.$$

We refer to these conditions (5.1) by speaking of an S-matrix. Thus we write the indeterminate S-matrix as $L(s) = \sum_j L_j s_j$.

The relationship

$$A = (E - S)(E + S)^{-1} = (E + S)^{-1}(E - S)$$

between finite and infinitesimal symplectic transformations, A and S, shows that the equation

$$(5.2) \qquad \frac{E - S}{E + S} = \frac{E - S'}{E + S'} \cdot \frac{E - S''}{E + S''}$$

defines an S-matrix S in terms of two indeterminate S-matrices S', S'' with the parameters s_j', s_j''. The parameters s_j of S are rational functions $\psi_j(s', s'')$. *In this sense the expression* $\frac{E - S}{E + S}$ *involving the* $\frac{1}{2}n(n+1)$ *indeterminates* s_j *constitutes a group.* Writing (5.2) in the form

$$2(E + S)^{-1} - E = (E + S')^{-1}(E - S')(E - S'')(E + S'')^{-1}$$

one finds at once

$$(E + S)^{-1} = (E + S')^{-1}(E + S'S'')(E + S'')^{-1}$$

or

$$E + S = (E + S'')(E + S'S'')^{-1}(E + S').$$

Hence the functions $\psi_j(s', s'')$ have the determinant

$$\Delta = | E + S'S'' |$$

as their common denominator, and

$$| E + S | = | E + S' | \cdot | E + S'' | \cdot \Delta^{-1}.$$

We derive the product matrix $\Pi_f A$ in the tensor space P_f from

$$A = (E - S)(E + S)^{-1}, \qquad S = L(s). \tag{5.3}$$

Evidently it is of the following form

$$\Pi_f A = R(s)/ \mid E + S \mid^f, \qquad R(s) = \sum_p C_p \sigma_p \tag{5.4}$$

where the numerator $R(s)$ is a polynomial of formal degree nf of the parameters s_j with matric coefficients C_p. Thus σ_p runs over the monomials of degree $\leq nf$ of the parameters s_j. The same holds for

$$\mid E + S \mid^f \cdot \Pi^{(f)} A = \sum C_p^{(f)} \sigma_p.$$

We maintain:

Theorem (10.5.A). *The linear closure of* (5.4) *in* κ, *i.e., the set* $\mathfrak{C}_f$ *of all linear combinations* $\sum \lambda_p C_p$ *with rational coefficients* λ_p, *is an algebra in the field* κ. *It contains the unit matrix and the transpose* C^* *of every one of its members* C.

Proof. The unit matrix is the coefficient C_o of the monomial $\sigma_o = 1$. Equation (5.2) entails

$$\Pi_f \left(\frac{E - S}{E + S}\right) = \Pi_f \left(\frac{E - S'}{E + S'}\right) \cdot \Pi_f \left(\frac{E - S''}{E + S''}\right)$$

or, because of (5.4),

$$\Delta^f \cdot \sum C_r \sigma_r = \sum_{p,q} C_p C_q \sigma'_p \sigma''_q.$$

After the substitution of $\psi_j(s', s'')$ for s_j each σ_r turns into a polynomial $\varphi_r(s', s'')$ divided by Δ^{nf}. Hence

$$\Delta^{(n-1)f} \cdot \sum_{p,q} C_p C_q \sigma'_p \sigma''_q = \sum_r C_r \varphi_r(s', s''). \tag{5.5}$$

We now apply the following trivial algebraic

Lemma (10.5.B). Let $\varphi = 1 + \cdots$ be a given polynomial of some variables $x_1, \cdots, x_l$ with the constant term 1. Then the coefficients of an arbitrary polynomial F of degree m may be linearly expressed by the coefficients of $\varphi F = G$.

Arranging the terms of F in ascending lexicographic order one obtains recursive linear equations for the unknown coefficients a of F, with the coefficients b of G as the known right members (only terms of G of a degree not exceeding m enter). Denote by F_μ the terms of degree μ in any polynomial $F = F_o + F_1 + \cdots$. Putting $\varphi = 1 - \omega$ and using the power series $(1 - \omega)^{-1}$ one finds more explicitly

$$F = \sum G_\mu \omega_{\mu_1} \omega_{\mu_2} \cdots$$
$$(\mu \geq 0; \quad \mu_1 \geq 1, \quad \mu_2 \geq 1, \cdots; \quad \mu + \mu_1 + \mu_2 + \cdots \leq m).$$

If φ has integral coefficients, then the coefficients of the linear combinations expressing the a in terms of the b are likewise integers.

This lemma is of immediate application to the equation (5.5), in which Δ begins with the constant term 1. The products $\sigma'_p\sigma''_q$ are monomials no two of which are equal, and thus we arrive at equations

$$C_pC_q = \sum_r \gamma^r_{pq}C_r$$

with rational coefficients γ^r_{pq}. They prove the first part of our statement. The arithmetician may care to observe that the γ are not only rational but even rational *integers*. The same equations hold for the matrices $C_p^{(f)}$ and thus the linear combinations $\sum \lambda_p C_p^{(f)}$ form an algebra $\mathfrak{C}^{(f)}$.

If S is an S-matrix, so is its transpose S^*. The following simple linear involutorial substitution of the parameters s_j,

$$(5.6) \qquad s(\alpha\beta) \to s(\beta\alpha), \qquad s(\alpha\beta') \to s(\alpha'\beta), \qquad s(\alpha'\beta) \to s(\alpha\beta')$$

changes the indeterminate $S = L(s)$ into S^*, hence A, (5.3), into A^*, $\Pi_f A$ into $(\Pi_f A)^*$, and since $| E + S | = | E + S^* |$, also $R(s)$ into $R^*(s)$. (5.6) induces an involutorial permutation among the monomials σ_p, $\sigma_p \to \sigma_{p^*}$. Our argument shows that

$$\sum C_p^*\sigma_p = \sum C_p\sigma_{p^*} = \sum C_{p^*}\sigma_p; \qquad C_p^* = C_{p^*}.$$

Thus the second part of our theorem is proved, and by means of (10.4.A) we conclude that the algebra $\mathfrak{C}_f$ in κ and hence $\mathfrak{C}^{(f)}$ are fully reducible.

From here on we can follow the same road as on p. 174 to prove Theorem (6.3.A). Still operating in the field κ we observe that a form $g(x, y, \cdots, z)$ that is invariant under the generic (5.3) is necessarily an infinitesimal invariant; apply the hypothesis to λS instead of S and neglect all but the first power of the parameter λ. Hence, according to Theorem (10.3.B), g is expressible as a polynomial depending on the skew products $[xy]$ of its arguments; and this means that the algebra described as $\mathfrak{A}^{(f)}$ on p. 174 is contained in the commutator algebra of the commutator algebra of $\mathfrak{C}^{(f)}$, *and therefore in* $\mathfrak{C}^{(f)}$. Indeed since $\mathfrak{C}^{(f)}$ contains the unit matrix and is fully reducible, Brauer's criterion (3.5.D) is applicable, to the effect that $\mathfrak{C}^{(f)}$ is the commutator algebra of its own commutator algebra. The algebra $\mathfrak{A}^{(f)}$ is defined by a system of linear equations with *rational* coefficients; hence the fact $\mathfrak{A}^{(f)} \subset \mathfrak{C}^{(f)}$ remains true in any field k of characteristic zero. (The inverse relation $\mathfrak{C}^{(f)} \subset \mathfrak{A}^{(f)}$ is trivial.) Any polynomial $\Phi(A)$ of degree f depending on the matrix $A = ||a_{ik}||$ with n^2 variable components a_{ik} proceeds from a linear form $\gamma(A^{(f)})$ of an arbitrary bisymmetric $A^{(f)}$ by the substitution $A^{(f)} = \Pi^{(f)}A$. If $\Phi(A)$ is annulled by the substitution (5.3) we must have $\gamma(C_p^{(f)}) = 0$, and therefore $\gamma(A^{(f)})$ vanishes for all matrices $A^{(f)}$ of $\mathfrak{C}^{(f)}$, hence of $\mathfrak{A}^{(f)}$. Thus results Theorem (6.3.B). In restating it I propose the following terminology for ideals of polynomials $\varphi(x_1, \cdots, x_l)$. Elements $\varphi_1, \cdots, \varphi_m$ of a given ideal

$\mathfrak{J}$, whose degrees are $f_1, \cdots, f_m$ respectively, are said to constitute a *form basis* of $\mathfrak{J}$ if any element φ of $\mathfrak{J}$ of degree f may be obtained in the form

$$\varphi = h_1\varphi_1 + \cdots + h_m\varphi_m$$

where h_i is a polynomial of degree $f - f_i$ (which implies, of course, the vanishing of h_i whenever $f - f_i \leq 0$). By introduction of homogeneous coordinates $\mathfrak{J}$ gives rise to an ideal $\bar{\mathfrak{J}}$ of homogeneous forms $\varphi(x_0, x_1, \cdots, x_l)$ such that $\varphi(x_0, x_1 \cdots, x_l)$ is in $\bar{\mathfrak{J}}$ if and only if $\varphi(1, x_1, \cdots, x_l)$ is in $\mathfrak{J}$. The statement "$\varphi_1, \cdots, \varphi_m$ constitute a form basis of $\mathfrak{J}$" means that the corresponding homogeneous *forms* $\varphi_1, \cdots, \varphi_m$ of degrees $f_1, \cdots, f_m$ constitute a basis of $\bar{\mathfrak{J}}$.

With any group Γ of linear transformations $A = \| a_{ik} \|$ there is connected an *ideal* $\mathfrak{J}(\Gamma)$: it consists of all polynomials $\Phi(A)$ of the n^2 variables a_{ik} which vanish for each element A of Γ. If Γ is the symplectic or the full or the proper orthogonal group we speak of the symplectic, the full and the proper orthogonal ideals respectively. We have arrived at the following proposition concerning the symplectic ideal, which holds in any field of characteristic zero:

THEOREM (10.5.C). (1) *Let $A = \|a_{ik}\|$ be the matrix with n^2 variable components a_{ik}. The components of the two matrices $A^*IA - I$ and $AIA^* - I$ constitute a form basis, the components of either of them a basis, of the symplectic ideal.*

(2) *The symplectic ideal is a prime ideal, and the expression* (5.3) *a generic zero; i.e., a polynomial $\Phi(A)$ vanishes for all symplectic transformations A if it is annulled by the substitution* (5.3).

6. The full and the proper orthogonal ideals

Whereas a form that is invariant under two substitutions A_1, A_2 is also invariant with respect to A_1A_2, it is by no means true for a polynomial $\Phi(A)$ that its vanishing for $A = A_1$ and $A = A_2$ implies its vanishing for A_1A_2. For this reason the observation that the expression $\dfrac{E - S}{E + S}$ constitutes a group was essential for proving the fact that it is a generic zero of the symplectic ideal.

In the orthogonal case the full and the proper orthogonal groups, O and O^+, and hence the corresponding ideals

$$\mathfrak{J}(O) = \bar{\mathfrak{o}}, \qquad \mathfrak{J}(O^+) = \mathfrak{o},$$

have to be distinguished. Let now $S = \| s_{ik} \|$ be the generic skew-symmetric matrix with $\frac{1}{2}n(n - 1)$ indeterminate components s_{ik} $(i < k)$. The two expressions

$$\frac{E - S}{E + S}, \qquad J_n \frac{E - S}{E + S} \tag{6.1}$$

jointly constitute a group. Hence by the same procedure as carried through for the symplectic group we arrive at the following result for the full orthogonal ideal $\bar{\mathfrak{o}}$ which holds in any field k of characteristic 0:

THEOREM (10.6.A). *The components of the two matrices $A^*A - E$ and $AA^* - E$ constitute a form basis, the components of either of them a basis, of the full orthogonal ideal $\bar{\mathfrak{o}}$. A polynomial $\Phi(A)$ vanishes for all orthogonal transformations A if it is annulled by the two substitutions*

$$A = \frac{E - S}{E + S} \qquad \text{and} \qquad A = J_n \frac{E - S}{E + S}.$$

The statements on p. 143, Supplement to Theorem (5.3.A), Theorem (5.3.B), the Corollary and its repetition in Lemma (7.1.B) on p. 177, have to be corrected accordingly: the vanishing of the polynomial Φ should be required not only for $\frac{E - S}{E + S}$ and J_n, but for (6.1). The ideal $\bar{\mathfrak{o}}$ is not prime, but the proper orthogonal ideal $\mathfrak{o}$ is because $\frac{E - S}{E + S}$ is its generic zero: Theorems (5.4.C and D) need no modification.

The two supplements A, B together contain a strictly algebraic approach to the theory of vector invariants and of the ideal of a given group Γ of linear transformations in an arbitrary field, as far as the most important "classical groups" $\Gamma = O(n)$, $O^+(n)$ and $Sp(n)$ are concerned.

C. SUPPLEMENT TO CHAPTER VIII, §§7–8, CONCERNING:

7. A modified proof of the main theorem on invariants

Hilbert's application of Cayley's Ω-process can be replaced by another purely algebraic process of a less formal nature that was discovered by Menahem Schiffer in 1933 and works *for any group Γ of linear transformations, all powers of which, $\Pi_f(\Gamma)$ ($f = 1, 2, 3, \cdots$), are fully reducible.* Mr. Schiffer, who never published his method, kindly gave me permission to have a description of it included in this book.*

Under the influence of the linear transformation $t \in \Gamma$ of the underlying n-dimensional vector space P any tensor of rank f (= vector in P_f) undergoes the transformation $t_f = \Pi_f(t)$, while an "anti-tensor" undergoes the contragredient transformation $\check{t}_f = \Pi_f(\check{t})$. Full reducibility of the tensor space P_f thus implies the same for the space of all anti-tensors of rank f. Let Σ be an invariant irreducible subspace of P_f under the group $\Pi_f(\Gamma)$; the vectors u in Σ are our *quantics*. Because of the basic assumption we may split P_f into Σ and a supplementary invariant subspace and thus obtain a linear operator I changing every vector x of P_f into a quantic u and reproducing every quantic u ($Iu = u$). We investigate homogeneous polynomials $P(u)$ of any degree m

* His unpublished MS. also contains an explicit statement of Theorem (10.4.A).

depending on a variable quantic u. Then $F(x) = P(Ix)$ is a polynomial of degree m depending on an arbitrary tensor x and may therefore be written as $\Sigma \alpha_{ik\ldots} x_i x_k \cdots$ with symmetric coefficients $\alpha_{ik\ldots}$. The multilinear symmetric form

$$F(x, y, \cdots) = \sum \alpha_{ik\ldots} x_i y_k \cdots$$

involving m independent tensors $x, y, \cdots$ may be used instead. Since every index $i, k, \cdots$ here stands for an f-uple $(i_1, \cdots i_f)$, $(k_1, \cdots, k_f)$, $\cdots$, the coefficients $\alpha_{ik\ldots}$ constitute an anti-tensor of rank mf that has the following two properties: (1) $F(x, y, \cdots)$ depends symmetrically on the m arguments $x, y, \cdots$; (2) $F(Ix, Iy, \cdots) = F(x, y, \cdots)$. Hence the polynomials $P(u)$ of degree m can be identified with the anti-tensors F of rank mf which possess these two properties, $P(u) = F(u, u, \cdots)$. They form an invariant subspace of the space of all anti-tensors of rank mf, and hence our basic hypothesis implies the fact that the space $\mathfrak{P}$ of homogeneous polynomials $P(u)$ of given degree is fully reducible. The argument carries over to polynomials depending with given degrees on several quantics $u, v, \cdots$ of the same or different types.

After this preliminary observation we are now going to construct a linear process Ω changing any homogeneous polynomial $P(u)$ into an invariant $\Omega P(u)$ of the same degree, a process which reproduces P, $\Omega P = P$, in case P is an invariant, and which carries a polynomial P that is divisible by an invariant J into a polynomial ΩP likewise divisible by J. Because the process is linear, $\Omega(P_1 + P_2) = \Omega P_1 + \Omega P_2$ for any two polynomials $P_1(u)$, $P_2(u)$ of equal degrees. Once such a process Ω is known, (8.7.1) yields

$$J(u) = \Omega(L_1 J_1) + \cdots + \Omega(L_h J_h);$$

but $\Omega(L_\alpha J_\alpha)$ is an invariant divisible by J_α, hence of the form $L_\alpha^* J_\alpha$ where L_α^* is an invariant, and thus the relation (8.7.1) may be replaced by

$$J(u) = L_1^*(u) J_1(u) + \cdots + L_h^*(u) J_h(u)$$

with new coefficients $L_\alpha^*(u)$ that are invariants; cf. (8.7.3).

Schiffer's construction of the desired process Ω is as follows. The homogeneous polynomials $P(u)$ of degree m form a vector space $\mathfrak{P}$ in which the invariants of degree m constitute an invariant subspace $\mathfrak{P}^i$. Hence we can decompose $\mathfrak{P} = \mathfrak{P}^i + \mathfrak{P}^s$ so that the supplementary space $\mathfrak{P}^s$ is likewise invariant; it contains no invariant P except 0. We define ΩP as the invariant part P^i in the decomposition $P(u) = P^i(u) + P^s(u)$ where $P^i \in \mathfrak{P}^i$ and $P^s \in \mathfrak{P}^s$. The only thing that remains to be shown is that if P is divisible by a given invariant J then ΩP is. This fact as well as the other that $\mathfrak{P}^s$ is uniquely determined is a consequence of the following

LEMMA (10.7.A). Let $\mathfrak{Q}$ be an invariant subspace of $\mathfrak{P}$ and perform a similar decomposition $\mathfrak{Q} = \mathfrak{Q}^i + \mathfrak{Q}^s$ for $\mathfrak{Q}$ as $\mathfrak{P} = \mathfrak{P}^i + \mathfrak{P}^s$ is for $\mathfrak{P}$. Then $\mathfrak{Q}^i \subset \mathfrak{P}^i$, $\mathfrak{Q}^s \subset \mathfrak{P}^s$.

This lemma is actually implied in our results on p. 94 concerning commutators of a fully reducible matric algebra. Here is a direct and independent proof. In terms of a basis of $\mathfrak{P}$ and $\mathfrak{Q}$ the elements of these two spaces appear as vectors x and y (columns of their components) respectively. Each vector y of $\mathfrak{Q}$ is identical with a certain vector $x = By$ of $\mathfrak{P}$, B being a constant rectangular matrix. Under the influence of any substitution t of the group Γ the vectors x and y undergo certain linear transformations

$$x^t = U(t)x, \qquad y^t = V(t)y,$$

hence

$$U(t)B = BV(t). \tag{7.1}$$

Use the bases adapted to the decompositions $\mathfrak{P} = \mathfrak{P}^i + \mathfrak{P}^s$, $\mathfrak{Q} = \mathfrak{Q}^i + \mathfrak{Q}^s$ into invariant subspaces and decompose the matrices accordingly:

$$U = \begin{Vmatrix} E & 0 \\ 0 & U_s \end{Vmatrix}, \qquad V = \begin{Vmatrix} E & 0 \\ 0 & V_s \end{Vmatrix}.$$

E stands for the unit matrix of as many rows as the dimensionalities of $\mathfrak{P}^i$ and $\mathfrak{Q}^i$ amount to. The elements of $\mathfrak{Q}^i$ are invariants; therefore $\mathfrak{Q}^i \subset \mathfrak{P}^i$, or B is of the form

$$\begin{Vmatrix} B_{ii} & B_{is} \\ 0 & B_{ss} \end{Vmatrix}.$$

This (is)- part of the equation (7.1) yields

$$B_{is}V_s(t) = B_{is}.$$

Hence every row b of B_{is} satisfies the equation

$$bV_s(t) = b$$

identically for $t \in \Gamma$. On the other hand the fact that no element of $\mathfrak{Q}^s$ is an invariant signifies that no column c except 0 can satisfy the relation $V_s(t)c = c$ (for all $t \in \Gamma$).

Lemma (10.7.B). Let $\mathfrak{B}$ be a fully reducible set of matrices V in an n-dimensional space. If there is no column c except 0 satisfying the relation $Vc = c$ for all V in $\mathfrak{B}$ then there is no row $b \neq 0$ satisfying the relation $bV = b$ for all members V of the set $\mathfrak{B}$.

Application of this lemma to the representation $V_s(t)$ in $\mathfrak{P}^s$ yields the desired result $B_{is} = 0$, $\mathfrak{Q}^s \subset \mathfrak{P}^s$.

Proof of Lemma B. Suppose a row $b \neq 0$ satisfies the equation $bV = b$ for all $V \in \mathfrak{B}$. The columns c for which $bc = 0$ form an $(n - 1)$ dimensional invariant subspace under $\mathfrak{B}$. Because of the assumed full reducibility there exists therefore a column c not in this subspace, $bc \neq 0$, such that $Vc = \lambda c$, the factor λ depending on V. This relation yields $bVc = \lambda \cdot (bc)$ whereas $bV = b$ gives $bVc = bc$, therefore $\lambda = 1$ in view of $bc \neq 0$. The equation $Vc = c$ $(c \neq 0)$ is contrary to the hypothesis.

The polynomials P of degree m which are divisible by a given invariant J form an invariant subspace $\mathfrak{Q}$ of $\mathfrak{P}$. Split such a polynomial P according to the decomposition $\mathfrak{Q} = \mathfrak{Q}^i + \mathfrak{Q}^s$ into an i- and an s-part, both belonging to $\mathfrak{Q}$. By Lemma A the i-part is in $\mathfrak{P}^i$, the s-part in $\mathfrak{P}^s$; hence the former is by definition ΩP, and ΩP is thus shown to lie in $\mathfrak{Q}$. This completes the proof of our

THEOREM (10.7.C). *Provided all powers of a given group Γ of linear transformations are fully reducible, the first main theorem holds for invariants $J(u, v, \cdots)$ under this group which depend on one or several quantics $u, v, \cdots$.*

The groups $SL(n)$, $O(n)$, $Sp(n)$ fall under this theorem, but not the non-homogeneous affine group treated in §8 by means of the adjunction argument. Even so the remark that wherever full reducibility prevails there the first main theorem holds, points out a fact of considerable interest.

According to Theorem (10.4.A) any group Γ of linear transformations in a real field k that contains the transpose t^* of any of its elements t satisfies the hypothesis of Schiffer's general proposition (10.7.C), and hence the latter implies E. Fischer's result (footnote 18 of Chap. VIII) concerning invariants of such groups.

D. SUPPLEMENT TO CHAPTER IX, §4, ABOUT EXTENSION OF THE GROUND FIELD

8. Effect of field extension on a division algebra

In IX, §4, the extension of the ground field k to a field K over k was studied under the aspect of multiplication of a given simple algebra $\mathfrak{a}$ by a (not necessarily commutative) division algebra. But the limitation to finite extensions which this procedure involves is only apparent. We repeat the argument here for any (commutative) field K over k, dropping at the same time the assumption that $\mathfrak{a}$ be normal. Heeding the warning sounded in the last lines on p. 290 we suppose the ground field k to be of characteristic zero.

The elements a of a simple algebra $\mathfrak{a}$ of order d in k are at the same time the vectors of a d-dimensional vector space P, and multiplication $x \rightarrow y = ax$ of a variable element x by a fixed element a of $\mathfrak{a}$ is a linear mapping $A = (a)$ of that space. The matrices (a) corresponding to the elements a of $\mathfrak{a}$ form the matric algebra $(\mathfrak{a})$ (regular representation). In restating Wedderburn's theorem (p. 91) let us denote by $\mathbf{M}_t\mathfrak{a}$ the algebra of all t-rowed matrices

$$\left\| \begin{matrix} a_{11}, & \cdots, & a_{1t} \\ \cdot & \cdots & \cdot \\ a_{t1}, & \cdots, & a_{tt} \end{matrix} \right\|$$

whose components a_{ik} are taken from $\mathfrak{a}$. The simple algebra $\mathfrak{a}$, its irreducible and its regular representations $\mathfrak{A}$ and $(\mathfrak{a})$ are then expressed by

$$\mathfrak{a} = \mathbf{M}_t\mathfrak{d}, \qquad \mathfrak{A} = (\mathfrak{d})_t, \qquad (\mathfrak{a}) = t\mathfrak{A} = t(\mathfrak{d})_t$$

in terms of a certain division algebra $\mathfrak{d}$ and a positive integer t.

The algebra $\mathfrak{a}$ in k is also an algebra $\mathfrak{a} = \mathfrak{a}_K$ of the same order in any field K over k. But simple in k it may cease to be so in K. In view of Wedderburn's theorem it is no essential restriction to assume $\mathfrak{a} = \mathfrak{d}$ as a division algebra in k. Let $\mathfrak{z}$ be the centrum of $\mathfrak{d}$ and d, ν, $m = d/\nu$ the orders of $\mathfrak{d}$, $\mathfrak{z}$ and $\mathfrak{d}/\mathfrak{z}$.

LEMMA (10.8.A). *In K the algebra $\mathfrak{d}$ becomes the direct sum of a number $\mu \leq \nu$ of simple algebras $\mathfrak{a}_1, \cdots, \mathfrak{a}_\mu$.*

The proof consists of three steps.

(1) The centrum $\mathfrak{z}$ of $\mathfrak{d}$ is a field over k and as such has a determining element z. The latter will satisfy an irreducible algebraic equation $f(z) = 0$ of degree ν in k, and the elements of $\mathfrak{z}$ are the polynomials $g(z)$ of z in k. Thus $\mathfrak{z}$ itself is isomorphic to the field of all polynomials of an indeterminate z in k modulo $f(z)$. In K the polynomial $f(z)$ of the indeterminate z will split into a number of prime factors $f_1(z), \cdots, f_\mu(z)$. The congruences

$$g(z) \equiv g_1(z) \pmod{f_1(z)}, \qquad \cdots, \qquad g(z) \equiv g_\mu(z) \pmod{f_\mu(z)}$$

establish a one-to-one correspondence between the polynomials $g(z)$ in K mod $f(z)$ and the μ-uples $(g_1(z), \cdots, g_\mu(z))$ each member $g_\lambda(z)$ of which is a polynomial in K mod $f_\lambda(z)$. This proves that $\mathfrak{z}$ as an algebra in K splits into a direct sum of $\mu \leq \nu$ division algebras,

$$\mathfrak{z} = \mathfrak{z}_1 + \cdots + \mathfrak{z}_\mu.$$

(2) Operating in k we found, (9.1.5), that for $\mathfrak{D} = (\mathfrak{d})(\mathfrak{d}')$ and the centrum $\mathfrak{Z}$ of $(\mathfrak{d})$ the relations hold

$$\text{(8.1)} \qquad \begin{aligned} \mathfrak{Z} &= m(\mathfrak{z}), \\ \mathfrak{D} &= (\mathfrak{z})_m, \end{aligned}$$

hence in K

$$(\mathfrak{d})(\mathfrak{d}') = (\mathfrak{z}_1)_m + \cdots + (\mathfrak{z}_\mu)_m.$$

$(\mathfrak{z}_1)_m, \cdots, (\mathfrak{z}_\mu)_m$ are irreducible matric algebras in K, and consequently the space P (over the field K) is split into μ subspaces $\mathrm{P}_1, \cdots, \mathrm{P}_\mu$ that are invariant and irreducible under the algebra $(\mathfrak{d})(\mathfrak{d}')$.

(3) As in the proof of Lemma (9.2.A) we pass from the algebra $(\mathfrak{d})(\mathfrak{d}')$ to $(\mathfrak{d})$ as follows. Let Σ be an invariant irreducible subspace of P_1 with respect to the matric algebra $(\mathfrak{d})$ and $\mathfrak{A}$ the algebra of transformations induced by $(\mathfrak{d})$ in Σ. Let $c_1 = \mathsf{l}, \cdots, c_d$ be a basis of $\mathfrak{d}$. Then $\Sigma c_1, \cdots, \Sigma c_d$ are equivalent subspaces of P_1 in the sense that in each of them $(\mathfrak{d})$ induces the same irreducible $\mathfrak{A}$. Hence we can pick out a certain number among them, say $\Sigma_1, \cdots,$

Σ_v, which are linearly independent and such that each Σc_i is contained in their sum $\Sigma_* = \Sigma_1 + \cdots + \Sigma_v$. Then Σ_* is invariant under $(\mathfrak{d})(\mathfrak{d}')$ and hence the total space P_1. There results a decomposition

$$(\mathfrak{d}) = v_1\mathfrak{A}_1 + \cdots + v_\mu\mathfrak{A}_\mu$$

of $(\mathfrak{d})$ into irreducible components $\mathfrak{A}_\lambda$, and thus $\mathfrak{d}$ in K appears as the direct sum of μ simple algebras $\mathfrak{a}$,

$$\mathfrak{d} = \mathfrak{a}_1 + \cdots + \mathfrak{a}_\mu. \tag{8.2}$$

After having proved Lemma (10.8.A) we make use of Wedderburn's theorem for the simple algebras $\mathfrak{a}_\lambda$ in K and thus find

$$\mathfrak{a}_\lambda = \mathbf{M}_{t_\lambda}\mathfrak{d}_\lambda, \qquad \mathfrak{A}_\lambda = (\mathfrak{d}_\lambda)_{t_\lambda}, \qquad (\mathfrak{a}_\lambda) = t_\lambda\mathfrak{A}_\lambda$$

where $\mathfrak{d}_\lambda$ are division algebras in K. Since (8.2) implies

$$(\mathfrak{d}) = (\mathfrak{a}_1) + \cdots + (\mathfrak{a}_\mu),$$

the explicit decompositions

$$\begin{cases} \mathfrak{d} = \mathbf{M}_{t_1}\mathfrak{d}_1 + \cdots + \mathbf{M}_{t_\mu}\mathfrak{d}_\mu, \\ (\mathfrak{d}) = t_1(\mathfrak{d}_1)_{t_1} + \cdots + t_\mu(\mathfrak{d}_\mu)_{t_\mu} \end{cases} \tag{8.3}$$

are obtained, and v_λ turns out to be the same number as t_λ.

Let $\mathfrak{z}_\lambda$, $\mathfrak{Z}_\lambda$; d_λ, ν_λ, m_λ have the same significance for $\mathfrak{d}_\lambda$ in K as $\mathfrak{z}$, $\mathfrak{Z}$; d, ν, m have for $\mathfrak{d}$ in k. The centrum of (8.3) is, in abstracto and in concreto,*

$$\mathfrak{z} = \mathfrak{z}_1 + \cdots + \mathfrak{z}_\mu, \tag{8.4}$$
$$\mathfrak{Z} = t_1^2\mathfrak{Z}_1 + \cdots + t_\mu^2\mathfrak{Z}_\mu.$$

But by (8.1)

$$\mathfrak{Z} = m(\mathfrak{z}), \qquad \mathfrak{Z}_\lambda = m_\lambda(\mathfrak{z}_\lambda).$$

The resulting relation

$$m \cdot (\mathfrak{z}) = t_1^2 m_1 \cdot (\mathfrak{z}_1) + \cdots + t_\mu^2 m_\mu \cdot (\mathfrak{z}_\mu)$$

is to be compared with the relation

$$(\mathfrak{z}) = (\mathfrak{z}_1) + \cdots + (\mathfrak{z}_\mu)$$

implied in (8.4) and thus yields

$$m = t_\lambda^2 m_\lambda \qquad (\lambda = 1, \cdots, \mu). \tag{8.5}$$

Note moreover the equations implied in (8.4) and (8.3):

* One readily sees that the $\mathfrak{z}_\lambda$ are the algebras denoted before by the same letters; this has, however, no bearing on the argument that follows.

$$\nu = \nu_1 + \cdots + \nu_\mu,$$
$$d = t_1^2 d_1 + \cdots + t_\mu^2 d_\mu, \tag{8.6}$$

[of which the second also follows from the first by means of $d = \nu m$, $d_\lambda = \nu_\lambda m_\lambda$ and (8.5)].

In IX, §4, the formula (8.6) was applied to a normal division algebra, in which case $\nu = 1$ and hence $\mu = 1$:

$$d = t_1^2 d_1.$$

Choosing K as an *algebraically closed field* over k ($d_1 = 1$), or constructing by successive algebraic adjunctions a splitting field K of $\mathfrak{d}$, we thus proved, Theorem (9.4.C), that the order d of a normal division algebra is a square number. However, $\mathfrak{d}/\mathfrak{z}$ and $\mathfrak{d}_\lambda/\mathfrak{z}_\lambda$ are normal division algebras; therefore we must have

$$m = n^2, \qquad m_\lambda = n_\lambda^2,$$

and thus the fundamental relation (8.5) is reduced to

$$n = t_\lambda n_\lambda.$$

The results are summarized in:

THEOREM (10.8.B). *The order of a division algebra $\mathfrak{d}$ in k relative to its centrum $\mathfrak{z}$ is a square number $d/\nu = n^2$. In a field K over k the following parallel decompositions take place*

$$\mathfrak{z} = \mathfrak{z}_1 + \cdots + \mathfrak{z}_\mu,$$
$$\mathfrak{d} = \mathbf{M}_{t_1}\mathfrak{d}_1 + \cdots + \mathbf{M}_{t_\mu}\mathfrak{d}_\mu,$$
$$(\mathfrak{d}) = t_1(\mathfrak{d}_1)_{t_1} + \cdots + t_\mu(\mathfrak{d}_\mu)_{t_\mu}$$

where $\mathfrak{d}_\lambda$ are division algebras in K and $\mathfrak{z}_\lambda$ is the centrum of $\mathfrak{d}_\lambda$. If $\mathfrak{d}_\lambda$ is of order n_λ^2 relative to $\mathfrak{d}_\lambda$, then

$$n = n_1 t_1 = \cdots = n_\mu t_\mu.$$

In an algebraically closed field K over k one must have $\mu = \nu$ and

$$n_1 = \cdots = n_\nu = 1,$$

and $(\mathfrak{d})$ decomposes into ν direct summands $n \cdot \mathfrak{M}_n$. The same will already happen in certain finite extensions K of k, the so-called splitting fields. In more explicit language, let $\mathfrak{c}_n^{(\nu)}$ denote the algebra consisting of all ν-uples $(C_1, \cdots, C_\nu)$ of n-rowed matrices C_λ in K and $\mathfrak{C}_n^{(\nu)}$ the corresponding matric algebra whose members are

$$\begin{Vmatrix} C_1 & 0 \cdots 0 \\ 0 & C_2 \cdots 0 \\ \cdot & \cdot\ \cdot\ \cdot\ \cdot \\ 0 & 0 \cdots C_\nu \end{Vmatrix}$$

Then $\mathfrak{d}$ becomes isomorphic to $\mathfrak{c}_n^{(\nu)}$ in the splitting field K and $(\mathfrak{d})$ equivalent to $n\mathfrak{C}_n^{(\nu)}$.

ERRATA AND ADDENDA

p. 96, after line 6: In formulating Theorem (3.5.D) the assumption that none of the irreducible parts of the matric set $\mathfrak{A}$ is the null-algebra (p. 85) should have been repeated. This additional assumption is certainly fulfilled if the *unit matrix* is a member of $\mathfrak{A}$.

p. 97, add at bottom of the page: The multiplication in a group ring does not fulfill the commutative law. Indeed it has become customary to apply the term ring (but not "field") also in the non-commutative case.

p. 143, Supplement to Theorem (5.3.A), Theorem (5.3.B) and its Corollary: The vanishing of the polynomial $\Phi(A)$ should be required not only for (3.2) and J_n, but for

$$A = \frac{E - S}{E + S} \quad \text{and} \quad A = J_n \cdot \frac{E - S}{E + S}.$$

Compare Chap. X, §6.

p. 177, Lemma (7.1.B): see correction to p. 143.

BIBLIOGRAPHY

(The number of each note is followed by the number of the page on which reference is made to it)

CHAPTER I

[1] (6) See e.g. Weyl, Raum Zeit Materie, 5th ed., Berlin 1923, p. 15.

[2] (13) Cf. D. Hilbert, Grundlagen der Geometrie, 7th ed., Leipzig 1930, Chapter 7.

[3] (14) We mention in particular: W. Burnside, Theory of Groups of Finite Order, 2nd ed., Cambridge 1911; G. A. Miller, H. F. Blichfeldt, L. E. Dickson, Theory and Application of Finite Groups, New York 1916; A. Speiser, Theorie der Gruppen von endlicher Ordnung, 3rd ed., Berlin 1937; H. Zassenhaus, Lehrbuch der Gruppentheorie I, Berlin 1937; the chapter on groups in v. d. Waerden, Moderne Algebra I, 2nd ed., Berlin 1937; and Chapter III, §§1–3, in Weyl, Gruppentheorie und Quantenmechanik, 2nd ed., Leipzig 1931.

[4] (14) Vergleichende Betrachtungen über neuere geometrische Forschungen, Erlangen 1872; also Math. Ann. *43*, 1893, p. 63, and Gesammelte mathematische Abhandlungen I, Berlin 1921, p. 460.

[5] (15) G. W. Leibniz, Initia rerum Mathematicarum metaphysica, in Leibnizens Mathematische Schriften, ed. C. J. Gerhardt, VII, Berlin 1848–63, p. 17; Zur Analysis der Lage, ibid. V, p. 178. How clearly Leibniz saw the problem of relativity is shown by his correspondence with Clarke, in particular his 3rd letter, Nos. 4 and 5, and his 5th letter, No. 47 (easily accessible in G. W. Leibniz, Philosophische Werke, ed. A. Buchenau and E. Cassirer, I, 2nd ed., in Meiner's Philosophische Bibliothek, Leipzig 1924).

[6] (17) Cf. Weyl, Raum Zeit Materie, 5th ed., Berlin 1923, p. 16.

[7] (18) For this notion and the foundations of the theory of invariants cf. B. L. v. d. Waerden, Math. Ann. *113*, 1936, p. 14.

[8] (22) I. Kant, Prolegomena zu einer jeden künftigen Metaphysik, die als Wissenschaft wird auftreten können, Kant's Werke, ed. Preuss. Akad. d. Wissensch., IV, Berlin 1903, p. 286.

CHAPTER II

[1] (27) Jour. reine angew. Math. *30*, 1846, p. 1. Coll. Math. Papers I, Cambridge 1889, p. 117.

[2] (27) Phil. Transact., vols. 144, 145, 146, 148, 149, 151, 157, 159, 169 (1854–1878). Coll. Math. Papers, vol. II, Nos. 139, 141, 144, 155, 156, 158; vol. IV, No. 269; vol. VI, No. 405; vol. VII, No. 462; vol. X, No. 693. The text refers to the first six numbers.

[3] (27) Math. Ann. *36*, 1890, p. 473; *42*, 1892, p. 313.

[4] (28) E. Galois, Oeuvres, Paris 1897, in particular his letter to Aug. Chevalier written on the eve of his death.

[5] (29) A modern book on the subject that will appeal to the mathematical reader is P. Niggli's Geometrische Kristallographie des Discontinuums, Leipzig 1919. A. Speiser's book, cited Chap. I[3], contains two interesting chapters on crystallographic classes and the symmetries of ornaments.

[6] (29) Lie himself synthesized his theories in the big three-volume work: S. Lie and F. Engel, Theorie der Transformationsgruppen, Leipzig 1893.

[7] (29) All of Frobenius' papers were published in the Sitzungsber. Preuss. Akad. A complete list of titles is to be found in Speiser's book, p. 143.

[8] (29) Thèse, Paris 1894, which is dependent on Killing's earlier but incomplete work, Math. Ann. *31*, *33*, *34*, *36* (1888–1890). E. Cartan, Bull. Soc. Math. de France *41*, 1913, p. 53.

[9] (29) W. R. Hamilton, Lectures on Quaternions, Dublin 1853. B. Peirce, Linear Associative Algebra, Washington 1870, and Am. Jour. of Math. *4*, 1881, p. 97. Th. Molien, Math. Ann. *41*, 1893, p. 83; *42*, 1893, p. 308.

[10] (29) J. H. M. Wedderburn, On Hypercomplex Numbers, Proc. London Math. Soc. (2) *6*, 1908, p. 77.

[11] (29) I. Schur, Trans. Amer. Math. Soc. (2) *15*, 1909, p. 159.

[12] (29) For the history of mathematics in the nineteenth century cf. F. Klein, Vorlesungen über die Entwicklung der Mathematik im 19. Jahrhundert I, Berlin 1926.

[13] (39) A. Capelli, Math. Ann. *29*, 1887, pp. 331.

[14] (47) H. Weyl, Math. Zeitschr. *20*, 1924, p. 139.

[15] (51) R. Weitzenböck, Komplex-Symbolik, Leipzig 1908; Invariantentheorie, Groningen 1923, III. Abschnitt.

[16] (52) The first main theorem for orthogonal vector invariants was first proved by E. Study, Ber. Sächs. Akad. Wissensch. 1897, p. 443. The following treatment according to Weyl, Math. Zeitschr. *20*, 1924, p. 136.

[17] (56) The group of Euclidean movements, i.e. the n-dimensional orthogonal group extended by a rim of width ν, and its vector invariants were treated by R. Weitzenböck. See his Invariantentheorie, Groningen 1923, XII. Abschnitt; also Wanner, Dissertation Zürich 1926, and the Author's mimeographed Notes on Elementary Theory of Invariants, Princeton 1935-36.

[18] (56) Cayley, Jour. reine angew. Math. *32*, 1846, = Coll. Math. Papers I, No. 52, p. 332. About the algebraic structure of the orthogonal group and the other classical groups in a field either of characteristic zero or of prime characteristic, consult v. d. Waerden, Gruppen von linearen Transformationen, Ergebn. d. Math. *4*, 2, Berlin 1935, and the literature cited there. The systematic study of the groups mod p was undertaken by L. E. Dickson; see his book on Linear Groups, Leipzig 1901.

[19] (66) Cf. E. Witt, Jour. reine angew. Math. *176*, 1937, p. 31; Satz 2.

[20] (70) E. Pascal, Mem. Accad. dei Lincei (4) *5*, 1888. B. L. v. d. Waerden, Math. Ann. *95*, 1926, p. 706.

CHAPTER III

[1] (81) I. Schur in his "Neue Begründung der Theorie der Gruppencharaktere," Sitzungsber. Preuss. Akad. 1905, p. 406.

[2] (87) The treatment here follows a paper by the author in Annals of Math. *37*, 1936, section 1, pp. 710–718. For the abstract approach see Deuring, Algebren, Ergebn. Math. *4*, 1, Berlin 1935, in particular Chapters I–IV. At the basis of the whole modern development of the theory of associative algebras is Wedderburn's paper Chap. II[10]. L. E. Dickson's books: Linear Algebras, Cambridge Tracts *16*, 1914; Algebras and Their Arithmetics, Chicago 1923, and its revised German edition "Algebren und ihre Zahlentheorie," Zürich 1927, are landmarks of the development.

[3] (92) Burnside, Proc. London Math. Soc. (2) *3*, 1905, p. 430. G. Frobenius and I. Schur, Sitzungsber. Preuss. Akad. 1906, p. 209.

[4] (99) Cf. Chapter V of my book Gruppentheorie und Quantenmechanik, 2nd ed., Leipzig 1931.

[5] (101) Whence follows the full reducibility of any of its representations. This fundamental fact was first proved by H. Maschke, Math. Ann. *52*, 1899, p. 363.

[6] (106) Cf. H. Weyl, Duke Math. Jour. *3*, 1937, p. 200.

[7] (111) In its application to the algebra of all bisymmetric transformations in tensor space our method (II) is closely related to the very first treatment of the representations of the full linear group by I. Schur in his Dissertation, Ueber eine Klasse von Matrizen, die sich einer gegebenen Matrix zuordnen lassen, Berlin 1901, while method (I) is applied to this problem in: I. Schur, Sitzungsber. Preuss. Akad. 1927, p. 58. Cf. van der Waerden, Math. Ann. *104*, 1931, pp. 92 and 800.

[8] (112) $a \to a^J$ is an involutorial anti-automorphism operating on the elements a of an algebra if

$$(a+b)^J = a^J + b^J, \qquad (\lambda a)^J = \lambda \cdot a^J, \qquad (ab)^J = b^J \cdot a^J; \qquad (a^J)^J = a \qquad \{\lambda \text{ any number}\}.$$

Our roof operation is of this type. Algebras with an involutorial anti-automorphism have been thoroughly studied by A. A. Albert; see in particular his paper, Ann. of Math. *36*, 1935, p. 886. They are important for the theory of the so-called Riemann matrices, cf. Albert, l.c., and Weyl, Ann. of Math. *37*, 1936, p. 709, and moreover for the algebraic construction of Lie algebras, cf. Ch. VIII[9].

[9] (114) Semi-linear transformations were first introduced by C. Segre, Atti Torino *25*, 1889, p. 276. In the field $K^\dagger$ of complex numbers one has the automorphism consisting in the transition to the conjugate-complex ("antilinear transformations"). The theory of representations of a finite group by semi-linear transformations was given by T. Nakayama and K. Shoda, Jap. Jour. of Math. *12*, 1936, p. 109. For the corresponding generalization of our theory see Weyl l.c.[6]; for semi-linear or anti-linear transformations in general: J. Haantjes, Math. Ann. *112*, 1925, p. 98; *114*, 1937, p. 293; N. Jacobson, Ann. of Math. *38*, 1937, p. 485; Asano and Nakayama, Math. Ann. *115*, 1937, p. 87; Nakayama, Proc. Phys. Math. Soc. Japan *19*, 1937.

CHAPTER IV

[1] (115) l.c., Ch. III[1].

[2] (120) A. Young, Proc. London Math. Soc. (1), *33*, 1900, p. 97; (1) *34*, 1902, p. 361. G. Frobenius, Sitzungsber. Preuss. Akad. 1903, p. 328. v. Neumann's simplified procedure in: van der Waerden, Moderne Algebra II, Berlin, 1931, §127.

[3] (127) A particularly neat way of carrying out such a construction was indicated by W. Specht, Math. Zeitschr. *39*, 1935, p. 696; see especially sections IV and V of his paper. It is based on I. Schur's earlier work, specifically Sitzungsber. Preuss. Akad. 1908, p. 64. The same goal is attained in A. Young's series of publications in the Proc. London Math. Soc. "On the Quantitative Substitutional Analysis" starting with those cited under[2] and followed by (2) *28*, 1928, p. 285; (2) *31*, 1930, p. 253; (2) *34*, 1932, p. 196; (2) *36*, 1933, p. 304. Cf. the seventh Abschnitt in J. A. Schouten, Der Ricci-Kalkül, Berlin, 1924. For the alternating group see G. Frobenius, Sitzungsber. Preuss. Akad. 1901, p. 303; concerning the octahedral group which will play a casual rôle in Chapter VII: A. Young, Proc. London Math. Soc. (2) *31*, 1930, p. 273; W. Specht, Math. Zeitschr. *42*, 1937, p. 120; for generalizations in another direction: W. Specht, Schriften Math. Sem. Berlin, *1*, 1932; Math. Zeitschr. *37*, 1933, p. 321.

[4] (130) The relation between the symmetric group and the full linear group was first discovered and applied to an analysis of the representations of the latter by I. Schur in his Dissertation, Berlin, 1901.

[5] (135) About the earlier history of the invariant theoretic expansions see R. Weitzenböck, Invariantentheorie, Groningen, 1923, p. 137.

CHAPTER V

[1] (137) This method is due to R. Brauer: On Algebras which are connected with the Semisimple Continuous Groups, Ann. of Math. *38*, 1937, p. 857.

[2] (141) R. Brauer, l.c.[1], p. 870.

[3] (141) Math. Zeitschr. *35*, 1932, p. 300.

[4] (149) See section 5 of R. Brauer's paper, l.c.[1]

[5] (158) By the infinitesimal method (Chapter VIII, B) E. Cartan constructed all irreducible representations of any simple continuous group and thus in particular of the orthogonal group $O(n)$ in his paper in Bull. Soc. Math. de France *41*, 1913, p. 53. It needed some further considerations supplied by H. Weyl, Nachr. Gött. Ges. Wissensch. 1927, p. 227 to show that the substrata of his representations are our subspaces $P_0(f_1 \cdots f_\nu)$, viz. the

proof of the following statement that lies in the direction of, but is essentially weaker than, the Theorems (5.2.B) and (5.3.A): Consider all polynomials $\Phi(x^1, \cdots, x^\nu)$ depending on ν arbitrary vectors $x^1, \cdots x^\nu$ which vanish whenever the ν^2 relations

$$(*) \qquad (x^\alpha x^\beta) = 0 \qquad (\alpha, \beta = 1, \cdots, \nu)$$

hold. They form an ideal of which the left sides of (*) constitute an ideal basis.

[6] (159) A. H. Clifford, Ann. of Math. *38*, 1937, p. 533.

[7] (164) The descent from the symmetric to the alternating group could be accomplished in similar manner as this transition from the full to the proper orthogonal group by virtue of A. Clifford's theorem. Cf. G. Frobenius, Sitzungsber. Preuss. Akad. 1901, p. 303.

CHAPTER VI

[1] (167) Weyl, Math. Zeitschr. *20*, 1924, p. 140. The group $Sp(n)$ with a rim of arbitrary width μ and its invariants depending on covariant and contravariant vector arguments is treated in R. Wanner, Dissertation, Zürich, 1926.

[2] (174) Cf. R. Brauer, Ann. of Math. *38*, 1937, p. 855; and H. Weyl, Math. Zeitschr. *35*, 1932, p. 300.

CHAPTER VII

[1] (184) For the fact that (2.10) imply $|A| \neq 0$ cf. H. Minkowski, Nachr. Gött. Ges. Wissensch. 1900, p. 90.

[2] (189) In order to prove the first main theorem for invariants (Chapter VIII B, in particular Theorem 8.14.A) the process was first introduced by A. Hurwitz, Nachr. Gött. Ges. Wissensch. 1897, p. 71, and applied to the real orthogonal group.

[3] (189) In three important papers in the Sitzungsber. Preuss. Akad. 1924, pp. 189, 297, 346, entitled "Neue Anwendungen der Integralrechnung auf Probleme der Invariantentheorie," I. Schur first applied the integration process to the representations of compact groups, especially of the real orthogonal group.

[4] (189) Math. Ann. *97*, 1927, p. 737.

[5] (193) About the whole subject and its literature consult: H. Bohr, Fastperiodische Funktionen, Ergebn. Math. *1*, 5. Berlin, 1932.

[6] (193) Ann. of Math. *34*, 1933, p. 147.

[7] (193) Transact. Am. Math. Soc. *36*, 1934, p. 445. Compare also W. Maak, Abh. Math. Sem. Hamburg, *11*, 1935–36, p. 240.

[8] (194) Ann. of Math. *37*, 1936, p. 57.

[9] (194) E. Cartan, Rend. Circ. Mat. Palermo *53*, 1929, p. 217. H. Weyl, Ann. of Math., *35*, 1934, p. 486.

[10] (194) L. Pontrjagin, Ann. of Math. *35*, 1934, p. 361.

[11] (198) In three papers on the Theorie der Darstellung kontinuierlicher halbeinfacher Gruppen durch lineare Transformationen, Math. Zeitschr. *23*, 1925, p. 271; *24*, 1926, pp. 338 and 377 (Appendix, p. 789) the author combined Lie-Cartan's infinitesimal with Hurwitz-Schur's integral approach. The first paper contains the determination of the class density and the characters of the unitary group by the integral method.

[12] (203) The formula, equating the two expressions (5.15) and (6.5), is originally due to G. Jacobi; see Muir, Theory of Determinants I (1906), p. 341. About the later work of Trudi, Naegelsbach and Kotska, ibid. III (1920), p. 135 and IV (1923), p. 145. A recent generalization by Aitken, Proc. Edin. Math. Soc. *1*, 1927, p. 55; *2*, 1930, p. 164.

[13] (203) First given by I. Schur in his Dissertation, Berlin, 1901.

[14] (208) The proof of Theorem (7.6.F), as of the corresponding theorems for the other classical groups, is a simplified version of the procedure I followed in: Acta Math. *48*, p. 255.

[15] (208) Sitzungsber. Preuss. Akad. 1900, p. 516. For other direct algebraic methods see I. Schur's Dissertation, Berlin, 1901, his papers Sitzungsber. Preuss. Akad. 1908, p. 664, and 1927, p. 58.

[16] (215) Am. Jour. of Math. *59*, 1937, p. 437. Cf. also F. D. Murnaghan, Am. Jour. of Math. *59*, 1937, p. 739; *60*, 1938, p. 44, and G. de B. Robinson, Am. Jour. of Math. *60*, 1938, p. 745.

[17] (215) Math. Zeitschr. *23*, 1925, p. 300.

[18] (215) I. Schur, Sitzungsber. Preuss. Akad. 1908, p. 664. Cf. the résumé in W. Specht, Math. Zeitschr. *39*, 1935, p. 696. Related investigations: A. Young, Proc. London Math. Soc. (1), *34*, 1902, p. 361. D. E. Littlewood and A. R. Richardson, Phil. Transact. Roy. Soc. (A), *233*, 1934, p. 99; Quart. Jour. of Math. (Oxford) *5*, 1934, p. 269. D. E. Littlewood, Proc. London Math. Soc. (2), *39*, 1936, p. 150; (2) *40*, 1936, p. 49; (2) *43*, 1937, p. 226.

[19] (220) Cf. the second of my papers on the theory of representations of semisimple groups, Math. Zeitschr. *24*, 1925, p. 328.

[20] (223) In Math. Zeitschr. *24*, 1925, p. 328, I treated the proper orthogonal group. What I added in Acta Math. *48*, p. 255, so as to cover the case of the full orthogonal group should be replaced by the development here given. A more algebraic deduction of the characters is the subject of R. Brauer's Dissertation, "Ueber die Darstellung der Drehungsgruppe durch Gruppen linearer Substitutionen," Berlin, 1925.

[21] (229) It is easy to carry over the combinatorial approach and the "row-wise" generating function to both the symplectic and orthogonal groups. Cf. F. D. Murnaghan, Nat. Ac. of Sciences *24*, 1938, p. 184. The formula for the number of invariants ($f_\alpha = 0$) is here as in the symplectic case, cf. (8.13), an immediate consequence of the first main theorem. I. Schur, in his papers Sitzungsber. Preuss. Akad. 1924, which inaugurated the application of the integration method to group theory, deduced from this equation the formulae (9.7), (9.15) for the volume measure on the orthogonal group which we secured by a direct geometric computation.

[22] (230) The formula for the orthogonal group was first given in R. Brauer's Dissertation, Berlin, 1925. Both of us have been aware for a long time that the same formula holds for any semisimple group; I give here my proof. Cf. Brauer's note, Comptes rendus, *204*, 1937, p. 1784. An explicit rule for the $\times$-multiplication of the two irreducible representations $\langle P(f_1 \cdots f_n)\rangle$, $\langle P(f' \cdots f_n)\rangle$ of the full linear group in: D. E. Littlewood and A. R. Richardson, Phil. Trans. Roy. Soc. (A), *233*, 1934, p. 99. Also F. D. Murnaghan, Am. Jour. of Math. *60*, 1938, p. 761.

[23] (234) Comptes Rendus *201*, 1935, p. 419. E. Cartan had guessed the correct result before: Ann. Soc. Polon. de Math. *8*, 1929, p. 181. See R. Brauer's own detailed account in the mimeographed notes of my lectures On the Structure and Representations of Continuous Groups II, Princeton, 1934–35.

CHAPTER VIII

[1] (239) The following textbooks are in the classic tradition: I. H. Grace and A. Young, The Algebra of Invariants, Cambridge, 1903. Glenn, The Theory of Invariants, Boston, 1915. L. E. Dickson, Algebraic Invariants, New York, 1913. A freer attitude as to the underlying group of transformations is taken in: R. Weitzenböck, Invariantentheorie, Groningen, 1923.

[2] (242) H. Weyl, Rend. Circ. Mat. Palermo *48*, 1924, p. 29.

[3] (250) See e.g. I. H. Grace and A. Young, The Algebra of Invariants, Cambridge, 1903, pp. 89–91, 96–97. The result was first derived by Cayley in his Memoirs on Quantics.

[4] (251) D. Hilbert, Math. Ann. *36*, 1890, pp. 473–534, = Gesammelte Abhandlungen II, Berlin, 1933, No. 16: "Ueber die Theorie der algebraischen Formen," Theorems I and II on pp. 199 and 211. van der Waerden, Moderne Algebra II, Berlin, 1931, pp. 23–24. The finiteness of an ideal basis for every ideal in R is equivalent to E. Noether's "Teilerkettensatz," cf. l.c., pp. 25–27.

[5] (254) The decisive facts are given in Hilbert's paper quoted under[4], including the theory of syzygies into which we did not enter here. A more detailed study of the ring of invariants and its quotient field aiming at a more finitistic construction of the integrity

basis is contained in Hilbert's later paper "Ueber die vollen Invariantensysteme," Math. Ann. *42*, 1893, pp. 313–373, = Gesammelte Abhandlungen II, No. 19, pp. 287–344. For a simpler proof of his "zero theorem" (p. 294) see A. Rabinowitsch, Math. Ann. *102*, 1929, p. 518, and van der Waerden, Moderne Algebra II, p. 11. The "zero manifold" consists of the sets of values $u, v \cdots$ for which all non-constant invariants $J(u, v, \cdots)$ vanish, and its construction as the intersection $J_1 = 0, \cdots, J_h = 0$ by means of a number of invariants $J_1, \cdots, J_h$ whose weights can be limited a priori plays an important rôle. Useful in this connection is a general criterion of finiteness due to E. Noether: Nachr. Gött. Ges. Wissensch. 1926, p. 28.

[6] (255) The idea of adjunction was emphasized by F. Klein, Erlanger program, passim.

[7] (258) This notion is due to O. Schreier, Abh. Math. Sem. Hamburg *4*, 1926, p. 15, and *5*, 1927, p. 233.

[8] (258) See H. Weyl, Die Idee der Riemannschen Fläche, Leipzig, 1913 (and 1923), §9. The idea of the universal covering manifold goes back to H. A. Schwarz and H. Poincaré (H. Poincaré, Bull. Soc. Math. de France *11*, 1883, pp. 113–114). For a genetic construction see P. Koebe, Jour. reine angew. Math. *139*, 1911, pp. 271–276. For the topological study of continuous groups in general see E. Cartan's two pamphlets: La théorie des groupes finis et continus et l'Analysis situs, Mem. des Sciences Math. *42*, Paris, 1930, and La topologie des groupes de Lie, Actual. Scient. *358*, Paris, 1936.

[9] (260) Lie-Engel, Theorie der Transformationsgruppen, 3 vols., Leipzig, 1893. More recent presentations: H. Weyl, Appendix 8 in Mathematische Analyse des Raumproblems, Berlin, 1923; L. P. Eisenhart, Continuous Groups of Transformations, Princeton, 1933; W. Mayer and T. Y. Thomas, Ann. of Math. *36*, 1935, p. 770. For a simplified treatment of the most important parts of E. Cartan's work on infinitesimal groups (cf. Chap. II[8]), see the author's papers in Math. Zeitschr. *23* and *24* (1925–26), and van der Waerden, Math. Zeitschr. *37*, 1933, p. 446. The construction of all (semi-) simple Lie algebras in $K^\dagger$ (or more generally in an algebraically closed field) has been the pivot of these investigations, as far as they deal with the structure rather than with the representations of groups. The same problem in an arbitrary field has recently been successfully attacked by N. Jacobson, Ann. of Math. *36*, 1935, p. 875; *38*, 1937, p. 508; Proc. Nat. Ac. of Sciences *23*, 1937, p. 240, and by W. Landherr, Abh. Math. Sem. Hamburg *11*, 1935, p. 41. Given an associative algebra $\mathfrak{a} = \{a\}$ with an involutorial anti-automorphism $J\colon a \rightarrow a^J$, its J-skew elements a satisfying $a^J = -a$ form a Lie algebra under the multiplication $[ab] = ab - ba$: this procedure of constructing Lie algebras has proved of paramount importance.

[9a] (260) For Lie groups this doubt has been settled by E. Cartan, Actual. Scient. *358*, 1936, p. 19.

[10] (267) I. Schur, Sitzungsber. Preuss. Akad. 1928, p. 96.

[11] (268) H. Weyl, Math. Zeitschr. *24*, 1926, pp. 348–353.

[12] (268) E. Mohr, Dissertation, Göttingen, 1933.

[13] (268) R. Brauer, Sitzungsber. Preuss. Akad., 1929, p. 3.

[14] (268) J. v. Neumann, Math. Zeitschr. *30*, 1929, p. 3. E. Cartan, Mémor. Sc. Math. *42*, 1930, pp. 22–24.

[15] (268) The author's original derivation of the connectivity of the classical groups in Math. Zeitschr. *23*, 1925, p. 291, and *24*, 1925, pp. 337 and 346, is more complicated. For arbitrary semi-simple groups see ibid. *24*, 1925, p. 380; E. Cartan, Annali di Mat. (4) *4*, 1926–27, p. 209, and (4) *5*, 1928, p. 253; Weyl, Mimeographed Notes on the Structure and Representation of Continuous Groups II, Princeton, 1934–1935, pp. 155–185.

[16] (269) E. Cartan, Bull. Soc. Math. de France *41*, 1913, p. 53. P. A. M. Dirac, Proc. Roy. Soc. (A), *117*, 1927, p. 610; *118*, 1928, p. 351. R. Brauer and H. Weyl, Am. Jour. of Math. *57*, 1935, p. 425. A detailed geometric study of the problem is contained in the mimeographed notes On the Geometry of Complex Domains by O. Veblen and J. W. Givens, Princeton, 1935–36.

[17] (270) The algebra was introduced by W. K. Clifford as early as 1878: Am. Jour. of Math. *1*, 1878, p. 350, = Math. Papers, p. 271. An interesting application of this algebra

was made by H. Witt for the study of quadratic forms in arbitrary fields, Jour. reine angew. Math. *176*, 1937, p. 31.

[18] (275) Attempts which miscarried were made by L. Maurer, Bayer. Akad. Wissensch. *29*, 1899, p. 147; Math. Ann. *57*, 1903, p. 265, and R. Weitzenböck, Acta Math. *58*, 1932, p. 231. Weitzenböck's paper contains a correct proof for any individual linear transformation. By an interesting direct algebraic approach, E. Fischer, Jour. reine angew. Math. *140*, 1911, p. 48, settles the question in $K^\dagger$ for each linear group which contains the transposed conjugate $\bar{A}^*$ of any of its elements A.

[19] (275) E. Noether, Math. Ann. *77*, 1916, p. 89. The same result for finite groups in a field of prime characteristic (dividing the order of the group): E. Noether, Nachr. Gött. Ges. Wissensch. 1926, p. 28. Projective invariants mod p were treated before by L. E. Dickson and his school: On Invariants and the Theory of Numbers, Madison Colloquium, 1913, and various papers during the following years in the Transact. Am. Math. Soc.

[20] (276) E. Cartan, Leçons sur les invariants intégraux, Paris, 1922; Ann. Soc. Polon. de Math. *8*, 1929, p. 181. E. Kähler, Einführung in die Theorie der Systeme von Differentialgleichungen, Leipzig, 1934. J. H. C. Whitehead, Quart. Jour. of Math. (Oxford) *8*, 1937, p. 220.

[21] (277) The coincidence of both definitions was proved by G. de Rham in his Thèse, Paris, 1931 (= Jour. Math. pures et appl. (9), *10*, 1931, p. 165), where he carefully lays the foundation of this theory.

[22] (277) J. W. Alexander, Ann. of Math. *37*, 1936, p. 698. The same idea was presented by A. Kolmogoroff at the Topological Conference in Moscow, Sept. 1935. Furthermore E. Čech, Ann. of Math. *37*, 1936, p. 681.

[23] (279) A direct topological approach: L. Pontrjagin, Comptes Rendus *200*, 1935, p. 1277.

CHAPTER IX

[1] (280) I follow my own method as expounded in Ann. of Math. *37*, 1936, pp. 743–745, and *38*, 1937, pp. 477–483. For the abstract treatment see: van der Waerden, Moderne Algebra II, pp. 172–177, 207–211. Deuring, Algebren, Ergebn. Math. *4*, 1, Berlin, 1935, and the literature cited there. Particularly important: E. Noether, Math. Zeitschr. *37*, 1933, p. 514.

[2] (282) First proved by Th. Skolem, Shr. norske Vid.-Akad., Oslo, 1927.

[3] (287) R. Brauer, Jour. reine angew. Math. *166*, 1932, p. 241; *168*, 1932, p. 44.

[4] (290) Cf. R. Brauer and E. Noether, Sitzungsber. Preuss. Akad. 1927, p. 221. Concerning E. Noether's related "verschränkte Produkte" and R. Brauer's "Faktorensysteme" see: H. Hasse, Transact. Am. Math. Soc. *34*, 1932, p. 171; R. Brauer, Math. Zeitschr. *28*, 1928, p. 677; *31*, 1930, p. 733; also Weyl, Ann. of Math. *37*, 1936, pp. 723-728, and Deuring, l.c.

[5] (290) van der Waerden, Moderne Algebra II, p. 174. J. H. M. Wedderburn, Ann. of Math. *38*, 1937, p. 854.

Supplementary Bibliography, mainly for the years 1940–1945

Important books: A. A. Albert, Structure of algebras, Am. Math. Soc. Coll. Publications *24*, New York, 1939. E. Artin, C. J. Nesbitt, R. M. Thrall, Rings with minimum condition, Univ. of Mich. Pubs. in Math. *1*, Ann Arbor, Mich., 1944. C. Chevalley, Theory of Lie groups, Princeton Math. Ser. *8*, Princeton University Press, 1946. W. V. D. Hodge, The theory and applications of harmonic integrals, Cambridge, Eng., 1941. N. Jacobson, Theory of rings, Am. Math. Soc. Mathematical Surveys *2*, New York, 1943. D. E. Littlewood, The theory of group characters and matrix representations of groups, New York, 1940. F. D. Murnaghan, The theory of group representations, Baltimore, 1938. André Weil, L'integration dans les groupes topologiques et ses applications, Paris, 1938.

On modular representations, which are mentioned in the footnote on p. 100, extensive work has been done by R. Brauer and his collaborators: R. Brauer, Act. sci. et industr. *195*, 1935. R. Brauer and C. Nesbitt, Toronto Studies, Math. Ser. *4*, 1937. T. Nakayama, Ann. of Math. *39*, 1938, 361–369. R. Brauer, Proc. Natl. Acad. *25*, 1939, 252–258; Ann. of Math. *42*, 1941, 53–61; 926–958. R. Brauer and C. J. Nesbitt, Ann. of Math. *42*, 1941, 556–590.

For Chapters VII and VIII compare now: D. E. Littlewood, Proc. Camb. Phil. Soc. *38*, 1942, 394–396; ibid. *39*, 1943, 197–199; Phil. Trans. Roy. Soc. London Ser. A, *239*, 1944, 305–365 and 387–417.

About Hodge's theory of harmonic integrals, which is related to the subject of Chap. VIII, §16, cf. H. Weyl, Ann. of Math. *44*, 1945, 1–6. Pontrjagin's method for determining the Betti numbers of compact Lie groups [see Bibliography, Chap. VIII,[23]] is more fully developed in: Rec. Math., New Ser. (Mat. Sbornik) (6) *48*, 1939, 389–422. Related papers: H. Hopf, Ann. of Math. *42*, 1941, 22–52, and H. Samelson, ibid., 1093–1137.

For the construction of all semi-simple Lie algebras compare, besides the papers mentioned in the Bibliography, Chap. VIII,[9]: E. Witt, Abh. Math. Sem. Hans. Univ. *14*, 1941, 289–322.

For Chap. X, Suppl. A and B, cf. H. Weyl, Amer. Jour. of Math. *63*, 1941, 779–784; for Suppl. D cf. N. Jacobson, Theory of Rings, Math. Surveys *2*, 1943, Chapter 5.

INDEX

Index to Chapter X

Texts and Readings in Mathematics

1. R. B. Bapat: Linear Algebra and Linear Models (Third Edition)
2. Rajendra Bhatia: Fourier Series (Second Edition)
3. C. Musili: Representations of Finite Groups
4. H. Helson: Linear Algebra (Second Edition)
5. D. Sarason: Complex Function Theory (Second Edition)
6. M. G. Nadkarni: Basic Ergodic Theory (Third Edition)
7. H. Helson: Harmonic Analysis (Second Edition)
8. K. Chandrasekharan: A Course on Integration Theory
9. K. Chandrasekharan: A Course on Topological Groups
10. R. Bhatia (ed.): Analysis, Geometry and Probability
11. K. R. Davidson: C* – Algebras by Example (Reprint)
12. M. Bhattacharjee et al.: Notes on Infinite Permutation Groups
13. V. S. Sunder: Functional Analysis — Spectral Theory
14. V. S. Varadarajan: Algebra in Ancient and Modern Times
15. M. G. Nadkarni: Spectral Theory of Dynamical Systems
16. A. Borel: Semisimple Groups and Riemannian Symmetric Spaces
17. M. Marcolli: Seiberg – Witten Gauge Theory
18. A. Bottcher and S. M. Grudsky: Toeplitz Matrices, Asymptotic Linear Algebra and Functional Analysis
19. A. R. Rao and P. Bhimasankaram: Linear Algebra (Second Edition)
20. C. Musili: Algebraic Geometry for Beginners
21. A. R. Rajwade: Convex Polyhedra with Regularity Conditions and Hilbert's Third Problem
22. S. Kumaresan: A Course in Differential Geometry and Lie Groups
23. Stef Tijs: Introduction to Game Theory
24. B. Sury: The Congruence Subgroup Problem
25. R. Bhatia (ed.): Connected at Infinity
26. K. Mukherjea: Differential Calculus in Normed Linear Spaces (Second Edition)
27. Satya Deo: Algebraic Topology: A Primer (Corrected Reprint)
28. S. Kesavan: Nonlinear Functional Analysis: A First Course
29. S. Szabó: Topics in Factorization of Abelian Groups
30. S. Kumaresan and G. Santhanam: An Expedition to Geometry
31. D. Mumford: Lectures on Curves on an Algebraic Surface (Reprint)
32. J. W. Milnor and J. D. Stasheff: Characteristic Classes (Reprint)
33. K. R. Parthasarathy: Introduction to Probability and Measure (Corrected Reprint)
34. A. Mukherjee: Topics in Differential Topology
35. K. R. Parthasarathy: Mathematical Foundations of Quantum Mechanics
36. K. B. Athreya and S. N. Lahiri: Measure Theory
37. Terence Tao: Analysis I (Second Edition)
38. Terence Tao: Analysis II (Second Edition)

39. W. Decker and C. Lossen: Computing in Algebraic Geometry
40. A. Goswami and B. V. Rao: A Course in Applied Stochastic Processes
41. K. B. Athreya and S. N. Lahiri: Probability Theory
42. A. R. Rajwade and A. K. Bhandari: Surprises and Counterexamples in Real Function Theory
43. G. H. Golub and C. F. Van Loan: Matrix Computations (Reprint of the Third Edition)
44. Rajendra Bhatia: Positive Definite Matrices
45. K. R. Parthasarathy: Coding Theorems of Classical and Quantum Information Theory (Second Edition)
46. C. S. Seshadri: Introduction to the Theory of Standard Monomials
47. Alain Connes and Matilde Marcolli: Noncommutative Geometry, Quantum Fields and Motives
48. Vivek S. Borkar: Stochastic Approximation: A Dynamical Systems Viewpoint
49. B. J. Venkatachala: Inequalities: An Approach Through Problems
50. Rajendra Bhatia: Notes on Functional Analysis
51. A. Clebsch (ed.): Jacobi's Lectures on Dynamics (Second Revised Edition)
52. S. Kesavan: Functional Analysis
53. V. Lakshmibai and Justin Brown: Flag Varieties: An Interplay of Geometry, Combinatorics, and Representation Theory
54. S. Ramasubramanian: Lectures on Insurance Models
55. Sebastian M. Cioaba and M. Ram Murty: A First Course in Graph Theory and Combinatorics
56. Bamdad R. Yahaghi: Iranian Mathematics Competitions, 1973-2007
57. Aloke Dey: Incomplete Block Designs
58 R. B. Bapat: Graphs and Matrices
59. Hermann Weyl: Algebraic Theory of Numbers (Reprint)
60. Carl Ludwig Siegel: Transcendental Numbers (Reprint)
61. Steven J. Miller and Ramin Takloo-Bighash: An Invitation to Number Theory (Reprint)
62. John Milnor: Dynamics in One Complex Variable (Reprint)
63. R. P. Pakshirajan: Probability Theory: A Foundational Course
64. Sharad S. Sane: Combinatorial Techniques